III International Scientific and Practical Conference "Transformation of the Labor Market: Risks, Trust, Prospects of International Communications" 2021

SHS Web of Conferences Volume 125 (2021)

Moscow, Russia
27-28 May 2021

Editor:

Alexander G. Tyurikov

ISBN: 978-1-7138-4078-7

Printed from e-media with permission by:

Curran Associates, Inc.
57 Morehouse Lane
Red Hook, NY 12571

Some format issues inherent in the e-media version may also appear in this print version.

For additional information, please contact EDP Sciences – Web of Conferences
at the address below.

EDP Sciences – Web of Conferences
17, Avenue du Hoggar
Parc d'Activité de Courtabœuf
BP 112
F-91944 Les Ulis Cedex A
France

Phone: +33 (0) 1 69 18 75 75
Fax: +33 (0) 1 69 28 84 91

contact-edps@webofconferences.org

Additional copies of this publication are available from:

Curran Associates, Inc.
57 Morehouse Lane
Red Hook, NY 12571 USA
Phone: 845-758-0400
Fax: 845-758-2633
Email: curran@proceedings.com
Web: www.proceedings.com

TABLE OF CONTENTS

INCLUSIVE WORKING DAY CULTURE: ABOUT WHAT IS AND WHAT SHOULD BE

RISK FACTORS OF LABOR MARKET DEVELOPMENT

EDUCATIONAL SERVICES MARKET: DEVELOPMENT TRAJECTORIES AND GROWTH POINTS

LABOR MIGRATION: REGULATORY MODELS AND PROSPECTS FOR ATTRACTING PEOPLE TO ECONOMIC ACTIVITY

Foreword

The modern stage in the development of socio-economic and cultural space is characterised by instability caused by the pandemic, the financial and economic crisis and the factors of the new reality. The sociological science is at the forefront of conceptualisation and research of the system of modern humanitarian relations, new socio-economic trends in the development of Russia and the world.

The Department of Sociology of the Financial University under the Government of the Russian Federation has been initiating professional communication of profile specialists for many years, organising conferences and other scientific and practical events reflecting the topical issues of sociology, economics, marketing and political science.

The III International Scientific and Practical Conference "Transformation of the Labour Market: Risks, Trust, Prospects of International Communications" (May 27-28, 2021) was a logical continuation of a series of scientific and practical events; it brought together over 100 representatives of the academic environment, small and medium-size business, specialists of state and municipal government bodies from more than 15 countries.

The collection of papers presents scientific reports of the participants of the conference organised on the basis of the scientific platform of the Department of Sociology at the Financial University under the Government of the Russian Federation.

During the plenary session of the conference, the proceedings of seven sections, three master-classes and the student discussion panel, the participants discussed the key problems of the 21st century, such as the system risks and factors of labour market development, the potential and ways to overcome the trust economy crisis, new gender features of the labour market, prospects and trends of labour migration, labour regulation models, the current state and prognostication of potential interaction of key stakeholders in the electronic and radioelectronic market, as well as the options for assessing employment, self-realisation and career development of persons with disabilities.

The results of doctoral students' scientific research on the problems of economic sociology and demography were also presented at the conference.

Owing to the coordinated work of the organising committee and the speakers, the conference was able to offer solutions for numerous complex scientific issues and problems of our time.

The broad international attention to the conference was accounted for by the importance of the discussed problems, the retrospective look into the past and the search for new, matching the current reality, ways of labour market development in the third decade of the 21st century, both in the Russian Federation and globally.

Conference Organising Committee

Statement of Peer review

In submitting conference proceedings to *Web of Conferences*, the editors of the proceedings certify to the Publisher that

1. They adhere to its **Policy on Publishing Integrity** in order to safeguard good scientific practice in publishing.
2. All articles have been subjected to peer review administered by the proceedings editors.
3. Reviews have been conducted by expert referees, who have been requested to provide unbiased and constructive comments aimed, whenever possible, at improving the work.
4. Proceedings editors have taken all reasonable steps to ensure the quality of the materials they publish and their decision to accept or reject a paper for publication has been based only on the merits of the work and the relevance to the journal.

Title, date and place of the conference

III International Scientific and Practical Conference " Transformation of the Labor Market: Risks, Trust, Prospects of International Communications», 27-28 May 2021, Moscow

Proceedings editor(s):

Alexander G. Tyurikov

Date and editor's signature

11.10.2021

Intercultural communication: English dominance

Natalya Vladimirovna Chernyishkova[1*], *Daria Valerievna* Sukhorukova[1], and *Marina Vadimovna* Aroshidze[2]

[1]Financial University under the Government of the Russian Federation, Department of Foreign Languages and Intercultural Communication, Moscow, Russia
[2]Batumi Shota Rustaveli State University, Department of Slavic Studies, Batumi, Georgia

Abstract. The importance of intercultural communication in today's globalised world is undeniable. Scholars from different countries have contributed to the development of this field and created educational materials – textbooks on intercultural communication devoted to classifications and features of different cultures – and textbooks on foreign languages that include exercises devoted to developing intercultural skills in everyday and professional contexts. Teachers of foreign languages are making an effort to use the available materials that provide a wide outlook on various cultures and help students prepare for being a part of intercultural community. However, there is an evident shortage of such materials in languages other than English. The article provides the results of the analysis of online sources that demonstrates that the majority of available resources that are in any way connected with intercultural communication are in English. The article dwells on the reasons for English dominance in intercultural communication study and suggests a global effort as a solution. Whether we are looking at textbooks purely on intercultural communication or foreign language learning resources, it is important to understand that the English language cannot be the only medium of intercultural competence formation – other languages can and should be a prism through which intercultural communication is studied.
Keywords: intercultural communication, intercultural competence, textbooks, educational materials, skills, foreign language learning

1 Introduction

Intercultural communication has become one of the integral parts of acquiring foreign languages regardless of the initial purpose – educational or professional. Today any person working in an intercultural context understands the value that comes with understanding representatives of other cultures not only in terms of their language but also in terms of their cultural background [1]. A person who can successfully communicate with people from other cultures becomes a competitive specialist in the labour market [2]. Such an

* Corresponding author: nvchernyishkova@fa.ru

employee will easily put a line between his/her culture and the other one, while also appreciating both similarities and differences between them [3].

When it comes to the term itself, "intercultural communication" can be confused with "cross-cultural communication", and these two terms are often used interchangeably. However, it seems to be a misconception that may be connected with only one term existing, for example, in French – "*interculturel*" [4]. In her article, Susan Fries, an Assistant Professor in the Département Langues et Formation Humaine at Institut National des Télécommunications in Evry, states that there cannot be any confusion about the English terms. "Cross-cultural" refers to something that is connected with several cultures, while "intercultural" – attention to the prefix 'inter' – presupposes interaction. This article looks at *intercultural communication*, and only this term has been analysed throughout the study.

The objective of the article is to study the availability of textbooks and educational materials on intercultural communication in different languages. We presuppose that there is a lack of such materials in languages other than English and attempt to identify possible reasons for this and look at what can be done to change it.

2 Methods

Being an important aspect of teaching foreign languages intercultural communication tends to be the subject of various textbooks and educational materials [5]. However, based on the personal experience of foreign language or intercultural communication classes at the university we can state that there is an evident shortage of educational materials in different languages excluding English. To analyse the actual situation we have studied the online sources – two book databases and an online shop: ISBN Database [6], Elsevier [7] and Amazon [8]. ISBN Database is a large collection of book data where one can do the search using ISBN, title, author, or publisher. It includes over 27 million ISBNs. Elsevier is a publishing company that grants access to various publications. Amazon is an online shopping platform that is available worldwide and can make a delivery to most countries. The study of the number of textbooks or other materials on intercultural communication was based on the search of the key words 'intercultural communication' and the specified language. The results for books based on these key words are presented in Table 1.

Table 1. Key words 'intercultural communication'.

	ISBN Database	Elsevier	Amazon
Overall	10,000	5,654	Over 3,000
English	10,000	5,566	Over 1,000
French	1292	19	-
German	1507	1	65
Spanish	230	61	50
Chinese	-	-	91
Russian	-	-	19
Italian	-	-	9
Japanese / Korean	-	-	17

3 Results

The issue of the lack of publications, may it be books, textbooks, articles or anything else, becomes undeniable if one looks at online databases or shops. The analysis of such sources has shown that the dominance of English as the language of intercultural studies is uncontestable. Table 1 reflects the current situation with books on intercultural communication; the figures are given as a result of search through these websites based on the key words 'intercultural communication' and the language criterion. The number of books available in English is much higher than for other languages.

ISBN database can offer more than 10,000 materials on intercultural communication, the majority of which are in English. The site does not even state the exact number of English sources and just shows the same figure – 10,000. However, more attention should be paid to the figures representing educational materials in other languages – French and German figures are still quite impressive, while the same cannot be said about Spanish.

The search for 'international communication' on Elsevier showed that the number of books amounts to 5,665, 98% of which are in English – a staggering figure compared to 19 – in French, 1 – in German and 61 – in Spanish. It should be noted that both ISBN database and Elsevier show search results for English, French, German and Spanish but not so much for other languages. Elsevier can offer two books in Dutch and one – in Portuguese.

The total number of books connected with 'intercultural communication' available on Amazon is more than 3,000 – more than 305 of them are in English. We are able to present figures for several languages, including Chinese, Italian, Russian, Japanese and Korean. Interestingly, there is no result for books connected with 'intercultural communication' in French.

It should be noted that we have not eliminated books or other educational materials from the search based on their topic. The presented results are purely for the key words 'intercultural communication' – ISBN database specifies that the search is done for books, on Elsevier we have chosen 'books' category, excluding journals, and on Amazon the search starts from choosing the section 'Books' and then entering the key words and the language. Therefore, the number of books that deal directly with intercultural communication can be even lower.

4 Discussion

The importance of creating textbooks of high quality is undeniable, and when it comes to textbooks on intercultural communication authors should understand how to choose the right approach to writing them. Safonova in her paper on intercultural communication educational materials outlines five methodological principles of such work:

1) correlation between global and national,
2) priority of intercultural dominant idea,
3) creation of problem-oriented methodological environment,
4) orientation towards professionally relevant bilingualism,
5) a multiple-level system [9].

These principles reflect how an author can work out an effective strategy for creating educational materials on intercultural communication; however, looking at textbooks that are currently available we can clearly see the dominance of the English language. The analysis of the online sources proves that English is perceived as the preferred language for creating textbooks or other educational materials on intercultural communication. On the one hand, it may seem logical since English is the "global" language – many people know it, take classes in it and use it for professional reasons [10]. Companies all over the world are becoming more culturally sensitive because of the foreign competition, and English is the language that serves as the basis for acquiring intercultural competence [11]. On the other hand, we live in the era of globalisation – cultures are intertwining, borders are being eliminated, that is why we are studying intercultural communication and aspiring to gain intercultural competence, and yet, to do that we mostly have to rely on educational materials in English.

The lack of educational materials in different languages may be also connected with the origin of scholars who have been studying intercultural communication because they are the ones who have made this topic widely known and contributed to people studying it more thoroughly. Most of them are Americans, for instance, Stella Ting-Toomey, William B. Gudykunst, Janet M. Bennett, Milton J. Bennett, Fred E. Jandt and Edward T. Hall – who is considered one of the "founding fathers" of the study of intercultural communication [12]. We should mention one Englishman – Richard D. Lewis – and Young Yun Kim – a scholar who was born and raised in Seoul, Korea but currently works at the University of Oklahoma, Norman. Last but not least are Dutch theorists – Fons Trompenaars and, most importantly, Geert Hofstede. Based on the fact that most of these academics' native language is English or they use it in their professional life, it is possible to assume that the dominance of the English language in educational materials on intercultural communication is logical. These scholars have created a theoretical basis for studying intercultural communication, their input is invaluable but it would be wrong to assume that this field is not developing [13]. Nowadays the concept of intercultural communication is well known and studied by people in different countries. People will fail if they rely only on the paradigms and frameworks outlined by Hall or Hofstede; people need to move further, to look not only for 'standard' examples of culture-induced behaviour but for anything that can be attributed to it [14]. Therefore, there is a need for more educational materials on this subject in different languages that would reflect what is happening in intercultural communication today and how it becomes more and more connected with other disciplines that use theories developed within the framework of intercultural communication [15].

Intercultural communication is now an integral part of teaching foreign languages, as we have understood that without proper skills a person cannot successfully communicate with representatives of other cultures. People may learn to speak a language but if they do not try to comprehend how culture influences the whole process of communication, they will fail to achieve the desired results, especially in professional context [16]. This is why it is essential for teachers to include intercultural aspect in their classes. Without proper educational materials it may be challenging because teachers often work with representatives of their own culture, and to dive into the intricacies of other cultures they do need books or other sources to rely on. Plus, it is important to develop new ways of teaching intercultural communication using the Internet as a way of interacting with representatives of other cultures, but this should be done properly through creating methodological approaches to this type of activity [17]. The opportunities of developing intercultural competence presented by the Internet have not been studied or discussed enough yet [18]. This requires a global effort that will lead to success thanks to an increasing number of online educational platforms. We can already speak about the integration of different online services into traditional approach, for instance, Google services have proved their effectiveness for teaching foreign languages [19]. Now the global community of teachers and learners need to take a step further and apply this specifically for improving intercultural skills.

The development of the world around us leads to changes in how people communicate, especially those who come from different cultures. People aspire to share their values and experiences with others and want to learn more about other cultures to understand them better [20]. Textbooks tend to describe certain frameworks of behaviour that may become partially stereotypical. It is a teacher's job to eliminate any type of stereotype formation about other cultures. If there are no stereotypes, no prejudice, individuals get a chance to become more interculturally conscious [21]. To achieve this, a teacher needs to look for educational materials that would follow certain standards, however, in reality to find one "ideal" textbook is difficult. That is why teachers have to analyse all the existing textbooks and choose the ones that follow these criteria:

1) sociocultural saturation – including enough information on various topics connected with intercultural communication;
2) system – achieving transparency of the presented information;
3) motivation – taking into account the age and professional needs of the students;
4) content – connecting the texts and other material to the professional field;
5) contrast – showing similarities and differences between cultures, which allows students to understand other cultures better;
6) information – including different viewpoints on the topics covered in the textbook,

and all this should be supported by exercises that lead to the formation and development of intercultural skills and competence [22].

If educational materials are chosen wisely, they will help to debunk stereotypes that are formed about most cultures, thus creating the basis for strong international communication skills [23].

Textbooks on intercultural communication should also include examples of 'realia' – elements of cultures that reflect the national character and historic events of a certain people because they will allow students to dive into another culture through words and things that constitute it on linguistic and mental levels [24]. In addition, the understanding of specific elements of a certain culture that can be partially connected with the language itself is also of utmost importance. For instance, in professional contexts people can face difficulties connected with different classifications of businesses [25]. The lack of understanding, in this case, may lead to negative consequences for both sides of professional intercultural communication.

5 Conclusion

The analysis of online sources has demonstrated the lack of educational materials on intercultural communication in different languages. The majority of textbooks are written in English because it is a "global" language and a native language for most scholars who have studied intercultural communication.

Creating educational materials in different languages is a necessity predetermined by intercultural communication itself – it should be studied through the prism of various languages, not only English. French, German, Spanish, etc. textbooks on intercultural communication will be of great importance for different specialists, including teachers of foreign languages.

Communicative language teaching that is quite popular today presupposes understanding other cultures [26]. Therefore, it may be more reasonable to create textbooks and educational materials on foreign languages with inclusion of intercultural aspects rather than focus on writing a textbook purely on intercultural communication. It is necessary to include a balanced cultural representation into the created language textbooks incorporating not only the aspects of a local culture [27]. By focusing on different cultures and looking for ways to explain them to students, educators face a challenge connected with the structure of these textbooks [28]. For example, neologisms should be seen as one of the integral aspects of developing language and culture [29]. They are directly connected with the formation of linguistic and intercultural competences. Therefore, exercises aimed at developing skills dealing with neologisms in business discourse should lead to acquiring knowledge about a specific language system and culture [30].

Currently, TESOL (Teaching English for Speakers of Other Languages)-textbooks include information on different cultures but it cannot be considered enough to help students develop intercultural competence and fully understand how to communicate effectively with representatives of other cultures [31]. To show a better aptitude for intercultural communication students should face concrete situations that may happen between specific groups, as it can provide insight into why they need intercultural competence in today's diverse world (professionally and globally) [32]. The emphasis on the cultural background and respect towards cultural heritage can be beneficial for students in terms of unique experiences they gain during the learning process [33].

It does not matter whether we are talking about textbooks on intercultural communication or textbooks on foreign languages that incorporate information on various cultures, the topicality of such educational materials is evident, and scholars around the world should become a part of a global effort to form students' outlook on other cultures through studying intercultural communication not only in English and acquiring intercultural competence while learning different languages.

References

1. E.A. Fedorova, et al., Int J Eng Adv Tech, **9(1)**, 4532-4535 (2019). https://doi.org/10.35940/ijeat.A1785.109119
2. I.I. Klimova, G.V. Klimova, S.A. Dubinka, XLinguae, **12(1)**, 207-218 (2019). https://doi.org/10.18355/XL.2019.12.01.16
3. E.A. Zueva, Higher Edu Today, **3**, 43-47 (2019). https://doi.org/10.25586/RNU.HET.19.03.P.43
4. S. Fries, J TESOL France, **10**, 5-17 (2003)
5. B. Urhan, *Intercultural Communication: A Literature Review*, in Metin I. (ed), İnönü Üniversitesi İletişim Fakültesi, 139-151 (2016)
6. ISBNdb: The World's largest book database. Accessed on: September 24, 2021. [Online]. Available: https://isbndb.com
7. Elsevier Wordmark. Accessed on: September 24, 2021. [Online]. Available: https://www.elsevier.com
8. Amazon. Accessed on: September 24, 2021. [Online]. Available: https://www.amazon.com/
9. V.V. Safonova, Moscow Univ Bul. Ser 19. Ling Intercult Comm, **2**, 65-76 (2018)
10. M.A. Antonyan, Pedagogy, **9**, 87-95 (2019)
11. F. Patel, M. Li, P. Sooknanan, Intercultural communication: Building a global community (2011). https://doi.org/10.4135/9781446270318
12. M. Baba, J App Behav Sci, **31**, 117-118 (1995). https://doi.org/10.1177/0021886395312001

13. T.N. Omelyanenko, *Problems of intercultural communication in teaching foreign languages, Language and culture in bilingual educational space*, in Materials of the III International Scientific and Practical Conference, 155-160 (2017)
14. M. Nguyen-Phuong-Mai, Intercultural Communication. An Interdisciplinary Approach: When Neurons, Genes, and Evolution Joined the Discourse (Amsterdam University Press B.V., Amsterdam 2017)
15. S.M. Croucher, S. Mélodine, R. Diyako, Comm Res Prac **1:1**, 71-87 (2015). https://doi.org/10.1080/22041451.2015.1042422
16. T.V. Sedova, *Integration of intercultural communication into teaching foreign languages as a factor of forming a successful professional, Theory and Practice of Teaching Languages and Cultures: Philosophical and Methodological Aspects*, in Materials of the II International Scientific and Practical Conference, 48-59 (2017)
17. Lugovaya AA, Selfmanagement, **2(24)**, 54-60 (2021)
18. V.V. Safonova, Eurasian Forum, **1(6)**, 6-20 (2014)
19. E.V. Savchenko, Izv: Herzen Univ J Human Sci, **199**, 99-107 (2021). https://doi.org/10.33910/1992-6464-2021-199-99-107
20. O.A. Azarova, et al., J Envir Manag Tourism, **10(6(38))**, 1376-1381 (2019). https://doi.org/10.14505/jemt.v10.6(38).20
21. M.Z. Altan, Bashkortostan Pedag J, **2**, 133-137 (2011)
22. T.M. Yozhkina, N.N. Sergeeva, Bul Ural Branch Rus Acad Edu, 123-127 (2006)
23. N.A. Kozlovtseva, N.N. Tolstova, Bakhtiniana, **14(1)**, 82-113 (2019). http://dx.doi.org/10.1590/2176-457338362
24. T.Y. Samuilik, The Rus Lang Cult Transl, **1**, 483-493 (2017)
25. E.A. Zueva, Z. Kralova, Int J App Ex Phys, **8-2**, 317-323 (2019). https://doi.org/10.30472/ijaep.v8i2.1.566
26. E.V. Morozova, Hist App Comm Lang Teach, **10-2**, 153-158 (2020)
27. H. Manzoor, A. Asma, S. Umrani, J Edu Soc Sci, **13(2)**, 1-9 (2019)
28. V.V. Levchenko, A.I. Oldmixon (Lager), T.N. Lubimova, Pedag Sci, **3(96)**, 39-46 (2019)
29. A.S. Mirzamova, Quest Theory Prac, **13(6)**, 284-290 (2020). https://doi.org/10.30853/filnauki.2020.6.5
30. E.M. Galishnikova, S.N. Khalevina, O.V. Meshcheryakova, XLinguae, **10(4)**, 357-366 (2017). https://doi.org/10.18355/XL.2017.10.04.29
31. E.S. Purgina, E.I. Purgina, Pedag Edu Rus, **11**, 31-36 (2015)
32. A.C. Kelly, Honors Prog Theses, 325 (2018)
33. Y.B. Novikova, et al., XLinguae, **10(4)**, 369-380 (2017). https://doi.org/10.18355/XL.2017.10.04.30

Intercultural communication and success in business

Svetlana Alekseevna Dubinko[1*], *Irina Iosifovna* Klimova[2], *Lizaveta Alehauna* Dubinka-Hushcha[3], and *Galina Vladimirovna* Klimova[2]†

[1]Belarusian State University, Department of English for Economics, Minsk, Belarus
[2]Financial University under the Government of the Russian Federation, Department of Foreign Languages and Intercultural Communication, Moscow, Russia
[3]Copenhagen Business School, Department of International Economics, Government and Business, Copenhagen, Denmark

Abstract. Effective business communication requires a certain degree of cross-cultural awareness and tolerance both on the ethnic and organizational levels. Communication patterns vary from culture to culture according to values, beliefs and traditional rules for interacting with others, which makes up a learned behavior. These deep-rooted factors lie inside cultural differences and change very slowly if ever as compared to those which are more likely to be on the surface, like food, tastes, art. The latter are more easily adopted or erased in today's global economy. The purpose of the study is to show that the internal traits compose different layers of metaphoricity in the language and, if understated, may lead to complexities or failures in business communication. Based on the experience of working in an international environment, which is analyzed by many scientists, and on the results of the study performed it can be noted that cross-cultural awareness and tolerance imply a variety of intelligences beyond (IQ), including emotional, cultural, social, and collective types of intelligence.
Keywords: collective intelligence, communication patterns, metaphoricity

1 Introduction

In a geographically structured organization, where divisions are represented by geography or country, understanding cultural complexities is critical to business success. The assumption as to the importance of fundamental cultural values in effective business communication lies on the intense studies in the field which have brought about ample information regarding the importance of cultural awareness in business communication. The role of cultural factors either in success or failure in doing business internationally is the subject of the current study. Such deep components of culture as values, beliefs, cognitive styles, relationship with the environment, attitude to social structure, time and

* Corresponding author: svetlanadubinko@gmail.com
† Died July 25, 2021.

communication patterns change very slowly or do not change at all, even in our global world. Values and beliefs are a key issue in understanding cultural differences. Their impact on the corporate climate within multinational companies is the object of the article. Most organizations have a vision and values that define their organizational culture. The vision and values provide a common foundation for the entire organization. They communicate what the organization is and how employees should conduct business and interact with each other and with the customers. They set the tone for behavioral and operational expectations. The hypothesis of the study based on the previous assumption implies that the process of globalization and the desire of multinational companies to standardize procedures will lead to some convergence between individualistic and collectivistic cultures in terms of organizational structure, which will lead to erasing some differences in cultural patterns in business communication, but the difference in deep-rooted values and worldviews will preserve organizational diversity for many years to come.

Based on the experience of working in an international environment, which is analyzed in the works published by many scientists, it can be noted that cross-cultural awareness and tolerance imply a variety of intelligences, including emotional, cultural, social, and collective types of intelligence [1-8].

The fundamental values of each culture are linked to the metaphorical structure of the most fundamental concepts of the language. Some values are deeply rooted in different cultures. The study of the levels of metaphoricity can significantly increase our competence regarding the communicative specifics of this or that culture.

1.1 Cultural traditions in the process of globalization

In today's global economy understanding cultural differences can make you successful in building business relationships and running business, while failing to recognize these differences can lead to misunderstandings, destroy business relationships, and damage billions of dollars' worth of transactions [9-13]. How do organizational, national, and ethnic values coexist? It is known that socially desirable, organizational values do not necessarily fully correspond to national or ethnic values [3, 14-19]. There may be some overlap, especially in the area of personal behavior towards other employees, but there are also certain differences. For example, the thesis "avoid conflicts, treat others with respect and dignity" seems quite harmless. However, ethnically, this can be interpreted in very different ways. It may suggest differences in such communicative behavioral tactics as directness and indirectness inherent in linear-active and multi-active cultures. To the members of the latter, American directness may seem offensive or even aggressive, while their linearly active partners may view indirection in business as insincere behavior that can destroy trust and lead to communication failure. Cultures vary in their egalitarian or authoritarian orientation; differ in building personal relationships and understanding their role in doing business.

In order to identify the value picture that is most favorable for building business relationships, it is necessary to compare socially desirable, organizational values and national, ethnic values inherent in representatives of this or that culture. The comparison, in our opinion, is possible and convenient to present on the basis of the main value orientations, for example, those presented in E. Hall's and G. Hofstede's scales. It should take into account the tasks assigned to the company and the complexity of its organization, including the cross-cultural factor of communication [9, 11, 20-22].

Can a large multinational organization have its own particular culture and spread it throughout its national branch? Multinational corporations are interested in promoting a common culture in their subsidiaries, because it is seen as a kind of connecting material that should help achieve integration and coordination [23]. However, subsidiaries often have an ingrained national culture that can resist what the headquarters would like to impose. This means that national values sometimes exceed corporate values. Citing research results, N. Adler clearly dispels the myth that corporate culture can erase national culture [24]. J. Hofstede, F. Trompenaars and other leading experts share the same opinion [21, 25]. If an organization is redefining and re-evaluating its values as a result of a new business strategy or leadership team, this is a great opportunity to learn from employees and their contributions-especially global contributions. From this point of view, collective intelligence is of particular importance as a form of universal, distributed intelligence resulting from the cooperation and competition of many individuals. Collective intelligence manifests itself in the areas of organizational behavior and leadership. Many organizations discover collective intelligence and put it into practice to classify, evaluate, and share information, as well as to make forecasts and solve problems [2, 18]. It seems that the increasing role of collective intelligence will be manifested not only in the choice of the company's business strategy, but also in the format of cross-cultural communication in order to find the most acceptable balance between organizational and ethnic values to achieve more effective results of the chosen business strategy.

1.2 Metaphorical Aspect of Business Communication

The linguistic metaphor is considered as a phenomenon that is secondary to the conceptual metaphor embedded in the conceptual system of a person and reflects his perception, thinking and activity [26]. A detailed classification of metaphors is presented in the works of G. Lakoff, M. Johnson, and M. Turner [26, 27].

The fundamental values of each culture are linked to the metaphorical structure of the most fundamental concepts and values. We talk in terms of metaphor in everyday life and business communication: we win or lose arguments, defend or lose our position and attack someone's point of view. A lot of what we do is structured with basic conceptual metaphors (like war / peace, etc.). And although there is no battle in the direct physical sense of the word, it is essentially transferred to the field of verbal battle [65]. Our logical arguments, direct or indirect tactics of speech behavior with the help of appropriate metaphors fixed in the cognitive base of a particular culture, reflect this: 'The discussion fell to the emotional level, but we raised it back up to the rational plane', 'The argument is shaky', The argument collapsed', The theory has no foundation', 'Let's gain an insight into the problem', 'Current work along these lines intends to shed some light on this issue'' 'Your argument won't hold water'.

In texts of an economic nature, we observe a wide application of metaphors that appeal to such conceptual issues in our daily life as time, movement, space, etc. So, money, financial affairs are often associated with liquid: 'After the merger, the new company will be able to attract new sources of resources', 'As new interest rates are high, money is flowing into Germany', 'There is a constant ebb and flow of money in and out of the system', 'Banks have been on the crest of a wave for much longer than anyone expected'. The following example metaphorically uses the concept of "time" in business: "Look, we've put a lot of money and effort into this project. We can't retreat now." A lot of metaphorical expressions refer particularly to money (spend, invest, budget, profitably, cost) or to limited resources (use up, have enough of, run out of) [26].

Metaphors reflect cultural coherence in the attitude to the concept of future. In the Western world, particularly in the United States, people tend to think of time as something fixed in nature, something from which one cannot escape: one thing at a time. This is the core issue in monochromic cultures. People who cannot plan events are not highly regarded. The word 'future' is used with the definite article in American English: 'in the future'. The American idea of the future is limited and shows the foreseeable future ('All upcoming events are listed in the paper', What's coming up this week?', 'I'm afraid of what's up ahead of us'). These different approaches to time and the idea of future show the difference in approaches to business appointments, agenda, contract, long-term and short-term planning, and deadlines.

The study of the levels of metaphoricity could significantly increase our competence regarding the communicative specifics of this culture, as well as shed light on the features of the conceptual structure of the English discourse [4, 28-30].

2 Methods

2.1 Study area and sample selection

The study employed a mixed-methods research design (quantitative – qualitative). The survey was conducted among the fourth-year students of the specialty "World economy" of the Belarusian State University, Faculty of International Relations in two stages: Part A and Part B. Part A was performed by the students after completing the theoretical course on intercultural communication, while Part B presupposed the application of students' practical skills in this area after completing a practical course of training in firms and companies working with their foreign counterparts for three months.

2.2 Questionnaire design and data collection

26 respondents participated in Part A, and 28 respondents – in Part B of the experiment. The response rate was 86.7%, or 26 out of 30 people for Part A, 93.3% or 28 out of 30 people for Part B. The majority of the respondents participating in part A (92.3%) were 23 years old, 7.7% of the participants were 22 years old at the time of the survey. 91.7% of the respondents in Part B were 23 years old, 8.3% were 22 years old. In the gender ratio, 82.2% of the respondents were women and 17.8% – men in Part A. Part B was represented by 74.5% of women, 25.5% – men. The participation of the respondents was completely voluntary. A cross-sectional survey questionnaire method was used to collect the required data. Two sets of tasks (one set for each part) were offered: a set of proverbs and a set of statements which reflect different cultural values as far as attitude to time, punctuality, deadlines, being-oriented or doing-oriented character of life are concerned.

Part A was aimed at evaluating the suggested proverbs and sayings in terms of cultural meanings and their accompanying behavioral patterns as individualistic or collectivistic norms, "both" or "puzzled", using the theoretical basis of the study course on intercultural communication.

The task in Part B was to relate this list of proverbs and statements to the students' cultural values and assumptions as "yes", 'no" or 'puzzled'. This task was given to the students after completing the course of practical training in multinational firms and companies working with their foreign counterparts. The answers were chosen for the analysis according to their predominance: those which had more than 50% (Table 1, Table 2). The survey was conducted by a research assistant anonymously in the Google forms application in the classroom. The data were collected using a questionnaire developed in

English. The questionnaires were personally distributed by the researcher, and the respondents were made aware of their rights to withdraw their participation at any time during the study.

C. Data analysis

The results of the study were presented using descriptive statistics.

Table 1. Stage A. Before completing the course of practical training in multinational firms and companies.

<table>
<tr><th></th><th colspan="4">Question 1</th><th colspan="2">Question 2</th></tr>
<tr><th>Proverbs</th><th>Ind.*, %</th><th>Col.**, %</th><th>Both, %</th><th>Puzzled, %</th><th>Yes, %</th><th>No, %</th></tr>
<tr><td>1. Cleanliness is next to godliness</td><td>32</td><td>67</td><td>≤1</td><td>≤1</td><td>64.3</td><td>35.7</td></tr>
<tr><td>2. God helps those who help themselves</td><td>87</td><td>12</td><td>≤1</td><td>≤1</td><td>74.3</td><td>25.7</td></tr>
<tr><td>3. A penny saved is a penny earned</td><td colspan="2">7</td><td>92</td><td>≤1</td><td>74.7</td><td>25.3</td></tr>
<tr><td>4. A man's home is his castle</td><td>95</td><td>4</td><td>≤1</td><td>≤1</td><td>6.4</td><td>93.6</td></tr>
<tr><td>5. Actions speak louder than words</td><td>74</td><td>25</td><td>≤1</td><td>≤1</td><td>83.6</td><td>16.4</td></tr>
<tr><td>6. It's not whether you win or lose but how you play the game</td><td>35</td><td>64</td><td>≤1</td><td>≤1</td><td>68.5</td><td>31.5</td></tr>
<tr><td>7. Winning isn't everything; winning is the only thing</td><td>97</td><td>2</td><td>≤1</td><td>≤1</td><td>24.7</td><td>75.3</td></tr>
<tr><td>8. Seeing is believing</td><td colspan="3">37</td><td>62</td><td>76.8</td><td>23.2</td></tr>
<tr><td>9. Time is money</td><td colspan="2">15</td><td>84</td><td>≤1</td><td>62.8</td><td>37.2</td></tr>
<tr><td>10. Better late than never</td><td>32</td><td>67</td><td>≤1</td><td>≤1</td><td>87.8</td><td>12.2</td></tr>
<tr><td>Statements</td><td></td><td></td><td></td><td></td><td></td><td></td></tr>
<tr><td>1. Everyone should have an equal chance</td><td>98.7</td><td>1.3</td><td>≤1</td><td>≤1</td><td>78</td><td>22</td></tr>
<tr><td>2. Group improvement is better than self-improvement; indeed, self-improvement is achieved through group improvement</td><td>24.2</td><td>75.8</td><td>≤1</td><td>≤1</td><td>87.6</td><td>12.4</td></tr>
<tr><td>3. Everyone needs breathing room</td><td>87.8</td><td>12.2</td><td>≤1</td><td>≤1</td><td>94.7</td><td>5.3</td></tr>
<tr><td>4. There is a solution to every problem; we just haven't invented it yet</td><td>75.8</td><td>24.2</td><td>≤1</td><td>≤1</td><td>36.7</td><td>63.3</td></tr>
</table>

5. Life is a lottery; it can be won with a little luck	26.2	73.8	≤1	≤1	86.7	13.3
6. One shouldn't mix business with pleasure. After all, life is short	12.6		87.4	≤1	65.4	34.6
7. The individual in a group is more important than the group	96.5	3.5	≤1	≤1	35.2	64.8
8. You shouldn't worry about the past or the future; the present is what really counts	25.6	74.4	≤1	≤1	84.7	15.3
9. You're as young as you act	67.8	32.2	≤1	≤1	77.4	22.6
10. The sky's the limit; you can do anything you wish if you just try	78.6	21.4	≤1	≤1	76.8	23.2

Table 2. Stage B. After completing the course of practical training in multinational firms and companies ≤1.

	Question 1				**Question 2**	
Proverbs	**Ind., %**	**Col., %**	**Both, %**	**Puzzled, %**	**Yes, %**	**No, %**
1. Cleanliness is next to godliness	32	67	≤1	≤1	64.3	35.7
2. God helps those who help themselves	87	12	≤1	≤1	74.3	25.7
3. A penny saved is a penny earned	7		92	≤1	74.7	25.3
4. A man's home is his castle	95	4	≤1	≤1	**76.2**	23.8
5. Actions speak louder than words	74	25	≤1	≤1	83.6	16.4
6. It's not whether you win or lose but how you play the game	35	64	≤1	≤1	26.8	**73.2**
7. Winning isn't everything; winning is the only thing	97	2	≤1	≤1	**64.4**	35.6
8. Seeing is believing	**73**	26	≤1	≤1	76.8	23.2
9. Time is money	15		84	≤1	62.8	37.2
10. Better late than never	32	67	≤1	≤1	87.8	12.2
Statements						
1. Everyone should have an equal chance	98.7	1.3	≤1	≤1	78	22

2. Group improvement is better than self-improvement; indeed, self-improvement is achieved through group improvement	24.2	75.8	≤1	≤1	36.7	**63.3**
3. Everyone needs breathing room	87.8	12.2	≤1	≤1	94.7	5.3
4. There is a solution to every problem; we just haven't invented it yet	75.8	24.2	≤1	≤1	31.8	**68.2**
5. Life is a lottery; it can be won with a little luck	26.2	73.8	≤1	≤1	86.7	13.3
6. One shouldn't mix business with pleasure. After all, life is short	12.6		87.4	≤1	65.4	34.6
7. The individual in a group is more important than the group	96.5	3.5	≤1	≤1	35.2	64.8
8. You shouldn't worry about the past or the future; the present is what really counts	25.6	74.4	≤1	≤1	84.7	**72.8**
9. You're as young as you act	67.8	32.2	≤1	≤1	77.4	22.6
10. The sky's the limit; you can do anything you wish if you just try	78.6	21.4	≤1	≤1	76.8	23.2

* Individualistic cultures
** Collectivistic cultures

3 Results

Part A. Proverbs 1, 6, 10 and statements 2, 5, 6, 8 in Part A were referred to collectivistic cultures by the majority of the respondents. Individualistic cultures were marked by points 2, 4, 5, 7 in proverbs and points 1, 3, 4, 7, 9, 10 in statements. Points 3, 9 in the list of proverbs and point 6- in the list of statements were qualified as belonging to both individualistic and collectivistic cultures, point 8 in the list of proverbs was marked as 'puzzled'. These points, as it appears, reflect the concepts which in our global world may be regarded as universal truths. The analysis of other data shows that the criteria which lie as the basis for differences reflected in the proverbs and statements determine the attitude to career and accomplishments, power distance, punctuality, efficiency, competition or social harmony, tendency to inductive or deductive thinking, the appropriate use of time in business communication. The analysis of students' belonging to this or that category according to their answers ('yes', 'no', 'both', 'puzzled') shows that 45% of the data analyzed go to the collectivistic patterns of behavior, 30%- to the individualistic ones, 15% to 'both' and 5% – to the category "puzzled'.

Part B. Points 1, 6, 10 in the list of proverbs and points 2, 5, 8 in the list of statements show the tendency towards collectivistic patterns. Answers in points 2, 4, 5, 7, 8 in the list of proverbs and those in points 1, 3, 4, 7, 9, 10 in the list of statements illustrate collectivistic behavior. Points 3, 9 among the proverbs and point 6 in the list of statements are regarded as 'both'. These points are more likely to reflect universal truths in business communication. The analysis of students' belonging to this or that category according to their answers ('yes', 'no', 'both', 'puzzled') shows that 20% of the data analyzed go to the collectivistic patterns of behavior, 65% – to the individualistic ones, 15% to 'both'.

4 Discussion

The analysis of the results obtained in Parts A and B shows a number of similarities and differences in theoretical understanding of categorizing cultures as individualistic and collectivistic ones before and after the students' completing a practical course of working with foreign counterparts in multinational companies as well as students' personal vision of their belonging to this or that cultural category. It may be noted that similarities embrace points 1, 2, 5, 9, 10 in the list of proverbs and points 1, 3, 5, 6, 7, 9, 10 – in the list of statements. It is worth mentioning that these points give the confirmation of students' understanding both the principles of categorizing cultures and their belonging to this or that category through the concepts expressed in proverbs and statements and their practical experience. The differences in the results of Part A and B of the experiment concern the student' change in their vision as belonging to the type of categories before and after the practical course due to the experience obtained while working with foreign counterparts which was a practical approbation of the theoretical principles of categorizing cultures. The students' answers in points 4, 6, 7, 8 in the list of proverbs and points 2, 4, 6 8 – in the list of sayings illustrate shifting from collectivistic to individualistic cultural patterns.

5 Conclusion

Judging by the results which show the similarities in the opinions as regards to the tasks in Part A and Part B the students position themselves as representatives of the individualistic cultures (answers in points 2 in the list of proverbs, 1, 3, 7, 9, 10 – in the list of statements) and as representatives of the collectivistic patterns (points 1, 10 in the list of proverbs and point 5 in the list of statements) both at stages A and B of the experiment. Points 9 (answers in the list of proverbs) and 6 in the list of statements are referred to the both cultural categories in Parts A and B of the study. Thus the concepts expressed in the latter may be qualified as universal truths. The differences in indices between stages A and B as far as students' vision of belonging to this or that category is concerned are due to practical experience and collective intelligence, which contributed to their assessment of internal values and beliefs in business communication.

The tendencies illustrated by the results of the study, in particular by the similarities and differences of the data in Parts A and B, show that there are some cultural patterns which may be the subject to changes in business communication and those behavioral norms which lie deep in the roots of the culture and remain unchanged. The study supports the previous assumption and the hypothesis that the process of globalization and the desire of multinational companies to standardize procedures will lead to some convergence between individualistic and collectivistic cultures in terms of organizational structure, but the difference in values and worldviews will preserve organizational diversity.

References

1. I. Alon, J.M. Higgins, Bus Horizons **48(6)**, 501-512 (2005). https://doi.org/10.1016/j.bushor.2005.04.003
2. L.A. Dubinka-Hushcha, *Augmenting collective intelligence through blended learning and open educational resources,* in Intercultural Communication and Professionally Oriented Teaching Foreign Languages: materials of the 13th international scientific conf., Minsk, October 30, 2019, 105-109 (Belarusian State University, Minsk, 2019)
3. Y.Y. Kim, The International Encyclopedia of Intercultural Communication (Wiley-Blackwell, 2017). https://doi.org/10.1002/9781118783665
4. I.E. Klyukanov, Principles of Intercultural Communication (Routledge, 2021)
5. D. Livermore, Leading with cultural intelligence: The real secret to success. Second edition (AMACOM, 2015)
6. K. Mchugh, et al., The Leadership Quart, **27(2)**, 218-241 (2016). https://doi.org/10.1016/j.leaqua.2016.01.001
7. M.F. Peters, Open Re Edu Res, **2(1)**, 259-266 (2015). https://doi.org/10.1080/23265507.2015.1084477
8. E. Smith, Toward Internationalization (Newbury House, Rowles, 1979)
9. L. Chen, Intercultural Communication (Handbooks of Communication Science, 2017). https://doi.org/10.1515/9781501500060
10. S.M. Croucher, Global Perspectives on Intercultural Communication (Routledge, 2017)
11. G. Ferri, Intercultural Communication: Critical Approaches and Future Challenges (Springer International Publishing, 2018). https://doi.org/10.1007/978-3-319-73918-2
12. R.E. Nelson, S. ScGopalan, Org Stud, **24(7)**, 1115-1151 (2003). https://doi.org/10.1177/01708406030247006
13. K. Sorrells, S. Sachi, Globalizing Intercultural Communication : a Reader (SAGE Publications, Inc, 2016)
14. I. Adizes, Global Crisis as Leadership Challenge (Bled School of Management, Bled, 2020)
15. N.V. Baryshnikov, *Plyuralisticheskaya didaktika yazykov i kultur* [Pluralistic didactics of languages and cultures], in Linguodidactic Problems of the Formation of Foreign Language Competencies: Collective Monograph Based on the Materials of the International Scientific and Methodological Symposium "Lempert Readings XIX", 16-22 (PSU Publishing House, Pyatigorsk, 2017)
16. X. Dai, Ch. Guo-Ming, Conflict Management and Intercultural Communication: The Art of Intercultural Harmony (Routledge, 2017)
17. M.D. Hills, Online Read Psy Cult, **4(4)** (2002). https://doi.org/10.9707/2307-0919.1040
18. I. Klimova, G. Klimova, S. Dubinka, XLinguae, **12(1)**, 207-218 (2019)
19. J. Santos, Globalisation is Dead. Long Live Globalisation! (Bled School of Management, 2017)
20. E.T. Hall, The Hidden Dimension (Anchor Books, 1969)
21. G. Hofstede, Culture's consequences. Comparing values, behaviors, institutions, and organizations across nations (Sage, Thousand Oaks, 2001)
22. S. Schwartz, Chapter 7 A Proposal for Measuring Value Orientations across Nations (2007)

23. S.C. Schneider, Human Res Manag, **27(2)**, 231-246 (1988). https://doi.org/10.1002/hrm.3930270207
24. N.J. Adler, International dimensions of organizational behavior (South-Western, Cincinnati, 1977)
25. Ch Hampden-Turner, F. Trompenaars, Group & Organization Management, 31(1), 56-63 (2006). doi:10.1177/1059601105276942
26. G. Lakoff, M. Johnson, Metaphors we live by (The University of Chicago Press, Chicago, London, 2003)
27. M. Turner, The literary mind (Oxford University Press, New York, 1996)
28. Research Methods in Intercultural Communication a Practical Guide (Wiley Blackwell, 2016)
29. G.J. Steen, Attention to metaphor: where embodied cognition and social interaction can meet, but may not often do so (Cambridge University Press, 2017)
30. 57 Metaphor Examples That'll Pack Your Prose with Persuasion (2021). Accessed on: September 24, 2021. [Online]. Available: https://smartblogger.com/metaphor-examples/

The role of the organizational culture of trust in the formation of regional social capital

Marina Alexandrovna Kolmykova[1*] and *Elena Nikolaevna* Korneeva[2, 3]

[1]Orenburg State University, Department of State and Municipal Administration, Orenburg, Russia
[2]Financial University under the Government of the Russian Federation, Department of Mass Communications and Media Business, Moscow, Russia
[3]Togliatti State University, Department of Master's Degree Programs (Business Programs), Togliatti, Russia

Abstract. The article discusses the problem of reducing the level of trust in various social institutions, including public authorities. It is shown that this problem is an obstacle to the formation of regional social capital. The research purpose is to determine the role of the organizational culture of trust in the process of formation and accumulation of regional social capital. Based on the study, it was found that the integrated level of trust is most observed in the social institution of the family, and the lowest one – in the social institution of the state, this confirms the tendency of a significant decrease in the level of trust in public authorities and, in turn, this prevents the accumulation of social capital in regions. It has been established that today regions face socio-economic challenges, namely: a pandemic, digitalization of social relations, changes in the labor market, the outflow of talented young people to the capital universities, brain drain, "anti-human factor" in enterprises. It is assumed that the most rational response to the challenges will be the development of regional social capital through the organizational culture of trust in enterprises. Moreover, regional social capital can be considered as a set of social capital of citizens with a pro-active attitude, inspiring political leaders, as well as public figures, and social capital of enterprises which organizational culture is based on trust. The article notes that the formation of the organizational culture of trust can contribute to a decrease in staff turnover, an increase in labor productivity, personnel involvement, an increase in the competitiveness of an enterprise, organizational commitment, an increase in business reputation.

Keywords: organizational culture, social capital, trust, regional social capital

1 Introduction

The process of social capital formation is complex and ambiguous, therefore, it is necessary to study it at the macro and micro levels. Social capital at these levels generally appears in

* Corresponding author: mari_kol@mail.ru

the form of public goods, which are produced and sold in a collective way. It is worth noting that each of them has distinctive features. It is advisable to refer to the opinion of foreign and domestic scientists who conducted research on the essence of the relationship between the categories of "trust" and "organizational culture". In this study, it is of certain interest, since it is possible to get a more detailed understanding of the role of the organizational culture of trust in the process of social capital formation.

It is necessary to pay attention to the research of Pescha, which focuses on the development phases of the element of trust in an organization. The research data contains a description of the calculation phase, the information-experimental phase and the identity phase. Any phase has its own characteristics. For example, the first phase is characterized by the slow development and disappearance in the event of a wrong action by an object of trust. The second phase contains components of regularity and systematicity of partners' meetings. The final stage of this phase implies a feeling of sympathy between partners. Moreover, both partners are already ready for mutual assistance, even if they have to incur damage to the assisting party [1].

At the very beginning, a team leader's trust in his subordinates, the working group is formed. Members of the working group simultaneously begin to build trusting relationships with each other. The result of this process is the formation of team members' trust in their leader. Honesty, professionalism, responsiveness, seniority and other aspects affect the strength of trusting relationships in an organization. As soon as the team leader is convinced that there is a certain level of trust in the organization, he will learn to trust himself.

Pescha's position is worthy of close attention, since the researcher analyzes the category of "trust" within the organizational culture, as well as its impact on performance. It should be noted that the above levels of trust, which are formed in the organization, are rationally appropriate for the theory of organizational culture by Schein, since he considered it as a tool of internal integration and external adaptation in the organization [2].

The interconnection of organizational culture and social capital is expressed in the fact that the formation of a positive corporate image and reputation of an organization through the establishment of trusting relationships with suppliers, partners and customers acts simultaneously as a tool of external adaptation and non-organizational social capital [3].

The successful activity of each enterprise is undoubtedly determined by the managerial actions of a team leader. He recruits employees to his organization, is directly involved in the creation of the working group. Consequently, the formation of the organizational culture of an enterprise occurs when the management of the enterprise is focused on people, their teamwork, progressive and stable activities [4].

It has been proven that a team leader is a subjective factor that predetermines the uniqueness of the organizational culture, which excludes the appointment of random persons (special education, experience in management activities in the management system, recommendations, personal qualities and personality traits, etc.) [5, 6].

In the process of choosing leadership styles, three important components can be distinguished that underlie the overall portrait of a team leader: personality traits, leadership style and authority, which affect the effectiveness of team leadership as a whole and, ultimately, the processes of forming a psychological climate [7].

Scientists have identified the main qualities of management culture associated with the successful enterprise performance. Primarily, this includes the support of the enterprise management of the organizational culture, which is aimed at increasing the value of each employee, as well as his creative potential [8]. This definitely forms the trust of subordinates in their leader and vice versa.

It should be borne in mind that an employee's satisfaction with his professional activities directly affects the performance of an entire organization [9]. The fact that job satisfaction is observed in the majority of employees may indicate a certain level of trust within the working group.

The research purpose is to determine the role of trust in the process of formation and accumulation of regional social capital.

To achieve this purpose, it is necessary to solve the following tasks:

1) to conduct an expert survey among the heads of enterprises and institutions of the Orenburg region to determine the integrated level of trust in various social institutions;

2) to conduct focus groups with representatives of the management of budgetary organizations and authorities of the Orenburg region, as well as with heads of departments and representatives of the top management of commercial organizations in the region;

3) determination of the role of trust in the process of formation and accumulation of regional social capital.

2 Methods

Based on the research purpose, in order to collect information, a decision was made to conduct an empirical study, consisting in an expert survey among heads of enterprises and institutions of the Orenburg region and conducting focus groups. The study used quantitative research methodology to collect and analyze the interpretations and values of the questionnaire.

When conducting focus groups, the authors proceeded from the need to analyze the opinions of heads of enterprises and institutions on the process of formation and development of regional social capital, as well as the determination of its role in the socio-economic development of the Orenburg region.

The first group included representatives of the management of budgetary organizations and authorities of the Orenburg region, the second one – heads of departments and representatives of the top management of commercial organizations in the region. The criterion for this selection is due to the fact that the heads of various enterprises and institutions in the region may have a special view and depth of understanding of the topic of social capital and the way it affects the regional development.

3 Results

In this study, it was decided to analyze the social relationships that are characteristic of various social institutions. It is also necessary to establish a possible level of trust as an integrated one.

It was found that this level of trust is most observed in the social institution of the family. Based on the opinion of representatives of Orenburg enterprises and institutions, a ranked list of manifestations of the integrated level of trust in the studied social institutions was formed.

Ranking the level of trust in social institutions:

1. Family;
2. Education;
3. Economy;
4. Health care;
5. Business;

6. State.

It can be observed that the experts attributed the lowest integrated level of trust to the social institution of the state. Within the framework of this study, team leaders were asked to assess the challenges facing each region that directly or indirectly affect the decrease in the integrated level of trust in various social institutions and hinder the accumulation of social capital. The main challenges were socio-economic ones: a pandemic, digitalization of social relations, changes in the labor market, the outflow of talented young people to the capital universities, "brain drain", "anti-human factor" in enterprises.

4 Discussion

In the authors' opinion, the most rational response to the challenges facing the socio-economic development will be the development of regional social capital through the organizational culture of trust in enterprises.

Human and social capital can be generally viewed as two human assets that exist within the development of a particular community. There are also assumptions that human capital includes "the skills and knowledge obtained in formal and informal learning". Consequently, the formation and development of these skills contribute to the broader social development of any of the communities [10].

At first glance, it may seem that the considered forms of capital are interconnected, they complement and reinforce each other, but there is a clear border between them. It should be borne in mind that human capital is embodied in people, while social capital is embodied in relationships between them [11].

In turn, social capital can be viewed from the perspective of a multidimensional system, which consists of structural, relational and cognitive aspects. That is why social capital often explains "how people perform better because they are somehow better connected to other people", and better networks predict business success [12].

The second dimension of social capital is a relational dimension. It has been suggested that this dimension of social capital is mainly related to ideas of trust, standards, identification and commitment, which facilitate coordination actions and reduce opportunism. Basically, the relational dimension is needed to acquire new knowledge. It also promotes social cohesion [13].

The third dimension of social capital is a cognitive dimension. It directly concerns language and narratives in the communication process, which create a common or mutual understanding in a relationship. A shared understanding and shared goals not only provide information to the parties, but also influence the perception of the relationship [14].

Currently, the above types of dimensions of social capital have become even more relevant, since there is an urgent need to determine the role of organizational culture and trust in the formation of social capital of a region as a whole. It should be noted that increasing the level of trust in a region is a rather problematic task.

It is necessary to pay attention to a certain paradox in society, on the one hand, trust in the authorities has been undermined and there is a decrease in social solidarity in society, on the other hand, the Russian mentality is a powerful tool that unites people. An example is the multiplicity of manifestations of social solidarity during the Great Patriotic War. Or the situation during the perestroika period, when people were not paid a salary for six months, but they remained loyal to the organizations they worked for, and were part of them despite everything [15]. This instills confidence that the accumulation of social capital in the regions will give the necessary impetus to the socio-economic development of the territories.

The modern world is characterized by new waves of changes in Russian society. There is an expansion of digitalization in the national economy simultaneously with the impact of digitalization on the life of each organization. This is the conclusion made by researchers of the leading international management consulting company "Oliver Wyman". In their research, they view organizational culture as a long-term system of beliefs, values and expectations. Researchers also argue that organizational culture is formed in the process of development of an organization when it adapts to the conditions of the external environment, as well as in the process of the organization's internal architecture [16].

Adapting to new realities for all enterprises and institutions is a paramount task, which is solved in different ways. In the authors' opinion, each organization in the region should form its own organizational culture of trust, which, in turn, should become a competitive advantage of an organization and will contribute to the accumulation of intra-organizational and extra-organizational social capital, which is also the basis for accumulating social capital in each region.

5 Conclusion

The formation of the organizational culture of trust can contribute to: reducing staff turnover, increasing labor productivity, staff involvement, increasing the competitiveness of an enterprise, organizational commitment, increasing business reputation.

The basic factor of the socio-economic development of a region is reflected in formal and informal relations between various organizations. The development of mutual trust in the region leads to an increase in the level of trust in the authorities as a whole. For this reason, there is an accumulation of social capital and an increase in the competitiveness of the region.

Regional social capital can be considered as the aggregate of social capital of individuals and legal entities. In other words, regional social capital, on the one hand, consists of social capital of citizens with a pro-active attitude, outstanding political leaders, as well as public figures who have gained authority among the public. On the other hand, the second element is social capital of enterprises, which organizational culture is based on trust. Therefore, the rational development of each of the elements is so important to get rich regional social capital as a result.

References

1. A.V. Pesha, Naukovedenie, **9(2)**, 100EVN217 (2017)
2. E. Schein, Organizatsionnaya kultura i liderstvo [Organizational culture and leadership], V.A. Spivak (ed) (Piter, St. Petersburg, 2002)
3. M.A. Kolmykova, Bul Samara Munic Inst Manag, **2**, 102-109 (2020)
4. Y.O. Da Silva, A. Firdiansjah, B. Triatmanto, Int J Adv Sci Res Eng, **5(3)**, 110-115 (2019). http://doi.org/10.31695/IJASRE.2019.33113
5. M.H. Rahman, T. Dey, M. Al-Amin, Asian Bus Rev, **9(1)**, 17-22 (2019). http://doi.org/10.18034/abr.v9i1.221
6. W. Cai, E.I. Lysova, S.N. Khapova, B.A.G. Bossink, Front Psychol **9**, 1767 (2018). http://doi.org/10.3389/fpsyg.2018.01767
7. T.A. Dukhina, Hist Soc-Edu Thought, **8(1(2))**, 104-144 (2016). http://doi.org/10.17748/2075-9908-2016-8-1/2-140-144

8. S.K. Tran, Int J Corp Soc Resp, **2**, 10 (2017). https://doi.org/10.1186/s40991-017-0021-0
9. B.A. Soomro, N. Shah, South Asian J Bus Stud, **8(3)**, 266-282 (2019). https://doi.org/10.1108/SAJBS-12-2018-0142.
10. J. Jackson, A. Carden, R.A. Compton, App Econ, **47(65)**, 5853–5867 (2015). https://doi.org/10.1080/00036846.2015.1058912
11. M. Moore, N. Recker, Loc Gov Stud, **41(6)**, 881–897, (2015). https://doi.org/10.1080/03003930.2015.1050092
12. R.S. Burt, K. Burzynska, Manag Org Rev, **13(2)**, 221–260 (2017). https://doi.org/10.1017/mor.2017.6
13. D.Z. Levin, J. Walter, M. M. Appleyard, R. Cross, Group Org Manag, **41(4)**, 415–457 (2016). https://doi.org/10.1177/1059601115574429
14. S. Madhavaram, S.D. Hunt, J Bus Res, **74(1)**, 38-46 (2017). http://doi.org/10.1016/j.jbusres.2017.01.007
15. A.G. Tyurikov, Human. Fin Univ Bul, **2**, 51-53 (2018). https://doi.org/10.26794/2226-7867-2018-8-2-51-53
16. O. Wyman, Corporate culture in Russia: new horizons of transformation (2019). Accessed on: October 8, 2021. [Online]. Available: https://www.oliverwyman.com/content/dam/oliver-wyman/europe/ru/files/corporate-culture-in-russia.pdf

Potlatch, absolute hospitality and metaphorization in constructing socioeconomic intercultural reality

Andrey Yakovlevich Bolshunov †, *Sofia Andreevna* Bolshunova[1], *Aleksandra Sergeevna* Proskurina, and *Aleksander Georgievich* Tyurikov

Financial University under the Government of the Russian Federation, Department of Sociology, Moscow, Russia

Abstract. The article reveals and substantiates the following thesis: the "economy of trust" exists exclusively in the context and sphere of specific social relations named potlatch. Potlatch is a traditional gift-giving feast that taken place among the natives of North America. Nowadays there are two points of view on the phenomenon of potlatch. On the one hand, potlatch is claimed to be a reckless waste. On the other hand, it is considered as a clear illustration of trust integration. At this point, the gift-giving process introduces the phenomenon of "absolute hospitality", the embodiment of which is the trust-based space of social integration. Absolute hospitality presupposes the free economic activity of the guest, which is based on unlimited trust and conditioned by nothing. The economy of trust can exist only in such space. Narrativization and metaphorization are essential features of gift-giving space and socio-economic relations developed in it. These are the main forms of construction and interpretation such spaces and relations.
Keywords: potlatch, absolute hospitality, metaphorization, cross-cultural communication, trust, narrative, economy of trust

1 Introduction

The subject of the article is special mechanisms of social integration in which frameworks of economic relations based on trust are possible. Trust is the significant condition of relations in this sphere because the degree of confidence in the ability to trust and trust behavior define interaction patterns [1]. *Kula* and *potlatch* are a clear illustration of trust integration.

Kula is a space outlined by symbolic gift-giving practice in which "natives conduct ordinary trade and make an exchange of different items between separate islands". "Kula unites a significant number of tribes and covers a wide range of actions that are related to each other and influence each other so that they form a single organic whole" [2]. As for the

[1] Corresponding author: ASBolshunova@fa.ru
† Died July 21, 2021.

potlatch, it has two images in modern scientific, economic, and political discourses. From the point of view of modern homo economicus potlatch is a reckless waste and even a "lustful investment" in social status, power, wealth, etc. Under this pretext, a merciless struggle was waged with the Indian potlatch from the time of the development of America until the second half of the twentieth century [3]. An alternative point of view is widely presented in the Italian- and Spanish-language literature. According to it the potlatch is a fundamental mechanism of socio-economic integration and even a total anthropological fact (fatto totale) [4-6].

The widespread view is that the circulation of gifts is the characteristic of for archaic economies only. But P. Sloterdijk emphasizes that "the modern world cannot function without the gifts-giving feast" [7]. J. Baudrillard also claims that "kula and potlatch have disappeared, but not their principle" and bases his "theory of objects" on it [8]. Moreover, he argues that "referential cost is destroying and giving way to a purely structural game of value" [9]. These examples point to gift-giving process to show the principles of constructing modern socio-economic and intercultural reality.

2 Materials and methods

Research methods include lifeworlds modeling (or social reality modeling) using the semantics of logically possible worlds, intensional and narrative logic. Discourse analysis including narrative analysis, self-understanding discourses in experimental groups, expert interview was also applied. The research was conducted in 2016-2021 inter alia as a part of the research project "Intercultural communication in the global world: modeling experience and application of new theoretical and methodological tools".

3 Results

The primary result of lifeworlds modeling is the activity-phenomenological model of intercultural communication. The model was applied for socio-economic analysis. So, the researchers highlights main parts of Constructing Socio-Economic Cross-Cultural Reality which are Potlatch, absolute hospitality and metaphorization.

Gifts-giving process introduces the *primary intensional* meaning (the intensional structure and context) of "*absolute hospitality*" [10]. The concept was introduced by J. Derrida and embodied in various forms of socio-economic integration [11]. The authors interpret embodiment through the prism of enactivism. The concept of "absolute hospitality" is not a rigid designation and has no reference but it has various embodiments or "objectification" in the terminology of E. Husserl. Kula and potlatch are examples of such embodiments. In a way, this is a "dream", "an attempt to provide hospitality in its semantic limit and completeness" [12]. "Nevertheless, we do not give up the dream of the pure gift, in the same way, that we do not give up the idea of pure hospitality. Even if we know it is impossible and that it can be perverse… If we try to draw a politics after of hospitality from the dream of unconditional hospitality, not only will that be impossible, but it will have perverse consequences. So, despite this perversion, despite this impossibility, we go on dreaming or thinking of pure hospitality, of pure gift" [13].

In economic terms, absolute hospitality presupposes *free economic activity of the guest, which is based on unlimited trust and conditioned by nothing*. "Absolute hospitality requires me to open my home" and to break up with traditional forms of hospitality based on laws and agreements [12]. It is no accident that during the Kievan Rus' period (approximately 9th to 13th century) guests called Russian merchants "*torovatyy*", which means generous and, at the same time, prone to possess luxuries, excessive riches. Such merchants were associated with gifts. By "guests" authors also mean tourists, investors, partners, and customers. In each of these cases, we are dealing with different embodiments of the intensional meaning of absolute hospitality.

The concepts of gift and absolute hospitality cannot be objectified or converted into secondary intensional meanings or propositions. A gift is something that has the meaning of a gift. Absolute hospitality is something that has the ultimate meaning of hospitality. Therefore, their embodiments require a "re-invention of language". This proccess carried out in the forms of new "narrative substances" deployment generated by it narratives and also metaphorization as a way of expressing the meaning of embodied gift and hospitality [14]. F. Ankersmit considers metaphor as a "linguistic device" by which narrative substances are introduced into the game and individualize the perspective in which reality should be perceived and analyzed. He agrees with S. Humphreys that "metaphors form the core of reality models". So, the narrative offers a metaphorical view of reality [14]. Even the product retains a touch of metaphor because the product is something that has the meaning of the product. This is manifested in the role that narratives play in economic consciousness and behavior [15, 16]. The meaning of the product is well objectified and has acquired the character of an "objective thought form" as Karl Marx said. It allows us to treat the concept of the product as a proposition. But, in addition to cost as a relationship in which goods are exchanged, there is also a "value" in the price of a good. Value is an expression of meaning of the production which carried out by branding, advertising, the context of demonstrative consumption and other modern forms of "commodity fetishism". In the context of these meanings, the goods in the transformed or as J. Derrida called it perverted forms restore the character of a gift. This fact allows J. Baudrillard arguing that *the law of cost has been replaced by the structural law of value*, which turns the modern economy into a "simulacrum of the third degree" [9]. In this regard "*turn to meanings*" and meaning-making in modern management and economic practices could be stated [17, 18].

One of the most operationalized approaches articulating the role of narratives in the construction of economic reality is the conviction narrative theory (CNT) [15, 19-21], a set of approaches that characterize decision-making under conditions of radical uncertainty, including the uncertainty caused by the fact that the concepts that actors use are primary intensionals, not rigid designatis. In this situation, the narrative becomes a model of development of the situation constructed by the participants of communication.

4 Discussion

The discussion of results can be made through discussing real practices which include described phenomenon. The following are two examples that reflect the results of pilot studies of the construction of socio-economic reality through metaphorization, the narratives of gift-giving and hospitality.

In archaic societies the so-called "exchange of women", constituting relations of kinship, fulfilled the function of social integration. *A woman in this "exchange" was not a commodity but a gift* (therefore it is in this case better to speak not of the exchange of women but of their circulation); the engagement and then the wedding as an apotheosis of potlatch, a contest in generosity, is a convincing evidence of this. Later, until the modern era, the identity of the gift often reflected the self-consciousness of women and was a product of gender socialization. Such self-consciousness can still be found today, but it is no longer dominant. Our pilot studies in the form of discourse and narratives of self-understanding have shown that modern young people tend to think of women in relationships as partners or as commodities, but not as gifts. They easily describe male-female relationships in categories of exchange (bargaining), violence, cooperation, manipulation, but relationships in which women present themselves as a gift are absent. Differences in the perception of marriage as a gift or as a commodity are clearly evident in religious (traditional) and legal understandings of the family. Religion (traditional culture) adheres to an existential interpretation of marriage as a sacrament; in the modern legal interpretation, marriage is increasingly acquiring the features of a contract. Family relations are the most striking confirmation of F. Fukuyama's statement: "legal institutions have replaced trust..." and "there is an inverse relationship between rules and trust" [22].

With this, traditional forms of social integration based on kinship relations, who implied absolute hospitality, are also disappearing. Specifically, this is reflected in the fact that kinship hospitality is increasingly becoming an object of irony, sarcasm (as, for example, in the famous phrase "hello, I am your aunt, I came from Kiev, I will stay with you" or in the post-covid jokes like "vaccination has drawbacks, now relatives can come to us whenever they want and stay as long as they want"). Of course, there is a search for new forms of being together based on an equal exchange, including gift-giving, in which the woman is more of a giver than a gift.

The second example is related to the transformation of the Russian economy. In traditional culture the economic consciousness of the Russians was not based on a binary (poverty/wealth), but on a trinary (poverty/wealth/"torovatost"); torovatyy is the one who is able to contribute to the common cause (for example, donating to the church), help those in need, meet the standards of hospitality, and so on. Russian proverbs prove this convincingly ("wealth is a sin before God, but poverty is a sin before other people", "it is not the rich who feed, but the torovatyy"). From a logical-linguistic point of view, poverty, wealth and torovatost are narrative substances which played a key role in the construction of the "house-building economy". A.P. Sumarokov (18th century) writes: "House-building consists in increasing the abundance... for everyone, not for one person"; "the one who gets rich by the exorbitant work of his subjects... must be called a robber", for he is "a hundred times more harmful than a thief to his country", he is "such a freak of nature, an ignorant..., illiterate creature who honours neither God nor mankind" [23]. Our pilot studies show that today the entrepreneur's torovatost is perceived as nonsense, and entrepreneurs do not understand the torovatost of a traditional merchant (they only perceive it as forms of irrational wastefulness and demonstrative consumption). However, the most convincing evidence of the transformation of Russians' economic consciousness is the fact that the word torovatost and related narratives have almost disappeared from the Russian language.

5 Conclusion

Modernity is characterized by "perverted" (J. Baudrillard) forms and degradation of mechanisms of social integration and intercultural communication based on the circulation of gifts. Examples of the transformation of family-marital behavior and the concept of torovatost prove that the narratives related to the circulation of gifts and absolute hospitality are partly out of use, and partly acquired a sarcastic and ironic connotation. Since economy and legal relations are not the areas of trust production per se, based on the analysis we can conclude that the degradation of the discussed mechanisms and forms of social integration is the cause of the crisis of trust in economic, political, social and everyday relations and intercultural communications.

The article was prepared based on the results of research carried out at the expense of budgetary funds under the state task of the Financial University under the Government of the Russian Federation.

References

1. E. Fehr, J Europ Econ Assoc, **7(2-3)**, 235-266 (2009)
2. B. Malinowski, Argonavty zapadnoi chasti Tikhogo okeana [Argonauts of the Western Pacific] (Center for Humanitarian Initiatives, 2015)
3. R. Butler, E. Adams, Potlatch Resistance: An Arendtian Analysis of Tina Loo's Theory of Rhetorical Legal Resistance (2018). Accessed on: September 25, 2021. [Online]. Available: https://www.academia.edu/37998274/%20Running_Head_Potlatch_Resistance_Potlatch_Resistance_An_Arendtian_Analysis_of_Tina_Loos_The%20ory_of_Rhetorical_Legal_Resistance
4. M. Esposito, Oikonomia: una genealogia della comunità. Tönnies, Durkheim, Mauss (Mimesis, 2011)
5. M. Gigli, L'economia del dono ai tempi della globalizzazione (2017). Accessed on: September 25, 2021. [Online]. Available: https://www.academia.edu/38827258/Leconomia_del_dono_ai_tempi_della_globalizzazione
6. Il Dono, Periodico di Ateneo (2013). Accessed on: September 25, 2021. [Online]. Available: https://www.academia.edu/6945921/Il_dono_fatto_sociale_totale
7. P. Sloterdijk, *What does a human have that he can give away? An Interview with Peter Sloterdijk by Sjoerd van Tuinen*, in J. Brouwer, S. van Tuinen (eds), Giving and Taking: Antidotes to a Culture of Greed (V2 Publishing, Rotterdam, 2014)
8. J. Baudrillard, For a Critique of the Political Economy of the Sign (Verso Books, 2019)
9. J. Baudrillard, Symbolic exchange and death (Sage, 2016)
10. J. Derrida, A. Dufourmantelle, Of hospitality (Stanford University Press, 2000)
11. M. Stapleton, T. Froese, *The enactive philosophy of embodiment: From biological foundations of agency to the phenomenology of subjectivity*, in Biology and subjectivity, 113-129 (Springer, Cham, 2016)
12. Yu.V. Vatolina, Bul SPb Univ. Intern Rel, **1**, 3-7 (2013)
13. J. Derrida, J.L. Marion, R. Kearney, *On the gift: a discussion between Jacques Derrida and Jean-Luc Marion*, in J.D. Caputo, M.J. Scanlon (eds), God, the gift, and postmodernism, 54-78 (Indiana University Press, 1999)
14. F.R. Ankersmit, Narrative logic (Nijhoff, Dordrecht, 1983)
15. A.D. Brown, E.R. Thompson, Bus Hist, **55(7)**, 1143-1167 (2013)
16. E. Vaara, J. Tienari, Organization Sci, **22(2)**, 370-390 (2011)
17. H. De Jaegher, E. Di Paolo, Emerging Comm, **10**, 33 (2008)
18. C. Naumer, K. Fisher, B. Dervin, *Sense-Making: a methodological persp*ective, in CHI '08: CHI Conference on Human Factors in Computing Systems, Proceedings of the SIGCHI Conference on Human Factors in Computing Systems, **8** (Florence, Italy, 2008)
19. R. Bronk, Erasmus J Philos Econ, **6(1)**, 82-107 (2013)

20. S.G. Johnson, A. Bilovich, D. Tuckett, PsyArXiv (2020). https://doi.org/10.31234/osf.io/urc96
21. D. Tuckett, M. Nikolic, Theory Psych, **27(4)**, 501-523 (2017)
22. F. Fukuyama, Trust: The social virtues and the creation of prosperity, **99** (Free Press, New York, 1995)
23. A.P. Sumarokov, *O domostroitelstve* [On Homebuilding], In O.A. Platonov (ed), Economy of the Russian Civilization (IRC, 2008)

The problem of social trust in the context of system methodology

Zoya Dmitrievna Denikina[1*], *Pirmagomed Shikhmagomedovich* Shikhgafizov[2], *Aleksandr Valentinovich* Sablukov[3], *Vyacheslav Leonidovich* Primakov[3], and *Valery Aleksandrovich* Lapshov[3]

[1]Financial University under the Government of the Russian Federation, Department of Humanities, Moscow, Russia.
[2]Financial University under the Government of the Russian Federation, Department of Sociology, Moscow, Russia
[3]Moscow State Linguistic University, The Department of Social Sciences at the Institute of International Relations and Social and Political Sciences, Moscow, Russia

Abstract. The article examines the epistemological status of the phenomenon of social trust. The research purpose is to explicate the concept of trust in connection with the fundamental transformation of socio-historical practice and social knowledge. The priority methodological task is to study the problem within the framework of the system methodology evolution and consider the parameters of trust in the intervals of non-classical and post-non-classical systems analysis. The study is based on a philosophical-scientific paradigm approach. Epistemological situations of autonomous and conventional application of different paradigms are modeled in the study of the phenomenon of social trust. In non-classical systems analysis, social trust is an internal characteristic of a society-system in the mode of its correct functioning. Social trust is one of the social order mechanisms. Social trust is also associated with subjective trust, ordering the interaction of subjects. Intersubjective interactions entail the legitimation of the social order. The transformation of social trust in the modern world is associated with trends in the reduction of the role of normativity, the fragmentation of socio-historical existence. There appears augmented reality, which contains to a certain extent the elements of pseudo-being. Subjective social trust can no longer support holistic meanings. In a situation of change in epistemological attitudes to determine the specifics of social trust, a systems analysis of weakly and highly non-equilibrium states of the intersystem environment is promising. Social order formulas are still inadequate for modeling intersystem environmental states (relations between states, blocs, unions, etc.). Within the framework of the post-non-classical systemic methodology, social trust refers to the signs of social holism. Social trust contributes to the emergence of stable intersystem conditions, the formation of a regulatory environment, which becomes an acting unit. Social trust in all philosophical-scientific paradigms is indicative of the rationality of social-historical existence.

* Corresponding author: zooden@mail.ru

Keywords: social trust, intersystem conditions, system environment, ontological trust, epistemological trust, value-based trust

1 Introduction

The problem of trust turns out to be a subject and methodological area where the interests of social philosophy, theoretical sociology, philosophy of history and philosophy of politics coincide thematically [1-3]. This general subject area can be designated as the systemic nature of the socio-political order. The initial hypothesis is that in post-non-classical social science, a systems analysis of any social phenomenon, including trust, is an alternative to postmodern thinking.

The systemic approach to social processes was developed by T. Parsons in the 1960s – 1970s [4]. His conceptual conclusions are still relevant today, but the proposed social landscape is not modeled within the framework of a super-complex, highly non-equilibrium, non-linear, baseless ontology. According to Parsons, intrasystem processes are effectively comprehended, and the interaction of the system with the external environment, the equilibrium environment is represented by the adaptation mechanism. As defined by Parsons, a social system is that constituted by states and processes of social interaction between acting units [5]. Accordingly, the key role is assigned to the interaction of individuals, their subjective ideas. In non-classical sociology, the phenomenon of trust is studied in the socio-psychological terms, as subjective meanings of actions. It is assumed that the systemic effect of trust arises at the moment of intersubjective meanings [6] common to interacting subjects (according to A. Schutz's phenomenological social theory [7], meanings are constituted in the stream of life experience) [8]. A subjective component of the phenomenon of trust is so great that it is impossible to secure public order without it.

The process of legitimizing social reality is another side of social trust. Legitimation [9] defines social reality as an objective reality [10]. According to M. Weber [11], a social subject gives reality the status of significance and universality. In any case, the social order, which includes subjective trust, is an intrinsic characteristic of society. It is necessary to highlight a non-classical idea that a social system functions correctly when there are social interactions based on comprehending the meaning of the actions of another individual. The system functions irrationally if social interactions are not mediated by trusting attitudes, conviction in appropriate response actions.

According to Parsons, the basic principle of the system is a long-term order. Any functional tasks within the system work for its integrity. In this case, the function determines "some necessary conditions for the preservation of independent existence of the system within some environment" [5]. An unpredictable change in the priorities of significance is dangerous for the system. Defending the system from a plurality of normative aspirations of subjects, Parsons uses the concept of "functional imperatives" – "universally shared rules". These rules are mainly drawn from practical (exemplary) administration tasks, and ensuring credibility of these rules depends on the ability of the system to socialize these rules. During the socialization, people receive a set of targeted landmarks that motivate the acceptance of the system policy [5]. Thus, a systems analysis of the phenomenon of social trust in the "pre-postmodern" period indicates the internal functional features of the system, its normativity. According to Parsons' system theory, the normativity of the system is determined by the ability to solve general problems of adaptation, integration, goal achievement and sampling [5]. Highlighting the simplest relationships in the system (belief, purpose, norm and value), Parsons points to the systemic ordering of the interaction of subjects, their trusting relationships.

2 Materials and methods

The methodological basis of the study consists of scientific works that reveal the problem of social trust in the 20th-21st centuries [7, 8, 10, 12-19]. The further epistemological history of the phenomenon of social trust is associated with two factors: the fundamental transformation of socio-historical knowledge and the evolution of systems analysis. The state of socio-historical existence can be expressed in a concentrated way in the word "loss". There is a loss of integrity and a threat of loss of economic, political sovereignty of national-cultural identity. There is a loss of systematicity, aggressive block thinking, getting out of contractual relations becomes the norm [20]. There is a loss of normativity, the prestige of international organizations and the role of international institutions decrease. There is a loss of responsibility as the reverse side of trust. Not only regularities, but even stable tendencies are not traced in the fragmented socio-historical existence. Communications between the subjects of local existence are phased out, the common ontological space is narrowed. In social knowledge, procedural practices are leveled: military expertise, environmental expertise, etc. Due to the impossibility of determining the degree of penetration of reality and pseudo-reality, subjective assessments lose their holistic meaning and are no longer a type of reflection [21]. Not only reasons, but even the conditions of existence are not collectively comprehended. The social contract within individual states and the outward-directed discourse are destroyed. "Structural and functional models, a systemic idea of sociocultural life were replaced by pluralism of the initial principles of organizing experience, the lack of a universal system of correlating the truth or falsity of statements... The postmodern worldview opposes the "dictatorship of the common"..." [22].

In the context of globalization, epistemological attitudes are undergoing significant changes. Currently, synergetics and the theory of autopoiesis are looking for sources and arguments for their conceptual models in a highly complex socio-historical world. In the authors' opinion, a systems analysis of weakly and highly non-equilibrium states of the intersystem environment is promising for determining the specifics of social trust.

3 Results

In modern systems analysis, it is necessary to take the next step "to admit that the environment for any system is also a higher-order system and not just a field of chaosogenic impacts... In organic integrity, any element is a bearer of qualities of the whole, only in a reduced potential state" [14].

One could hardly disagree with U. Beck's idea of the need to abandon "methodological nationalism", which considers the state in national isolation, and the transition to "methodological cosmopolitanism", which makes it possible to talk about human society as a whole, about the connection of local society with all humanity. In his interdisciplinary analysis, U. Beck comes to the conclusion that "within the national-state modernity it was assumed that in the world of national players there are only two paths to stability: equilibrium (balance of intimidation) or hegemony. On the contrary, in the age of globalization the following alternative is fair: the loss of national sovereignty or transnational cooperation" [12].

4 Discussion

In the authors' opinion, it is methodologically productive to analyze modern socio-historical ontology as the interaction of the system and the environment. The specificity of this process lies in the fact that the mode of a highly non-equilibrium state is inherent only in some national-state formations (systems), while the external environment with intersystem interactions does not comply with an equilibrium ontological description. Even synergetic scenarios are not correlated with this existence of the intersystem environment, the conceptual formula "order-chaos-order" is not yet adequate for modeling intersystem environmental states, in which bifurcation diversity is not observed and stable attractors do not appear.

The ontology of a stable world cannot be unipolar: on the one hand, equilibrium crisis-free autonomy of the state (system) and, on the other hand, a fragmented highly non-equilibrium external environment with a minimum of common intersystem interactions. If the state wants to remain a legal actor, a subject of political-economic decision-making, it is involved in the mechanism for creating intersystem interaction and maintaining trust.

Trust is related to the category of ontological signs of the common, which is manifested in the single. If a group of states recognizes international law, applies the recommendations of international organizations, it has a similar reflection on historical and current events and there appears the effect of socio-historical trust. The emergence of conditions favorable for interstate relations serves as an indicator of the resumption of linear functional and causal determination. This means that in the intersystem environment, one can trace stable trends that the goals lead to predictable results, while the identified reasons entail unambiguous consequences. Thus, according to F. Fukuyama [13], a similar perception of values leads to predictability.

The conditions make the environment normative and value-significant for national states (systems-subjects). To a large extent, intersystem conditions are "man-made", they are formed through special efforts of the subjects of international communication. In this sense, trust is an attributive characteristic of intersystem conditions. Trust as a parameter of intersystem conditions also contributes to the dynamics of intrasystem interactions, the emergence of new functions, the renewal of structures, and the desired choice is made in intersystem interactions. Conditions are the environment for cause-and-effect relationships. The presence of an element of trust in them removes the uncertainty in cause-and-effect relationships. It becomes possible to predict a range of consequences, and the conditionality of political-economic events becomes linear.

Stable conditions lead to ontological equality and balance, prolongation of agreements without absolutizing the current tasks. Cause-and-effect relationships are constantly transforming, and intersystem conditions set the long-term background of these changes. It appears that the conditions are always essentially determined, but their existence is discontinuous: they may or may not be, the environment always exists, and it can be weakly or highly non-equilibrium.

In contrast to the conditions, the intersystem environment has the following ontological meaning. It can be an actor, and its ontological carriers are social groups, parties, elites, states, economic and political unions, etc. According to N. Luhmann, a system always delimits itself from the environment, preserving its essential values [23-25].

5 Conclusion

It can be suggested that in the systemic autopoiesis the environment is also essentially defined and ontologically not aggressive and, in this case, intrasystem decisions "heal" the systemic states.

Weakly non-equilibrium geopolitical processes are favorable for intersystem and intrasystem communications. Trust between the system and the environment is transformed into a self-organization mechanism. Trust as a parameter of the intersystem environment contributes to the preservation of the identity of each system, in political terminology – national security [26, 27]. That is why the role of international law and the normative intersystem environment is great. Trust is a sign of the rationality of social-historical existence, trust helps to restore essentialism, logic and holistic values [28]. Ontological trust is expressed in relation to the past (historical reality); in relation to the present (socio-political and economic reality); in relation to the future (global virtual reality). Epistemological trust abandons simulacra, returns to expert knowledge, to the recognition of coherent and relative truth. Axiological trust appeals to the norm, to the recognition of atomistic values (the role of an individual and the sovereignty of the state) and holistic values within the framework of universal human ideas and supranational values [29].

Thus, social trust as a scientific category is actively included in the process of complex socio-philosophical comprehension. The problem of trust is one of the most demanded topics of interdisciplinary knowledge [30]. The study of trust in the perspective of modern systems analysis provides an idea that there is a fundamental determinism behind the fundamental indeterminism of social interactions [31, 32], the existence of a system in an irrational environment threatens its identification and limits the ways of reproduction and self-regulation.

References

1. P. Sztompka, Socjologia. Analiza społeczeństwa [Sociology. Analysis of society] (Wydawnictwo Znak, Kraków, 2002)
2. P. Sztompka, Trust: A Sociological Theory (Cambridge University Press, Cambridge, 1999)
3. P. Sztompka, Zaufanie fundament społ eczeństwa [Trust the foundation of fellowship] (Wydawnictwo Znak, Krakow, 2007)
4. T. Parsons, *The Concept of Society: The Components and Their Interrrelations*, in T. Parsons (ed), Societies: Evolutionary and Comparative. Perspectives (Prentice-Hall, Englewood Cliffs, 1966)
5. T. Parsons, The system of modern societies (Prentice-Hall, Inc., Englewood Cliffs, 1971)
6. A. Schütz, G. Walsh, F. Lehnert, The Phenomenology of the Social World (Northwestern University Press, 1967)
7. A. Schütz, The Structures of the Life-World (Northwestern University Press, Evanston, 1973)
8. A. Schütz, On Phenomenology and Social Relations: Selected Writings, in H. Wagner (ed) (University of Chicago Press, Chicago, 1970)

9. M. Weber, Gesammelte Aufsätze zur Religionssoziologie [Collected essays on the sociology of religion] (Mohr Siebeck GmbH & Co, Tubingen, 1920)
10. P. Berger, L.T. Luckmann, The Social Construction of Reality. A Treatise on sociology of Knowledge (Penguin Press, 1966)
11. M. Weber, Wirtschaft und Gesellschaft [Economy and Society] (Tübingen, 1972)
12. U. Beck, Chto takoe globalizatsiya? [What is globalization?] (Moscow, 2001)
13. F. Fukuyama, Konets istorii i Poslednii chelovek [The End of History and the Last Man] (Free Press, 1992)
14. S.V. Shachin, Zakony dialektiki i printsipy posledovatelnosti: opyt novogo obosnovaniya [The laws of dialectics and the principles of consistency: The experience of a new justification] (LENAND, Moscow, 2014)
15. M. Woolcock, Theory Soc, **27**, 151-208 (1998)
16. P.C. Bauer, Socius: Soc Res Dynam World, **4**, 1-14 (2018)
17. M.J. Brandt, G. Wetherell, P.J. Henry, Polit Psychol, **36**, 761-768 (2015)
18. A.K. Clark, M.A. Eisenstein, Soc Sci Res, **42**, 361-375 (2013)
19. C. Dawson, Sociology, **53**, 590-599 (2019)
20. R. Inglehart, The Silent Revolution in Europe: Changing Values and Political Styles among Western Publics (Princeton, New York, 1977)
21. J. Alexander, Europ J Soc Theory, **5(1)**, 5–85 (2002). https://doi.org/10.1177/136843100200500100l
22. M.V. Bakhtin, Epistemologiya sotsialno-istoricheskogo znaniya (modeli istorii i filosofsko-antropologicheskie predstavleniya) [Epistemology of socio-historical knowledge (models of history and philosophical and anthropological representations)] (VNIIgeosystem, Moscow, 2013)
23. N. Luhman, Okologische Kommunikation: kann die moderne Gesell-schaft sich auf okologische Gefahrdungen einstellen? [Ecological communication: can modern society adapt to ecological hazards?] (Opladen, 1986)
24. N. Luhmann, Soziologische Aufklärung 6: Die Soziologie und der Mensch, 113-124 (1995)
25. N. Luhmann, *Die Gesellschaft der Gesellschaft* [The Society of society], in Gesellschaft als soziales, v.1 System, 11-189 (Suhrkamp Verlag, Frankfurt am Main, 1997)
26. J. Delhey, K. Newton, Eur Soc, 5, 93-137 (2003)
27. P.T. Dinesen, R. Bekkers, The foundations of individuals generalized social trust: a review, in P.A.M. Van Lange, B. Rockenbach, T. Yamagishi (eds), Trust in Social Dilemmas, 77-99 (Oxford University Press, 2017). https://doi.org/10.1093/oso/9780190630782.003.0005
28. V. Ilyina, SHS Web Conf, **88**, 01007 (2020). https://doi.org/10.1051/shsconf/20208801007
29. P.T. Dinesen, K.M. Sønderskov, Am Socio Rev, 80, 550-573 (2015)
30. F. Fukuyama, Doverie. Sotsialnye dobrodeteli i put k protsvetaniyu [Trust. Social virtues and the path to prosperity] (LLC "Publishing House AST", Moscow, 2004)
31. A. Giddens, Consequences of Modernity (Stanford, 1990)

32. R. Sztompka, Trust: A Soziological Theory (CUP, Cambridge, 1999)

Dynamics of the gender structure of teaching staff in modern Russia

Lyudmila Alekseevna Brushkova[1]*, *Ivan Andreevich* Vladimirov[2], and Niginakhon Arslanovna Shermukhamedova[3]

[1]Financial University under the Government of the Russian Federation, Department of Sociology, Moscow, Russia
[2]Financial University under the Government of the Russian Federation, Faculty of Social Sciences and Mass Communications, Moscow, Russia
[3]National University of Uzbekistan named after Mirzo Ulugbek, Department of Philosophy and Fundamentals of Spirituality, Tashkent, Uzbekistan

Abstract. The subject of the article is the dynamics of the gender structure of university staff in modern Russia. The purpose of the article is to reveal the changes that have occurred in the gender structure of university teachers over the past 20 years. Research methods include the analysis of statistical and sociological data, cross-tabulation and calculation of the feminization index. The authors note that over the twenty-year period, the feminization of Russian universities has intensified. It captures increasingly higher levels of the university job hierarchy, which is manifested in an increase in the number of women in the positions of associate professors, professors, heads of departments, vice-rectors, etc. The significant predominance of women in the positions of assistants, teachers and senior teachers is the basis for the further feminization of university staff in Russia. The analysis of the gender structure of postgraduate and doctoral students shows that women lag behind men only in the youngest age categories of postgraduate students (up to 27 years old) and doctoral students (up to 39 years old) and the oldest category of doctoral students (over 60 years old). This is explained by the fact that at these ages, women are likely to perform their reproductive and educational functions. Women dominate in all other age cohorts of postgraduate and doctoral students. The authors conclude that, despite the existing manifestations of discrimination against women in higher education (gender pay gap, low representation of women at the highest level of university administration, the glass ceiling effect in promoting women up the career ladder, etc.) In general, women are making progress in higher education, making the industry increasingly feminized.
Keywords: gender, employment, education, universities, feminization, teachers, discrimination, Russian society.

* Corresponding author: lbrushkova@fa.ru

1 Introduction

Today, gender issues are being actively developed by representatives of various sciences, including sociology.

Despite the fact that there is a lot in common in gender issues of different groups of countries, there is also significant specificity due to the level of their economic, political and national-cultural development. Foreign research focuses on studying the issues of positioning men and women in popular culture, including advertising [1, 2]. Special attention is paid to the study of gender issues in the black community [3, 4]. A fairly large number of works are devoted to the problem of women leaders [5], women on the board of directors [6], women in non-traditional female positions [7] and women entrepreneurs [8].

The transitional countries, including Russia, are still under the influence of the socialist legacy, despite the fact that the period of socialism is increasingly receding into the past (this year it has been exactly 30 years since the collapse of the USSR and the socialist system). Thus, in the works of the famous Russian sociologist Sillaste, modern gender processes are considered in the context of the formation of a new gender order that comes from the West and does not always correspond to Russian cultural and national traditions [9].

The basis of the gender order in modern society is the gender structure of employment – the distribution of men and women by various spheres and employment levels. It is not only part of the social division of labor, but also one of the manifestations of social inequality within the population. Over the past thirty years, Russia has been going through a complex and painful transformation of all spheres of society, moving onto the "rails" of capitalism. Certainly, this also applies to gender relations in the labor market.

The gender segregation of the Russian labor market and the professional mobility of men and women are well studied in the works by Maltseva and Roshchin [10], the career strategies of men and women in the context of the gender order reproduction are considered in the works by Tartakovskaya [11]. The works by Sillaste are devoted to the gender aspects of poverty [12]. Self-employment in terms of gender is analyzed in the works by Brushkova [13].

The gender structure of employment in any country is characterized by a certain level of segregation of men and women, a manifestation of which is masculinization or feminization of professions and employment spheres. The main negative consequence of gender segregation is the gender pay gap and hence the life opportunities of men and women [14]. In Russia, according to the Federal State Statistics Service (Rosstat), women's average salary in 2019 was 67.3% of men's salary (in 2015 – 73%), which indicates an increase in the gender pay gap [15].

According to Rosstat data for 2019, female industries in Russia include as follows:

• education (82% of women);
• health care and social services (80%);
• hotels and catering facilities (74%);
• finance and insurance (69%);
• culture, sports, entertainment and leisure (66%).

The list of male industries looks completely different:

• construction (87% of men);
• mining (82%);
• transportation and storage (78%);
• water supply, sewerage, waste collection and disposal (67%);
• information and communication (65%).

The sphere of higher education and its transformation is the object of close attention of Russian sociologists. Thus, the age peculiarities of the structure of teaching staff in Russia, as well as the features of the dynamics of mass higher education in Russia, were analyzed in the works by Pugach [16].

The purpose of this article is to analyze the dynamics of the gender structure of teaching staff in modern Russia (over the past 20 years), its main trends and mechanisms. The choice of higher education as a subject of analysis is deliberate, since, according to statistics, the permanent process of feminization is recorded in this sphere.

2 Materials and methods

The latest statistical collection "Women and Men of Russia" prepared by Rosstat was used as the main source of information. Unfortunately, today there is no representative sociological data on the gender structure of teachers in Russian universities. The research mostly concerned only individual universities.

The cross-tabulation method was used in the work for the secondary analysis of official statistical data. To visualize data on age cohorts of postgraduate and doctoral students, the feminization index was calculated for each group.

3 Results

Russia is one of the few countries in the world where the education sector is characterized by a very high level of feminization. It reaches its maximum values at the level of preschool and school education, where it is often possible to find an almost entirely female composition of working groups.

According to the latest data from Rosstat, the ratio of women and men in professional positions of teachers in preschool institutions is 99% versus 1%. The share of women among teachers providing training in primary, basic and secondary education programs is also very high – 88% [15].

Over the past 20 years, there has been a gradual but steady increase in the number of female university teachers. Currently, 58% of women and 42% of men work in the country's universities. In 2000, the ratio was almost parity – 48% of women versus 52% of men, and the difference was a little over 10 thousand people [15].

Feminization is taking over increasingly higher levels of the hierarchy of Russian universities. Currently, women dominate not only in the positions of assistants, teachers and senior teachers, but also among associate professors of universities – more than 60%. The number of women among deans and heads of departments is growing – 45 and 47%, respectively, as well as among professors – 35% [15].

By comparison, in 2000, women lagged behind men even at the level of associate professors (only 44% were women), while at the level of professors, deans and heads of departments women were less than 30% [15].

Currently, 70-90% of the positions of assistants, teachers and senior teachers are occupied by women, which is definitely the basis for further feminization of teaching staff of Russian universities.

In Russia, as in most developed and transitional countries, girls steadily dominate the environment of university students. As for postgraduate and doctoral studies, the situation is not so obvious (see Table 1).

Table 1. Gender structure of students, postgraduates and doctoral students, % [15].

Student category	Women	Men
Students in undergraduate, specialist and graduate programs (students)	53	47
Students in academic and teaching staff training programs (postgraduate students)	44	56
Students in academic staff training programs (doctoral students)	46	54

Among those enrolled in universities the number of girls is 53%, while among postgraduate students in all sciences – 44% on average, and in engineering and general scientific disciplines – about 25%. This is most likely explained by the fact that after graduation, a certain part of the girls switches to performing reproductive and family functions, which forces them to interrupt their careers for a certain period and postpone the implementation of their scientific and professional plans.

However, as for the gender ratio of postgraduate and doctoral students by age, it can be seen that the lag in the number of women is observed only in the youngest (among postgraduate and doctoral students) and the oldest (among doctoral students) age groups, which, as noted above, is more likely due to their involvement in childbearing, childcare and upbringing of children and grandchildren. But women recover their positions and begin to confidently outpace men at older ages – after the age of 30 among postgraduate students and after the age of 40 among doctoral students. Moreover, the gender gap was established at the level of 10-12% (see Tables 2 and 3).

Table 2. The number of postgraduate students by age group, % [15].

Age group	Women	Men
Up to 26 y.o.	36	64
27-29 y.o.	50	50
30-34 y.o.	56	44
35-39 y.o.	55	45
40 y.o. and older	56	44

Table 3. The number of doctoral students by age group, % [15].

Age group	Women	Men
Up to 39 y.o.	38	62
40-49 y.o.	56	44
50-59 y.o.	56	44
60 y.o. and older	24	76

To reflect the degree of feminization of different age groups of postgraduate and doctoral students, the authors calculated the feminization indices (Fig. 1 and Fig. 2), where value 1 is gender parity, value greater than 1 is dominance of women, value less than 1 is dominance of men.

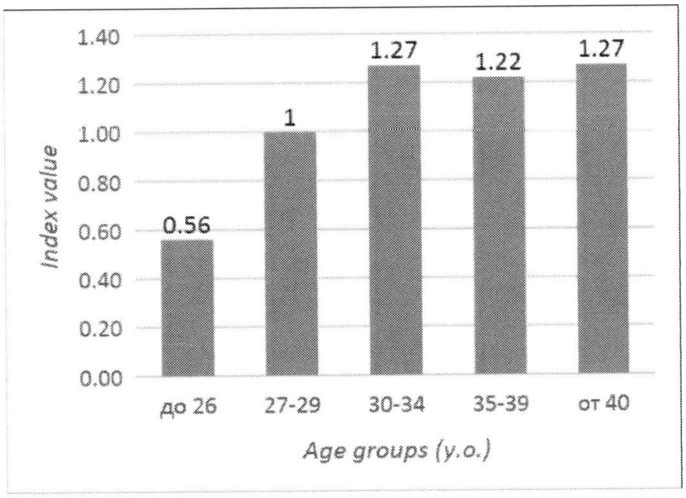

Fig. 1. Postgraduate feminization index. *Source:* Compiled by the authors.

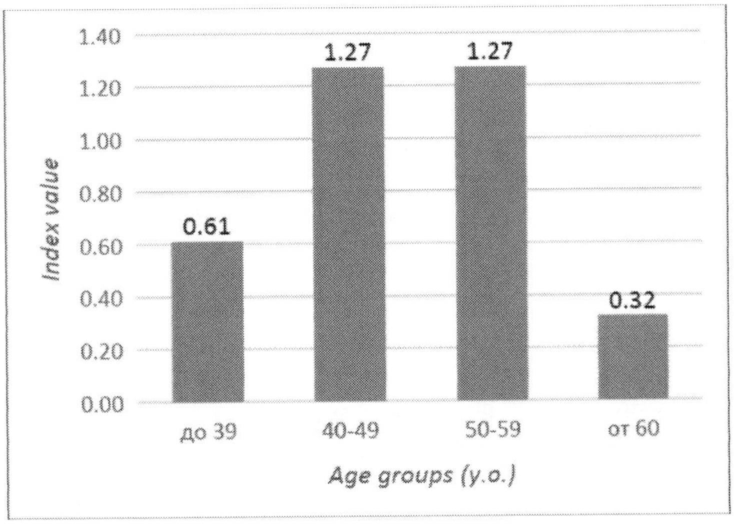

Fig. 2. Doctoral feminization index. *Source:* Compiled by the authors.

The postgraduate feminization index clearly demonstrates that women confidently dominate among the older age groups of postgraduate students, and only in the 27-29 age groups gender parity is observed.

The doctoral feminization index shows that there is no gender parity in this category of doctoral students. In the groups aged under 39 and over 60, men predominate among doctoral students, while in the middle age group (aged 40-59) there is a noticeable asymmetry in favor of women, which is similar by its values to the situation at the postgraduate level.

According to the international monitoring, despite much housework and childcare, Russian women are more likely than men to obtain doctoral degrees. Thus, in the 2015-2016 academic year, 55% of all doctors of sciences were women and 45% – men [17].

Over the past 20 years, the qualitative characteristics of the Russian scientific female community have significantly improved. In 2019, 42% of researchers with PhD degrees were women (an increase of 8% since 2000), doctors of sciences – 27% (an increase of 8%) (see Table 4).

Moreover, since 2015 (among PhDs) and 2018 (among doctors of sciences), these indicators have not increased, which indicates a certain stagnation in this area, the reasons for which need to be specially studied.

Table 4. The number of women with PhD and doctoral degrees, % [15].

Year	PhD	Doctor of Science
2000	34	19
2005	36	21
2010	39	23
2015	42	26
2016	42	26
2017	42	26
2018	42	27
2019	42	27

4 Discussion

The development of higher education in Russia, including postgraduate and doctoral studies, is explained by some researchers by the market laws of supply and demand for this resource [18]. However, within the framework of this approach, it is impossible to explain the gender disparities that have formed in the structure of academic and teaching staff in Russian universities.

In the authors' opinion, women's increased interest in academic and teaching career can be considered in the context of the human capital theory by Schultz, Becker and the multiple capital theory by Bourdieu.

The authors think that since the 1960s, in industrialized countries, including Russia, women have deliberately begun to develop human (according to Bourdieu, cultural) capital, both their own capital and their daughters' as well. This is evidenced by statistical data on the better school and university performance of girls and young women (in comparison with boys and young men), the numerical dominance of girls in universities, their greater involvement in extracurricular activities, various courses of supplementary education, etc. The authors' study of parents' investments in supplementary education for children, conducted in 2019, also supports this hypothesis.

5 Conclusion

The analysis of the gender structure of teaching staff in Russia over the past 20 years allows concluding that this sphere of activity is feminized. Currently, women dominate at all levels of the higher educational hierarchy except for the highest positions (professors, heads of departments, vice-rectors and rectors).

In the future, the feminization of this sphere will only intensify, since already now women prevail in the field of training teaching and academic staff (postgraduate and doctoral studies), as well as in lower teaching positions (assistants, teachers and senior teachers).

This state cannot be considered optimal, since the lag of men in this sphere can cause negative socio-economic and cultural consequences, including long-term ones. It is optimal to maintain gender balance in higher education, when both genders and society as a whole benefit equally.

The study found alarming tendencies towards stagnation in the proportion of female PhDs and doctors of science. Finding out the reasons for this requires additional research.

Despite the obvious progress in promoting women in employment in Russian universities, there are still certain "zones" of gender discrimination. Women are still underrepresented in the senior management of universities, as well as in the bodies regulating higher education (ministries, etc.). The predominance of women in lower and middle teaching positions is one of the reasons that women's average salaries in universities are lower than those of men.

The provisions and conclusions of this study can be useful to everyone involved in the management of higher education in Russia. The results obtained can also be used in further research on gender problems in modern society, as well as in the process of teaching sociological and managerial disciplines.

The study was carried out within the framework of the scientific project "Challenges and risks of the formation of a new gender order and the social mechanism of its management in Russia" No. 20-011-00519 with the financial support of the Russian Foundation for Basic Research.

References

1. S.L. Grau, Y.C. Zotos, Int J Adv, **35(5)**, 761-770 (2016). https://doi.org/10.1080/02650487.2016.1203556
2. L.T. Zayer, M.A. McGrath, P. Castro-Gonzalez, Europ J Market, **54(1)**, 238-260 (2020). https://doi.org/10.1108/EJM-07-2018-0502
3. M. Ong, J.M. Smith, L.T. Ko, J Res Sci Teach, **55(2)**, 206-245 (2018). https://doi.org/10.1002/tea.21417
4. L. Rosenthal, M. Lobel, Ethn Health, **25(3)**, 367-392 (2020). https://doi.org/10.1080/13557858.2018.1439896
5. M.H. Elmagrhi, C.G. Ntim, A.A. Elamer, Q. Zhang, Bus Strat Environ, **28(1)**, 206-220 (2018). https://doi.org/10.1002/bse.2250
6. X. He, S. Jiang, Bus Strat Environ, **28(7)**, 1341-1356 (2019). https://doi.org/10.1002/bse.2319
7. J. Harris, S.G. Naoum, J. Rizzuto, C. Egbu, J Manag Eng, **36(2)** (2020). https://doi.org/10.1061/(ASCE)ME.1943-5479.0000731
8. R.T. Harrison, C.M. Leitch, M. McAdam, J Econ Geog, **20(4)**, 1041-1067 (2020). https://doi.org/10.1093/JEG/LBZ035
9. G.G. Sillaste, Woman Rus Soc, **2**, 3-16 (2019). https://doi.org/10.21064/WinRS.2019.2.1
10. I.O. Maltseva, S.Yu. Roshchin, Gendernaya segregatsiya i mobil'nost' na rossiiskom rynke truda [Gender segregation and mobility in the Russian labor market]. (Higher School of Economics Publishing House, Moscow, 2006)
11. I.N. Tartakovskaya, Soc Res, **5**, 84-93 (2015)
12. G.G. Sillaste, Digest Sci Life Fin Univ, **5**, 6-9 (2018)
13. L.A. Brushkova, Self-govern **3**, 27-30 (2020)
14. C. Collins, L.C. Landivar, L. Ruppanner, W.J. Scarborough, Gender, Work Org, **28(S1)**, 101-112 (2021). https://doi.org/10.1111/gwao.12506

15. Rosstat opublikoval sbornik "Zhenshchiny i muzhchiny Rossii" – 2020 [Rosstat has published the collection "Women and men of Russia" – 2020]. Accessed on: October 4, 2021. [Online]. Available: https://rosstat.gov.ru/folder/70843/document/115497
16. V.F. Pugach, Higher Edu Rus, **1**, 47-55 (2017)
17. Status and Trends in the Education of Racial and Ethnic Groups. Accessed on: October 4, 2021. [Online]. Available: https://nces.ed.gov/programs/raceindicators/
18. V.F. Pugach, Higher Edu Rus, **29(2)**, 74-82 (2020). https://doi.org/10.31992/0869-3617-2020-29-2-74-82

Family entrepreneurship as a social resource for increasing employment of Russian families

Ulyana Alekseevna Agapova[1], *Galina Georgievna* Sillaste, and *Svetlana Aleksandrovna* Yushkova

Financial University under the Government of the Russian Federation, Department of Sociology, Faculty of Social Sciences and Mass Communications, Moscow, Russia

Abstract. The subject of the article is the poorly studied to date social phenomenon of family entrepreneurship as a type of joint labor activities of family members, which improves the family's standard of living. The phenomenon of family entrepreneurship is considered from the standpoint of the sociological approach in four perspectives: as a social process, as a type of collective work activity, as a form of group economic behavior of family members as a small group, and as a form of educational work on the formation of children's needs and working skills. From the perspective of gender sociology family entrepreneurship is considered as a type of gender-neutral entrepreneurship, in which all family members can participate, regardless of gender attribute, but with different degrees of involvement in a family business. A positive prerequisite for the development of family entrepreneurship in the Russian market is the active integration of women in business and the development of gender-oriented entrepreneurship (male and female) in Russia. Each of these types, with its own experience, specifics, and dynamics of development, has a positive impact on the spread of family entrepreneurship. Based on the method of content analysis the degree of actualization of public and scientific attention to the problem of family entrepreneurship, scientific approaches to its study and the potential for further spread are revealed. The findings of the content analysis are supplemented by the results of narrative interviews with entrepreneurs revealing the degree of their understanding of the essence of family entrepreneurship, its strengths, and weaknesses, risks, and potential, the attitude of entrepreneurs to increasing economic activity of family members. The assessment of factors hindering the development of family entrepreneurship, its prospects, and expectations of forms of government assistance to this process is provided. The social potential of family entrepreneurship, its educational role in the formation of children's labor skills, and intergenerational continuity of family business are emphasized in the research.

Keywords: family entrepreneurship, social resource, employment, the standard of living, gender neutrality, new gender order, social functions

[1] Corresponding author: agapova.ulyana2016@yandex.ru

1 Introduction

Family entrepreneurship is a relatively new subject of sociological study, which is increasingly attracting the attention of sociologists, economists, psychologists, and lawyers in the context of the development of an inclusive economy and the formation of a new gender order in the world. On the one hand, it is traditionally an economic activity, representing an important segment of the economy, related to the development of households and their contribution to the social and economic development of regions, to the improvement of the material situation of families, and to the reduction of poverty. Economists pay consistent attention to this area, tracking the real contribution of small and medium-sized enterprises to the economy. According to Rosstat, it increased to 32% in 2020 [1]. The rate in Europe is 60%.

The novelty of the author's presentation of the problem is in the study of family entrepreneurship as a social phenomenon from the perspective of sociological, specific socio-gender approach with the support of the theory of gender sociology as a private sociological theory [2, 3], the theory of gender equality and gender neutrality, on the one hand, and the sociology of entrepreneurship [4] as a field of economic sociology [5] and general sociological laws on the other. The category of family entrepreneurship is based on the indicator of "household", which is understood as relatives living together (related and not related by filiation), jointly providing themselves with everything necessary for life, combining and spending their funds. The family as a small group and workable collective, uniting representatives of two gender communities (male and female), of different ages and social experience, is considered as the subject of the action. These gender groups may differ in dynamics and style of economic behavior but are united by common goals and objectives. The goal is to organize a successful business for the well-being of the whole family. The common interest is the quality results of collective work, which can bring income and improve the quality of life of all family members. From the perspective of the social gender approach family entrepreneurship can be studied in four trajectories: as a social phenomenon, as a type of collective labor activity, as a type of group economic behavior, and as a form of child labor education. Both objective and subjective factors contribute to the expansion of family entrepreneurship [6]. Among the objective factors are extreme economic conditions created under the influence of the Covid infection that require more flexible forms of business, and reliability and trust of partners. Among the subjective are the positive experience of integrating women in business, accumulated over the previous decades.

Thus, if, according to the first survey of public opinion of women [7], only 1% of women participated in entrepreneurship in 1991, then by 2015 the percentage was 27% [8] and in 2019 it reached 46% for the first time. In total, 5 million 602 thousand 738 Russian entrepreneurs (men and women) are involved in entrepreneurship, and this community is gender asymmetrical with 54% of men.

The experience and skills of spontaneously formed youth and children's entrepreneurship as new social phenomena of Russian reality can be considered an additional factor contributing to the development of family entrepreneurship. Thus, in 2019, the most active entrepreneurial groups were men and women aged 18 to 24 years [9]. At the same time, the restraining factor for family entrepreneurship is objectively the lack of legal interpretation of this category and of relevant legal regulators of the family business in the legislation of the Russian Federation. Whereas the main emphasis of the Federal Law on self-employment is made on gender-equal participation of men and women in entrepreneurial activity [10]. There is no established position on the conceptual definition of the "family entrepreneurship" phenomenon in the modern scientific literature [11]. However, women over the years of market economy development In Russia have actively mastered entrepreneurship as a new type of economic activity and successfully advanced in it. The average age of a Russian woman entrepreneur is 38 years old. In the EU countries, the age is 35 years old [12]. Social functions of family entrepreneurship are economic (through specific forms of participation of family members in the development of the economy), monetary (through profit and its distribution), production (depending on the direction of activity), educational (related to the formation of children independence and valuable attitude to work), teaching (the formation of professional skills entrepreneurial activity), management, and communication. In the conditions of the formation of a new gender order as part of the new social world order, family entrepreneurship will play an increasing role as a type of gender-neutral entrepreneurship, which expands the boundaries of gender equality in the economy.

2 Materials and methods

Problematic situation lies in a social contradiction that has developed between the interest of families in ensuring their employment, increasing their income and quality of life, on the one hand, and the lack of necessary support from the state, hindering the inclusion of families in entrepreneurial activities, as well as of knowledge of social practices of family entrepreneurship, on the other. The social contradiction is exacerbated by the fact that the legal acts of the Russian Federation still lack the definition of "family entrepreneurship".

A combination of three methods for collecting primary sociological information including the traditional method of analysis of documents (statistical, legislative, scientific), the content analysis of publications (scientific and newspaper) of the past 10 years (2010-2020), and narrative interviews with entrepreneurs of the metropolis was used to study the problem. Taking into account that there is practically no sociological research of family entrepreneurship essence an attempt was made to determine the frequency of reflection of this issue in scientific publications and newspaper articles of the last 10 years (2010-2020).

To what extent does the topic of family entrepreneurship enjoy the attention of the media and finds its reflection in it? The authors turn to the results of the publications dynamics analysis, which are typologized into three types of groups: scientific theses, newspaper articles, and scientific publications (Fig. 1).

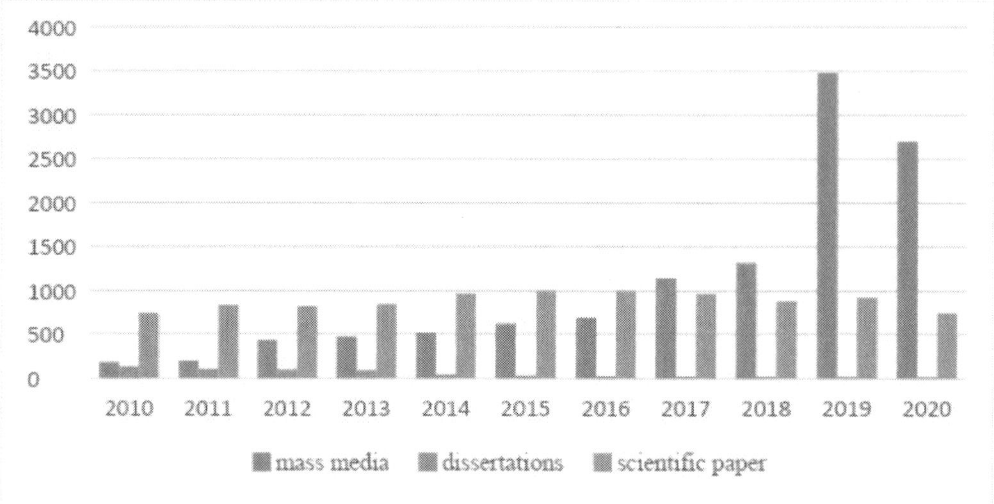

Fig. 1. Dynamics of the number of publications on family entrepreneurship in the media in 2010-2020.

The analysis of *defended theses* devoted to the study of family entrepreneurship has shown that there are very few of them. The minimum that was defended refers mainly to the 2010-2015 period as an inertial continuation of the period of active interest in this topic, which is typical for the 90s.

As for *newspaper articles* in the media as channels of operational information, family entrepreneurship is a subject of consistent interest of journalists. Therefore, the number of articles on the subject grows year by year (from 181 in 2010 to 3,478 in 2019) and declines to 2,697 by the end of 2020. The drop is understandable, given the introduction of the pandemic isolation regime and the reduction in contacts between journalists and real entrepreneurs.

As for *scientific articles*, the trend has been relatively stable over all 10 years: from 738 in 2010 to 743 in 2020. This suggests that the interest in the problems of family entrepreneurship is stable and not conjectural among scientists.

The change in the dynamics of attention to the studied subject of analysis suggests that entrepreneurship [13] and specifically family entrepreneurship is of steady interest to both journalists and scientists. This interest will persist in the coming years. The catalyst for the sustainability of such attention was the May Decrees of the President of the Russian Federation (2012), which defined the prospects for the country's development until 2024 and the place of entrepreneurship development in them. Thus, the decrees set the task of creating conditions for their comfortable functioning and development, such as tax benefits, a simplified system of lending and reporting, as well as the preferential procedure for leasing premises and equipment, and much more. In general, it can be noted that the measures of entrepreneurship support in Russia peaked in 2015-2016. Benefits for family entrepreneurship are legislated, and this will support both the development of family entrepreneurship and entrepreneurship in general.

The second task of the content analysis is to determine the prevalence of entrepreneurial topics of publications depending on the scientific branches of knowledge. Ten branches of science were analyzed ranging from linguistics to economics and business (Fig. 2).

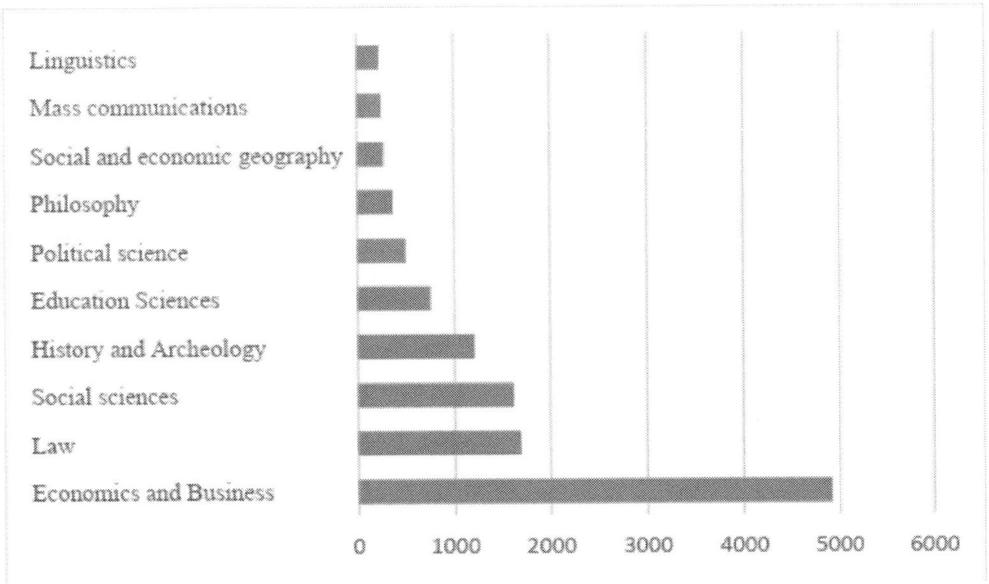

Fig. 2. Areas of scientific knowledge showing interest in the topic of family entrepreneurship. *Source:* Compiled by the authors.

According to the bar chart, the economics and business industry has the greatest number of publications on entrepreneurship (4,916). This confirms the prevalence of the economics approach to family entrepreneurship research. The second place (1,689) is taken by the field of legal sciences. Sociological sciences are in third place (1,616). These are the three leading fields of scientific knowledge in which the problems of entrepreneurship, including family entrepreneurship, are most actively studied. Least of all this problem is raised in such branches as linguistics (232) and mass communications (251).

Four other fields of knowledge that show attention to the problems of entrepreneurship have been analyzed: history, archaeology, education, and political science. The interest of these sciences in the entrepreneurial topic is most likely due to the development of markets, the functioning of which is inextricably linked to entrepreneurship in its various forms and volumes.

As for the attention to entrepreneurship from sociological science, it simultaneously reflects the prevalence of research in the field of sociology of entrepreneurship [14] and the expansion of interest to family entrepreneurship specifically. However, it should be recognized that the vast majority of the analyzed publications are devoted to entrepreneurship in general as a type of economic activity and economic behavior. As for family entrepreneurship, they account for a minimal share of research so far only emphasizing the relevance of the scientific demand for sociological study.

3 Results

The purpose of the study conducted by narrative interview of entrepreneurs of the capital megalopolis [15] is to determine their ideas about the essence, opportunities, and reserves of family entrepreneurship. Five social indicators were measured: understanding of the essence of family entrepreneurship, the degree of its development, involvement of friends and acquaintances in family entrepreneurship, assessment of measures required from the state for the active development of family entrepreneurship, and its impact on the lives and future plans of entrepreneurs themselves.

1) What are the entrepreneurs' ideas about *the meaning of "family entrepreneurship"?* The positions vary:

"It is a business that involves all family members and close relatives. It mostly fits family dynasties, where, for example, there are dynasties of doctors".

"A business where the core capital, the main investment comes from family members replacing the top management positions, and where everyone is interested in making money and getting the business on its feet".

"...People who are also in an informal relationship, that is a husband and wife who do business together, being each other's business partner. ...It is when the people closest to each other do business together. It involves the first circle of people – parents and children".

"In family entrepreneurship, the concept of the conventional company is greatly altered and participants in the top management or board of directors are family members".

"It's doing business together, involving several family members, who are in charge of the business and make decisions collectively".

In other words, the essence of "family entrepreneurship" as understood by entrepreneurs is that it is a business in which all family members, close relatives, participate, and where the capital base, the major investments come from all family members, and decisions are made on the basis of collective opinion. These are quite competent positions in ideas about the essence and features of family entrepreneurship.

The second indicator is the degree of development of family entrepreneurship. A number of interviewees emphasized agriculture as an important branch of family entrepreneurship development:

"agriculture is a very developed sphere of family entrepreneurship because this industry is quite capacious. Despite the education, each family member can find a use for himself. Farming is a high-profile economic practice where specialists in different areas are needed. It includes economics, law, construction, mechanics, cattle breeding – you need to understand everything".

"The development of family entrepreneurship is not particularly different from entrepreneurship in general, because there is no particular difference in whether you are doing business on your own, with a friend, or with your family. When you decide to do business with the people closest to you, there's no question of trust. There's more loyalty, more hope that everyone will really and consciously do their job. Overall, the prospects are great. Especially in difficult times, when you create your own business and prefer reliable, trusted people who can also take some part of the responsibility".

"The development of family entrepreneurship in our country is still at a low level because young people in Russia are strange – they try to live as interesting as possible, but go against the opinion of their elders; the only exception, perhaps, is the youth of the Caucasus region". "Family entrepreneurship is relevant, but it's not properly developed".

"Entrepreneurship is not in the best condition today. If we talk about family entrepreneurship, I consider a large part of family entrepreneurs as fictitious, shares of many organizations sit on the "family shoulders", while serious managerial decisions are made by one or two people".

Respondents' opinions were divided. While recognizing the importance of family entrepreneurship, more than half note negative trends in its development, stagnation, and its low level. *The third* indicator notes the influence of family entrepreneurship on the life and activity of the entrepreneur. The results of the study built a pyramid of opportunities for family entrepreneurship (Fig. 3), according to which the three most important and stimulating of the seven measured are the opportunity for self-development and the opportunity to work together with the family. In other words, self-development, trust, and family improvement.

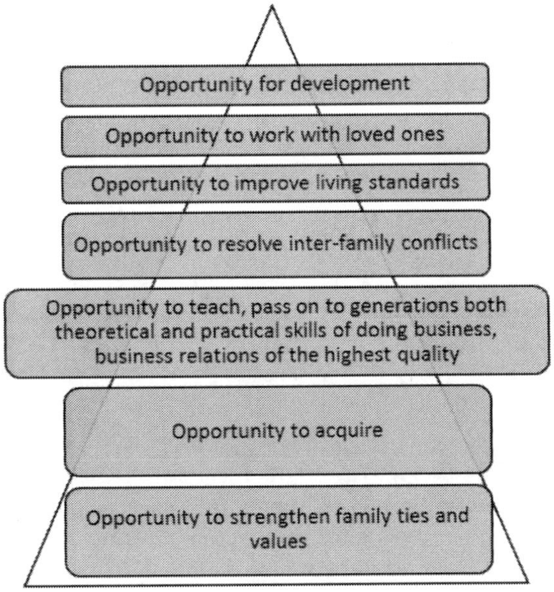

Fig. 3. Pyramid of family business opportunities. *Source*: Compiled by the authors.

The fourth indicator: pros and cons of family business (Table 1).

Table 1. Pros and cons of family entrepreneurship.

Pros	Cons
Complete trust; interest; all family members devote themselves fully and give even more than they can; you won't be cheated; there are no and can't be any financial leaks – this is a big advantage.	Human Factor. Running a successful family business requires that all members have an equal stake in it. It takes family members to learn how to interact not only on family matters but on many other things as well (which takes time). While the adjustment period lasts, conflicts may arise

The diagnosis of positive and negative aspects of family entrepreneurship reflects mainly their subjective nature, depending on the personal qualities of the entrepreneur and his family.

The fifth indicator – in which areas of the economy may the development of family entrepreneurship be most promising? According to the study, those are handicraft production, agriculture, catering and medicine, floristry, sports. These sectors allow the development of the so-called "spot" (small) business (pharmacies, flower stores, small gyms, bakeries, nail salons, etc.).

What do entrepreneurs see as the risks of family entrepreneurship? Let us refer to Table 2.

Table 2. Risks of family entrepreneurship and measures to neutralize them.

Risks and obstacles to development	Measures of overcoming

Excessive control of the market by the state Tough competition between small business and big companies People's mentality High level of taxation	Concessional lending Organization of special training events Reduce state control

According to the study, the main risks for small businesses are connected with the fact that there are no established (even if unspoken) rules for fair business practices in Russia. Ferocious "no rules" competition is the most dangerous risk factor, making it impossible for both entrepreneurship in general and family business, in particular, to develop properly. Such competition is supplemented by unfavorable taxation conditions, excessive control by the state, and insufficient resources.

This means that family entrepreneurship is forced to exist and develop under the same conditions as an ordinary business due to the lack of its legal formalization, which is very unprofitable for family entrepreneurship. Therefore, the legal stipulation of the status of family entrepreneurship in the Russian Federation will create more favorable conditions for its qualitative development and will promote the formation of family entrepreneurship culture in Russia.

For this purpose entrepreneurs propose a number of measures on the part of the state that would promote active development of family business, such as state participation in regulation and legislative definition of "family entrepreneurship", reduction of various tax payments, and revision of family business taxation system in general with its specific features being considered. Specific suggestions to the state made by entrepreneurs are as follows:

"To reconsider the tax component of family entrepreneurship"

"To regulate, help and develop any sphere of activity from the standpoint of legislation, and then control family entrepreneurship in compliance with the established norms"

"To introduce a general tax for the family. What is happening now? – The owner pays taxes, then, if family members are registered as employees, they also pay income tax. So it turns out that they pay tax from the general business, plus each of them pays individually. This is unfair".

"To introduce certain privileges for family entrepreneurship, because family entrepreneurship is a good idea, but the giant companies in the market "devour" everything. Look at what Yandex is doing? And Sberbank? This is a pure monopoly".

"To oblige large market players to act as investors in family projects in order to help implement them, but not destroy them".

"To introduce a scheme under which maternity fund could be invested in your business".

"To introduce economic programs, to change the law of inheritance, so that there would not be two owners of family businesses; to educate heirs in family entrepreneurship, to develop approaches to the economic management of family businesses, to introduce training programs".

These and many other suggestions expressed reveal the need for effective programs, projects, educational directions to support and develop family entrepreneurship. It is indicative that the majority of respondents consider their business to be a family business and feel much unprotected.

Entrepreneurs' proposals indicate an urgent need for a specific law on family entrepreneurship in Russia that would clarify its procedure for development and status.

4 Discussion

The development of family entrepreneurship in the context of the development of an inclusive economy and the formation of new gender order raises many new questions about the integration of family as a small consolidated group into the development of entrepreneurship and the economy as a whole. On the one hand, family members united by common interest and trust in each other are a reliable starting point for the development of "family entrepreneurship". On the other hand, participation in a risky financial and organizational enterprise requires serious educational and practical preparation of family members, especially the parents, to ensure the organizational and financial management of the family business. How do the entrepreneurs assess the pros and cons of family entrepreneurship (see Table 3)?

Table 3. Pros and cons of family entrepreneurship according to entrepreneurs.

Pros	Cons
"Total trust, personal interest; all family entrepreneurship participants are fully committed to the joint business in all possible ways; there is no cheating and no financial leakage – which is a big advantage"	"Great dependence on the human factor. Running a family business successfully requires all members to have the same interest in it. Family members must learn to interact beyond family matters (which takes time). Family conflicts can arise during the adaptation period"

Obviously, both positive and negative aspects of family entrepreneurship identified in the study belong to the group of subjective factors. Their influence depends primarily on the personal qualities of the entrepreneur and his family members. In this context, the question arises about the family's choice of certain sectors of the economy in which, in their opinion, family entrepreneurship is most promising. According to the results of the narrative interview, such spheres are small businesses of handicraft production, agriculture, catering, beauty and health. i.e., which provide an opportunity to develop a "spot business". Examples are pharmacies, flower stores, small gyms, bakeries, nail salons.

Stiff competition, unfavorable taxation conditions, excessive control by the state, lack of resources, blurred rules of the "game" in the organization of family business are largely due to the lack of a well-thought-out legal framework for the development of family entrepreneurship in Russia. This is why addressing this issue can be regarded as a prerequisite for expanding the entrepreneurial involvement of the family institution in the development of the national economy.

Based on entrepreneurial estimations, legislative consolidation of the family entrepreneurship status can also open new opportunities for its expansion on the scale of regional economies and contribute to the formation of a "business culture", a "culture of entrepreneurship" as a new social norm in Russia.

5 Conclusion

The results of the study confirm the conceptual validity of the author's idea of the project. The sociological approach to the study of family entrepreneurship from the perspective of gender sociology and the theory of gender equality in the conditions of a new gender order formation significantly deepens the analysis of family entrepreneurship in society, where women are a stable majority in all spheres of life.

Family entrepreneurship, on the one hand, expands economic opportunities for integrating women into entrepreneurship without "leaving the family". On the other hand, it contributes to the restoration of "generational entrepreneurship" in Russia. In contrast to developed Western countries, post-Soviet society has formed "first-generation" family entrepreneurship, compared, for example, with the fourth or even fifth generation of family entrepreneurship in the U.S. In this regard, the creation in Russia of a legislative platform as a pillar of its development is extremely important.

The conducted study belongs to the exploratory type of research, which can serve as a reference point for further comparative studies, the results of which can be compared with the obtained pilot empirical base and constructively develop further scientific search.

The authors are grateful to the Russian Foundation for Basic Research, which supported the sociological research within the framework of the scientific project No. 20-011-00519 A "Challenges and risks of formation of a new gender order and the social mechanism of its management in Russia".

References

1. Maloe i srednee predprinimatelstvo v Rossii [Small and medium-sized enterprises in Russia]. Accessed on: October 2, 2021. [Online]. Available: https://rosstat.gov.ru/folder/210/document/13223
2. G.G. Sillaste, Sci. Cult. Soc, **4**, 108-123 (2016)
3. G.G. Sillaste, Gendernaya sotsiologiya i rossiiskaya deistvitelnost [Gender Sociology and Russian Reality]. (Infra-M, Moscow, 2012)
4. A.Y. Chepurenko, Sotsiologiya predprinimatelstva [Sociology of entrepreneurship]. (Publishing House of the Higher School of Economics, Moscow, 2007)
5. Y.V. Veselov, et al., Economic sociology: theory and history. (Publishing house of St. Petersburg University, 2012)
6. Zakon Gruzii. O predprinimatelyakh [The Law of Georgia. On Entrepreneurs]. Accessed on: October 2, 2021. [Online]. Available: https://matsne.gov.ge/en/document/download/28408/49/ru/pdf
7. G.G. Sillaste, Women and Democratization: Women's Public Opinion on Current Socio-Political Issues. (Moscow, 1991)
8. Indeks zhenskoi predprinimatelskoi aktivnosti (WBI), 4 volna [Women's Entrepreneurial Activity Index (WBI), 4th wave]. Accessed on: October 2, 2021. [Online]. Available: https://nafi.ru/projects/predprinimatelstvo/indeks-zhenskoy-predprinimatelskoy-aktivnosti-wbi-4-volna/
9. O.R. Verkhovskaya, et al., Natsionalnyi otchet. Globalnyi monitoring predprinimatelstva. Rossiya 2019/2020 [National report. Global monitoring entrepreneurship. Russia 2019/2020]. Accessed on: October 2, 2021. [Online]. Available: https://gsom.spbu.ru/images/cms/menu/otchet_2020-red-2n-04-06.pdf
10. Soyuz Sovetskikh Sotsialisticheskikh Respublik. Zakon ob individualnoi trudovoi deyatelnosti [Union of Soviet Socialist Republics. The Law on Self-Employment]. Accessed on: October 2, 2021. [Online]. Available: https://docs.cntd.ru/document/9024308
11. T.G. Kasyanenko, A.M. Vorotilkina. *Semeinyi biznes: opredelenie ponyatiya i klassifikatsiya* [Family business: definition of the concept and classification], in

"Culture. The science. Education: modern trends", Ch. 2 Science, 175-188. (Sreda, Cheboksary, 2020)

12. Kratkosrochnye initsiativy dlya dostizheniya dolgosrochnykh tselei. Mezhdunarodnoe issledovanie semeinogo biznesa — 2019 [Short-term initiatives to achieve long-term goals. International study of family business – 2019]. Accessed on: October 2, 2021. [Online]. Available: https://www2.deloitte.com/content/dam/Deloitte/ru/Documents/tax/family-business-2019.pdf
13. M. Munkert, et al., Founding a company: Handbook of legal forms in Europe. (Springer, 2015)
14. R. Goffee, R. Scase (eds), Entrepreneurship in Europe (Routledge Revivals). The social process. (Routledge, 2016)
15. R. Lisowska, Adv Res Entrep, **4(3)**, 7-9 (2016). https://doi.org/10.15678/EBER.2016.040307

The specificity of adaptation of future engineers to pandemic conditions on labor market: a gender perspective

Ekaterina Igorevna Makarenko[1*], *Irina Gennadievna* Prokhorova[2], and *Asem Hekimoglu* Naushabaeva[3]

[1]Moscow Automobile and Road Construction State Technical University (MADI), Department of Sociology and Management, Moscow,
[2]Financial University under the Government of the Russian Federation, Faculty of Social Sciences and Mass Communications, Moscow, Russia
[3]Bitlis Eren University, Faculty of Economics and Administrative Sciences, Bitlis, Turkey

Abstract. The paper is based on the advance studies of the adaptation of MADI students, future engineers, to the conditions of a forced pandemic associated with the COVID-19 virus. At the first stage of the study (spring 2020), using the content analysis method, the main directions of adaptation and the patterns of future engineers were identified: educational activity, social responsibility, social communication, financial constraints that arose during the period of a sharp change in learning activity, and also the specificity of changes in the place and living conditions, including moving to another region of Russia. At the second stage (spring 2021), using the semi-formalized interview method, a survey of students of the same technical university was conducted about the difficulties faced by boys and girls at the beginning of the pandemic, based on the previously identified directions of adaptation. The results of the study have shown a gender difference in understanding the difficulties of forced adaptation to the conditions of a modern pandemic: girls adapted better to school. The specificity of gender adaptation will help to predict the behavior of graduates of technical universities on labor market in the future.
Keywords: gender, gender sociology, economic sociology, future engineers, adaptation, pandemic, distance learning, labor market

1 Introduction

Engineering education at all stages of its development plays an extremely important role for the economic development of the country, since the production base and national security of states depend on the quality of training highly qualified technical personnel. Klyucharev rightly notes that the field of engineering education plays an important role in the development of the national economy [1]. Technical education has its own peculiarities when training experts: it requires more practical skills, a serious laboratory base and

* Corresponding author: makarenko_madi@mail.ru

opportunities for students to directly conduct experiments and master complex fundamental disciplines. In this paper, the object of research is students of technical universities studying under the specialty training program, future engineers; they are representatives of the Russian technical intelligentsia.

Recently, the concept of gender has been increasingly used in the Russian socio-political and scientific field. By itself, this term is not new, but it has a different interpretation from the point of view of Russian and foreign theoretical positions. Thus, Professor Sillaste, the head of the national scientific school "Gender and Economic Sociology" believes that this term can be identified as the category of "gender" and he notes that "the social content of the concept of "gender" is immanently inherent in Russian sociological science" [2]. The modern Western scientific approach often divides these concepts: "gender" is of a biological, physiological nature, and the category of "gender" is social and cultural: Giddens and Sutton believe that the main social understanding of gender lies in its construction: "gender is an unstable social category that can accommodate many variations and can change quite radical" [3]. In particular, scientists believe that once you can change your body, you can also change the socio-cultural grounds for its formation. The Western understanding of gender refers us more to the sociobiological and psychological approaches. Whereas the authors of the paper will adhere to the domestic scientific tradition, considering the specificity of the adaptation of future engineers to the conditions of a pandemic from the point of view of the socio-gender approach, which highlights precisely the social aspects in this complex process.

Turning to the problems of social adaptation in general, it is worth highlighting that it can be voluntary and forced. Voluntary means the direct desire of an individual or a group to change their lifestyle, place of residence, status, and other social indicators. Forced adaptation occurs under the influence of certain conditions that force various societies to leave their usual way of life. We attribute the pandemic (COVID-19) to a forced form of adaptation. The population of the whole world and citizens of many countries find themselves in this difficult situation. In particular, such a social group as students. Russian students, who numbered 4.068.3 thousand people by the beginning of the pandemic (spring 2020) [4], also faced the problem of forced adaptation to the abrupt transition to distance learning. This also affected students of technical universities, who have a long tradition of gender asymmetry towards men. Similar processes took place not only in MADI. This problem is also characteristic of the Financial University. The results of the study are reflected in the article "Gender Equality in Science: Achievements and Problems" [5].

2 Methods

The authors have conducted an advanced sociological study on this issue. It was carried out in 2 stages. At the first stage (April-May 2020), 45 student essays were collected by sending an email to one of the authors. Then, 36 works were analyzed by the content analysis method (9 were rejected). These were the reflections of students of the Moscow Automobile and Road Construction State Technical University (MADI), studying under the specialty programs, i.e. future engineers. The units of analysis were the following areas of adaptation: (highlighted in order of importance for students)

1. Learning activity
2. Social responsibility
3. Changes in communication
4. Financial and other issues
5. Place and conditions of residence.

Stage 2: it was conducted during the sociology classes by one of the authors of the paper.

A year later, in April-May 2021, the method of semi-formalized interviews was used with MADI students (N = 140; among them were: 104 men and 36 women). The respondents were students of the faculties: design and mechanical, road construction, road transport and management of MADI.

3 Results

The first stage of the signal research has shown the expected priority of students in adaptation: *learning activity*. In this regard, in the context of a "sudden" pandemic, students were primarily concerned with the following issues: mastering complex technical disciplines; the impossibility for some students of access to the Internet, the unwillingness of many teachers to work within the framework of distance learning [6].

Social responsibility was associated with the following indicators:

- organization of medical care, work of doctors with coronavirus patients;
- wearing masks and using protective measures;
- volunteer movement among young people, which began to gain momentum at the beginning of the pandemic.

The peculiarities of social communication in these conditions include the fact that it has almost completely moved into virtual space. *In financial matters*, students turned to new forms of earnings – food delivery and distance working. However, it is worth noting that most of the students in their essays turned to the help of their parents and relatives. It is also necessary to indicate that such a parameter as "changing the place and living conditions" was associated, first, with the departure from the hostel in Moscow. At stage 1, the gender preferences of students were not analyzed.

At the second stage, during a semi-formalized interview, students were asked to assess the importance of the problems that young people face when adapting to a pandemic, as well as to identify it. Immediately, we note that in a year we can state the same priority of adaptation directions for future engineers as it was in 2020. However, the gender preferences of future engineers in terms of adaptation were analyzed.

Table. 1 Preferences in the directions of adaptation to the conditions of the COVID-19 pandemic by gender (in %).

No.	Indicator	Men	Women
1	Learning activity	75%	95%
2	Social responsibility	81%	70%
3	Social communication	90%	65%
4	Financial constraints	95%	52%
5	Conditions and place of residence	75%	90%

What do these data say?

In the educational activity, the adaptation of women was more successful. We believe this is due to the fact that women enter universities for technical specialties much more consciously than young men, so their motivation works to achieve success. In turn, one of the authors of the paper, who has been working at a technical university for many years, can confirm this feature. Kolesnikova and Kudenko also argue about this examining the processes of vocational guidance in terms of gender [7]. Reflecting about the specificity of the transition of Russian students to the distance learning in recent years, Zborovsky and Abramova consider it as an increase in the number of student failures and a decline in the overall quality of education [8]. It is interesting that in order to improve the quality, foreign scientists (in particular, Tsiatsos Th. from Greece) believe that it is necessary to use gamification to support engineering education [9].

In the article "The importance of constructivism in the teaching social and humanitarian disciplines" [10], these problems were also confirmed based on the results of the study of using new, constructivist elements in pedagogical activity. The need to introduce new elements of constructivism into learning activities is determined by the peculiarities of modern social life, in which virtual digital reality and IT technologies play the main role in the formation of everyday reality.

Let us note that in percentage terms, the problems of *social responsibility* are of almost equal importance for both girls and boys. Gender parity in this regard is extremely important and testifies to the opportunities for young people, regardless of gender, to express themselves in society. This positive trend reflects the tradition of domestic higher technical education – training a broad educated one who sees and analyzes social problems, who understands what technology and weapons are created for [11]. Modern Western experts are in solidarity with this position of teaching engineering disciplines in Russia. Thus, Zandvoort (Netherlands) directly points out that technical education should be based on such principles as humanitarian and ethical responsibility, since they determine the "social face" of an engineer; and modern engineers who want to be realized in their profession should be ready to build "social design" and the ability to carry out their ideas and mission in public policy [12]. Castro and Sancristobal believe that the training of an engineer in the current context is accompanied by a social understanding of the essence of engineering [13].

Turning to such a parameter as *social communication*, we understand that we are dealing with a new generation (Z as it is often called), which uses information technologies in a completely new way than representatives of older ages. "Modern children communicate and learn "online". In addition to the advantages, there are obvious problems here: there is less human communication, they are poorly versed in human emotions and behavior. The Internet comes to the fore as an authority among children, not parents. Virtual communication prevails over real: people communicate more on the Internet, but communicate less in real life" – this is how this generation characterizes… [14] The modern English scientist Mason calls it even the "network generation" and believes that young people, having access to countless information resources, develop completely new values, rules and norms of behavior [15]. It can be seen from the above table that men adapted better under conditions of limited real communication. Young women want more communication with their peers outside of information devices.

Considering such an indicator of adaptation as *financial constraints*, we expect an increased interest in these issues among male students. Domestic gender stereotypes that a man "bears the burden of responsibility" for financial support is also manifested at the level of students. In addition, during the pandemic, men had more opportunities to earn money. During the interview, the guys were happy to talk about how they were actively involved in the process of making money (unfortunately, sometimes to the detriment of their studies).

The place of residence. Most of the girls have left the hostel and returned home.

The change in living *conditions has affected the students living* in the hostel to a greater extent. In the spring of 2020, they were all forced to return home. For the girls, this opportunity seemed more attractive. They noted during the interview that communication with loved ones and cooking classes went to their benefit and joy. We also note that women were more disciplined – they "came to the sessions of classes" on time and passed exams better.

4 Conclusion

It should be noted that the study of students of technical universities in the issues of adaptation to the difficult conditions of the pandemic caused by the COVID-19 virus has revealed some difference in these problems in terms of gender. Girls have adapted better in terms of learning activities and its forced transition to online forms and to changes in the place and living conditions; men have found themselves in the implementation of social activity and solving question of finances.

References

1. G.A. Klyucharev, Soc Res, **3**, 51-59 (2020). https://doi.org/10.31857/S013216250008820-5
2. G.G. Sillaste, Zhenshchiny v mirovoi sotsiologii [Women in world sociology]. (Prometheus, Moscow, 2021)
3. E. Giddens, F. Sutton, Osnovnye ponyatiya v sotsiologii [Essential Concepts in Sociology]. (Higher School of Economics, Moscow, 2019)
4. Federalnaya sluzhba gosudarstvennoi statistiki [Federal State Statistics Service]. Accessed on: October 4, 2021. [Online]. Available: https://rosstat.gov.ru/
5. I.G. Prokhorova, L.A. Brushkova, Self-government, **3(125)**, 16-19 (2021)
6. E.I. Makarenko, Automobile. Road. Infrastructure, **2(24)** (2020)
7. E.M. Kolesnikova, I.A. Kudenko, Bul Peoples' Friendship Univ Rus, **21(2)**, 239-252 (2021)
8. G.E. Zborovsky, P.A. Ambarova, Soc Res, **3**, 17-27 (2021). https://doi.org/10.31857/S013216250012904-7
9. Th. Tsiatsos, Int J Eng Pedag, **10(2)** (2020). https://doi.org/10.3991/ijep.v10i2.13771
10. B. Leyla, et al., SHS Web Conf, **103**, 01025 (2021). https://doi.org/10.1051/shsconf/202110301025.
11. E.I. Makarenko, Regionology, **28(2)**, 322-349 (2020). https://doi.org/10.15507/2413-1407.111.028.202002.322-349
12. H. Zandvoort, *Engineering Education for a Sustainable, Just and Peaceful Society*, in TEK Sustainable Development Seminar. 19 November, 2009. (Helsinki, 2009). https://doi.org/10.13140/2.1.2754.3042
13. M. Castro, E. Sancristobal, Int J Eng Pedag, **10(1)**, (2020). https://doi.org/10.3991/ijep.v10i1.12927
14. D.O. Nikitina, Sociology, **3**, 136-140 (2021)
15. P. Mason, Postkapitalizm: rukovodstvo k nashemu budushchemu [Postcapitalism: a guide to our future]. (Ad Marginem Press, Moscow, 2016)

Capability of the Russian women in science: support required

Aleksandr Mikhailovich Egorychev[1], *Anna Gogievna* Akhtyan[1*], *Irina Aleksandrovna* Fedoseeva[2], *Svetlana Nikolaevna* Fomina[1], and *Aleksandr Maksimovich* Tyutchenko[3]

[1]Russian State Social University, Social Work Department, Moscow, Russia
[2]Novosibirsk State Pedagogical University, Department of Practical and Special Psychology, Novosibirsk, Russia
[3]Moscow City Pedagogical University, Department of Special Psychology and Psychosocial Technologies, Moscow, Russia

Abstract. The article considers the problems related to the development of the capability of Russian women in science, necessity to optimize the scientific capability of the students of a higher educational institution. The objective of the research: a scientific grounding of the focused support of the woman's scientific capability in Russia and the development of the perspective directions of its optimization. The methods of information collection: theoretical (analysis, synthesis, comparison, correlation and generalization of the author approaches in the scientific literature) and empiric (observation, conversation, interview, sociological poll) – regarding the state and perspectives of the development of the scientific capability of the Russian women. The conditions and reasons were revealed that do not allow realizing the capability of the Russian women in science completely; the conditions of optimization of the development of the scientific capability of the students of a higher educational institution were discovered taking into account the gender aspect; the conclusions were obtained that are related to the optimization of the woman capability in the scientific activity in all the fields of the country. The novelty of the research performed is the following: the historical and social role of a woman in science and education was shown, the necessity of development of the woman scientific capability in the country was grounded; the fields of scientific interests of the students of a higher educational institution were revealed and grounded taking into account the gender aspect; the possibilities for optimization of the scientific capability of the students were determined.
Keywords: woman scientific capability, vocational education system, gender aspect, Russian women

1 Introduction

* Corresponding author: ahtyananna@mail.ru

The present problem is urgent not only for Russia; it is typical for all countries in the world. Thus Khamdamova believes that "… there is lack of women-scientists is still observed in the world as well as the absence of women-researchers and scientists taking the high managerial positions" [1].

This problematics attracts national researchers: *Russian women-scientists: historical aspect* [2]; *gender aspects in science and education* [3, 4]; *women in science and education* [4-9]; *student science: gender aspects* [10, 11], self-development of women [12].

This problem attracts also the attention of the foreign authors [13-20].

Analysis of the scientific literature allows discovering the availability of certain nuances in the present problem:

- There is an interest in the gender problem and a woman role in the development of society all over the world;

- Solution of the problem of woman's capability support in science relates to the activity of the entire Russian educational system taking into account the gender approach.

Thus Enloe in his article "Gender Makes the World Go Round" discusses in detail and thoroughly the problems of the world's existence within the frameworks of the gender approach [15].

In their research Holman, Stuart-Fox and Hauser raise a clear question: how much time will it take before the women are equally represented in science? [19].

Burdukovskaya and Rvacheva notice: "The introduction of the category "gender" into the pedagogical process has a principal significance as in the higher educational institution many ideas of the professional identity are formed as well as of the life strategy, access to the resources and power based upon the socio-gender orientation" [21].

The history of Russia has many witnesses that show the significant contribution of women to the development of their Motherland. Let us take as an example those women scientists who contributed significantly to the development of Russian science (Romanskaya, Kovalevskaya, Sklodowska-Curie, Karavaeva, Zeigarnik, Oleinik, etc.). Even in the 18th century, Russian women were actively involved in scientific and educational activities. Ekaterina II supported the woman education in the country. Valkova notices: "Her Majesty and the high society following her looked quite favorably at the exercises of women in the sciences" [2].

The present tendency was supported in Russia after the 1917 October Revolution. The Council of People's Commissars of RSFSR prepared and issued a DECREE dated on August 02, 1918 "Of the rules of admittance into the higher educational institutions" that clearly stated: "Every person regardless his citizenship and gender who has reached the age of 16 years can join the listeners of any higher educational institution without producing the diploma, certificate of secondary education or any school-leaving certificate".

Pushkareva writes: "If until 1917 only 10% of Russian women had access to the scientific work, for a short period of fifty years a number of women-scientists grew 5 times in our country. … The process of feminization of science in our country in comparison with Europe and the USA was always progressive and was not limited, and on the contrary, it was encouraged all the time" [7].

Due to the positive historical processes by the beginning of the 21st century, a share of Russian women studying in the system of higher education was akin to the men and exceeded them. Thus according to the data of the All-Russia census 2010 (collected volume of Russian Federal State Statistics Service (Rosstat) "Men and women of Russia") – 58% women have higher education (15975 thousand people) and 42% men (11556 thousand people). Meanwhile, a number of women engaged in the scientific fields was significantly lower in comparison with men. In Russia according to the data of Rosstat 136.4 thousands of women-researchers work in the field of research and development, which is 39.2 of the total number of researchers [22].

Nevertheless, in Russian science there is a bigger number of women in comparison with other countries – about 50%; in Japan, they are 16%, in Great Britain – 23%, in Germany – 20%, in the USA – 36%, in France – 28% [6]. If in 1991 a number of women in science was 51%, by 2000 it was 63% of the total number of scientific researchers [5].

There are so-called "dark sports" regarding the carrier growth of a woman in science. Women are weakly represented in the management bodies of science and higher educational institutions of the country, they rarely are at the head of the large research groups, there are only 19% women among the heads of the laboratories, and among the deputy directors – 4%, and among the heads – 2% [7].

2 Materials and methods

Objective of the research: the grounding of the necessity to support the capability of the Russian women in science and development of the perspective directions of its optimization. Tasks: the grounding of the necessity to develop the women's scientific capability in the country; revealing of the sphere of scientific interests of the students of higher education institutions taking into account the gender component. Research methods: *theoretical and empirical*. Research tools included: a questionnaire (12 questions) for polling of the students of the Russian State Social University – 14 faculties. *Sampling procedure*: age of interviewed students is 18-25 years; sampling population: ± 894 respondents, including 484 – girls, 410 – young men. Time of research: 04.04.-21.12.2020.

3 Results

3.1 Theoretical analysis of the research problem

The reasons were revealed that do not allow to implement the capability of the Russian women in science completely:

- The standard historical socio-cultural tradition related to the positive attitude to the male status and its activity in the affairs of state and society, belittling of the woman status, giving her the secondary roles in all social spheres.
- The main concept of the woman's scientific capability is a socio-humanistic sphere and to a lesser extent – physico-mathematical, natural scientific, technical directions. The reason is the stereotype established in the Russian consciousness- a woman has a quite different set of mind that does not correspond to the exact sciences.
- Absence of enough time for women as it is required by the scientific type of activity, their involvement in the problems of creation, maintenance and preservation of the family (household worries, birth of children, their upbringing, etc.).
- Absence of proper upbringing in the higher education system regarding the working out of the sustainable motivation of the students to the scientific activity taking into account the gender component.

3.2 Empiric analysis of the research problem

The information was collected that characterized the students' scientific capability at a higher educational institution taking into account the gender aspect. We shall represent the results obtained on a selective basis and comment them in brief.

Question: What do you think, does the educational process at the university promote the formation of researcher skills among the students?

A half of all respondents (54.8%) believe that the educational process at the university promotes the formation of researcher skills among students; 19.2% do not agree; 26.0% have no answer. The distribution of the responses of the young men and girls does not have many deviations according to the gender, they are equal.

Question: How do you estimate the level of your involvement in the research work of the university?

The estimation was not at the high level: only 0.8% of the respondents estimated it as "quite high"; 13.9% estimated it as "high"; one third (37.7%) estimated their participation at the "medium" level; 19% estimated it at the "low" level; 13.9% – at the "very low" level; 11.7% had no answer.

Question: What could increase your motivation for the research activity?

The students evaluated rather critically their mechanisms of development of motivation to active participation in the research work. Surprisingly one-third of students (31.3%) was undecided. The "introduction of the material reward" was rated first (26%); and the "active and accessible organization of research work on the Chairs of the faculty" was rated second (16.2%); the "improvement of the informational and material support of the research activity" was rated third (12%).

The obtained results show that there is no certain gender domination in the responses of the young men and girls, they have almost equal values according to the choice of reasons interfering with the involvement in the research activity.

4 Discussion

The analysis showed the multivalued position of the national scientists regarding the problem of development and support of the woman's scientific capability in Russia. Meanwhile, the percent of involvement of Russian women in science and education gets ahead of the values of many developed countries of the world. The tendency related to the involvement of the Russian women in the scientific and educational activity arisen in the 18th century continued during the whole 19th century and expressed brightly in the 20th century (Soviet period), and was supported in the new time (beginning of the 20th century).

Following the results of the performed *theoretical analysis* the certain conclusions can be made:

- Possibility of joining the world of science and education by a Russian woman happened much earlier in Russia than in other developed countries of the world and it was more efficient;
- Despite the maintaining of the historical positive tendency a particular "imbalance" is observed in the country in favour of men (minimal number of the Russian women-scientists who joined the higher spheres of the scientific hierarchy);
- The scientific capability of a Russian woman is not realized completely and this requires the creation of the corresponding conditions, development of the socio-cultural mechanisms of its optimization and support.

Empiric method of research allowed revealing the peculiarities of the organization of the student science in the higher educational institution, its advantages and disadvantages.

The advantages include:

- most of the students in the higher educational institution (84,8%) are ready for active participation in the student scientific activity when satisfying all necessary conditions.
- the results of the poll showed the gender unit of the students in the sphere of motivation to the scientific activity.

The disadvantages include:

- low level of motivation of students of both genders to the participation in the scientific activity of the higher educational institution. In terms of amount (0.8%+13.9%=14.7%) estimate their participation in the student science as "quite high" (0.8) and "high" (13.9), and 37.7% estimate is as "medium", 13.9% as "low";
- a disturbing indicator characterizing "disengage" or "low level of understanding" of the role of science by the students, one-third of respondents (31.3%) could not formulate their attitude to it.

5 Conclusion

The results of the research performed allow us to formulate the following conclusions:

1. The woman's scientific capability serves as the most important component of the innovative resource of the country; its optimization has a direct impact on the development of the country, all its social institutions and structures, democratization of the Russian society.
2. The support and development of the woman's scientific capability is connected to the history of the development of our state and society, reflects its best traditions, senses and values of culture.
3. The Russian system of higher vocational education serves the most important factor and condition of formation of the sustainable motivation of students to be engaged in research activity taking into account the gender component.

References

1. F. Khamdamova, Sci Rev, **10(27)**, 29–32 (2019). https://doi.org/10.31435/rsglobal_sr/30122019/6865
2. O.A. Valkova, Woman Rus Soc, **1**, 89-98 (2018)
3. A.G. Allakhverdyan, *Izmeneniya v strukture personala sotsialnykh nauk: gendernyi analiz* [Changes in the structure of staff of social sciences: gender analysis], in O.A. Valkova (ed), Materials of the International scientific conference "To the Century anniversary of the Great Russian revolution: women faces of the Russian science – inheritance", November 07-08, 2017, Moscow, 372-378. (Publishing House: "Yanua-K", Moscow, 2017)
4. S.M. Vinogradova, Yu.G. Dunaeva, D.F. Ziatdinov, Bul St. Petersburg Univ. Polit Sci. Int Rel, **10(3)**, 260-268 (2017). https://doi.org/10.21638/11701/spbu06.2017.306
5. E.Z. Mirskaya, E.A. Martynova, Bul RAS, **63(8)**, 693-700 (1993)
6. I.S. Bogomolova, S.V. Grinenko, E.K. Zadorozhnyaya, Eng Bul Don, **2-2(36)**, 118 (2015)
7. N.L. Pushkareva, Woman Rus Soc, **3(56)**, 24-35 (2010)

8. I. Emelyanovich, Zh. Komarova, Sci Innov, **3**, 18-37 (2018)
9. T.K. Rostovskaya, A.M. Egorychev, Man World Cult, **2**, 31-38 (2016)
10. I.L. Sizova, N.V. Kladova, T.M. Khusyainov, Woman Rus Soc, **1(78)**, 60-68 (2016)
11. A.M. Egorychev, A.G. Akhtyan, L.V. Mardakhaev, S.N. Fomina, A.K. Bykov, AD ALTA: J of Interdiscipl Res, **10(2)**, Spec. Iss. 14, 61–65 (2020)
12. A.M. Mityaeva, S.N. Fomina, N.A. Tsvetkova, A.I. Rybakova, N.S. Gedulyanova, E.A. Khovanskaya, Turkish Online J of Design Art & Communication, **7**, 1–17 (2017)
13. M.W. Rossister, Soc Stud Sci, **23(2)**, 325–341 (1993)
14. O.S. Rose, What is gender history? (Polity Press, Cambridge, 2010)
15. C. Enloe, *Gender Makes the World Go Round*, in S. Chan, C. Moore (eds), Theory of International Relations. Vol. IV: Contemporary Reflexive Approaches in International Relations, 89-105 (SAGE Publications Ltd, London; New Delhi, 2006)
16. S.J. Ceci, W.M. Williams, PNAS, **108(4)**, 3157–3162 (2011)
17. R.E. Nisbett, et al., Amer Psychol, **76(2)**, 130-159 (2012)
18. N. Hudgins, Hold still, Madame: wartime gender and the photography of women in France during the Great War. (Univ. of St Andrews, 2014)
19. L. Holman, D. Stuart-Fox, C.E. Hauser, PLoS Biol, **16(4)**, e2004956 (2018). https://doi.org/10.1371/journal.pbio.2004956.
20. K. Cavoukian, N. Shahnazarian, *Armenia: Persistent Gender Stereotypes*, in S. Franceschet, M.L. Krook, N. Tan (eds), The Palgrave Handbook of Women's Political Rights (2019)
21. E.A. Burdukovskaya, E.V. Rvacheva, Bul Amur State Univ, **42**, 72-74 (2008)
22. Gendernyi sostav nauchnykh kadrov [Gender composition of scientific staff]. Accessed on: October 3, 2021. [Online]. Available: https://issek.hse.ru/news/341451906.html

Gender aspect in psychological support of Russian teachers of foreign schools

Alina Aleksandrovna Pozdnyakova[1*], *Elena Mikhailovna* Markova[1], *Darya Aleksandrovna* Dmitrieva[1], *Elena Aleksandrovna* Baychuk[1], and *Larisa Yurievna* Morozova[2]

[1]The Kosygin State University of Russian, Institute of International Education, Moscow, Russia
[2]Moscow Pedagogical State University (MPGU), Institute of International Education, Moscow, Russia

Abstract. "Staff feminization" of the Russian schools is a general tendency of the modern education, however, the problems that female teachers face when performing their professional activity in a foreign cultural environment are rather specific and therefore they require the most careful consideration and further correction through the system of psychological support. The study of the mechanism of psychological support, as well as the methods of its implementation, is the objective of the present article. The tasks of the research are the following: 1) determination of the stress factors, accompanying the activity of the female teachers of the Russian school abroad; 2) selection of the contemporary psychological techniques required for neutralization of the stress factors; 3) formation of a cluster of measures required for psychological support of teacher of Russian as a foreign language; 4) modelling of the educational content directed to the formation of the psychological competencies of the Russian teacher of a foreign school. The system of psychological support has two levels and includes the following measures: 1) prevention of the impact of negative factors and 2) liquidation of the competency-based lacunes in the field of psychological knowledge. Taking into consideration the specific character of a foreign cultural environment allows determining the individual trajectories of the professional development of a teacher. The system of psychological support is not closed; it should be updated and improved constantly taking into account and under the influence of the demands of the time to become a part of the mechanism of the general training of a teacher of the Russian school abroad.
Keywords: educational process, psychological support, additional professional training, Russian school

1 Introduction

The organization of work of Russian schools abroad raises a wide range of issues, including those related to the psychological support of female teachers. The gender aspect in this sense is relevant as 98% of teachers of the Russian language and Literature and 95.4% of

* Corresponding author: apozdnyakova@live.ru

teachers of national languages and literature in the Russian Federation are women [1], and it is they who for the most part form the staff for foreign schools.

The problems that female teachers face in foreign cultural environment are quite often gender-specific. Therefore it seems correct to talk about the importance and systematic approach of the psychological support of the teachers working abroad.

The objective of this article is to study the mechanism of psychological support of the female teachers of foreign schools as well as the methods of its implementation. The tasks of the research are the following: 1) determination of the stress factors, accompanying the activity of the female teachers of the Russian school abroad; 2) selection of the contemporary psychological techniques required for neutralization of stress factors; 3) formation of a cluster of measures required for psychological support of the teachers of Russian as a Second (Foreign) Language (RSL); 4) modelling of the educational content directed to the formation of psychological competencies of the Russian teacher of the foreign school.

2 Methods

The following methods are applied for implementation of the objective and solution of the tasks set in the work:

1) theoretical analysis of the literature of psychological and pedagogical, sociological, culturological focus related to the specified problems (the most useful for research there were the "basic" works of Rosen [2], Dolton and Makepeace [3], Court [4], Anker [5, 6], experimental works of the foreign authors dedicated to the study of the various aspects of the educational milieu [7-9], works of Baskakova and Soboleva [10-12], Sillaste [13, 14], Efimova [15]);

2) analysis of the current programs of professional development of teachers of RFL;

3) studying the practice of organizing courses of psychological support for teachers;

4) bulk collection of materials for analysis (questionnaire survey, testing, interviewing Russian teachers of the foreign schools);

5) analysis of summarized statistic data related to the study of the teaching staff regarding teaching and learning [1, 16-18].

3 Results

The analysis of the results of interviewing of the female teachers who performed their professional activity in the Russian schools abroad in the different periods of time allowed distinguishing the four groups of factors being to a certain extent stress-based and causing the problems of psychological character.

The first group is factors related to the selection of candidates and the specific character of the educational institution that the teacher joins.

The concept "Russian school" combines the various types of educational institutions: with different ethnic composition, different structure, various demands of students [19]. These demands are not differentiated and often are not taken into account while forming the educational programs as they cannot be included in the qualification requirements specified for the selection of candidates for work in a certain organization. A teacher who got the job offer for work in the "Russian school" does not often have any idea what he should do and what student body he should work with and therefore quite soon he is in the situation of psychological discomfort. Often a woman cannot cope with it by herself. 27% of teachers pointed out the non-conformity of expectations and more than 40% pointed out the incomplete conformity of expectations. Moreover, most teachers (82%) state that without the preliminary awareness of the specific character of the "Russian school" where they have to work, it is difficult to make an own strategy of the professional development and it is a negative psychological factor that does not allow getting complete satisfaction from the teaching process.

The second group is the factors related to the objectives of the taught student body. Evidently, the people (children and adults) rarely start learning the Russian language "for no reason" and even if it is so with time the objective is always formed. These can be: 1) practical goals (a wish to know the Russian language to be able to read (books, newspapers, Russian fairy tales, etc.); striving to get to Russia for study; a wish to work in a Russian speaking team, etc.), 2) cultural goals (interest to the Russian history, culture; a wish to travel around Russia; love for the Russian music, art, etc.), 3) specific goals (a wish to "maintain the language of ancestors"; a wish to learn an "exotic" language, etc.). However, the achievement of the goal can be realized only in case of the positive motivation available, formation and maintenance of which is possible only when two factors interact in the educational process: the authority of a teacher and the popularity of the language under studies. In case when the Russian language is not popular in the country it is rather difficult to maintain the motivation of a student and if the teacher does not have authority it is impossible.

In the opinion of the majority of respondents, the loss of authority (even partial) is the main reason of failure (or termination) of the pedagogical activity. According to the scale of significance, the loss of authority is above such factors as the lack of interest in the profession (35%; the lack of sufficient working experience abroad (17%); a negative attitude to the certain student group (5%); the "professional burnout" (2%).

The third group is the factors associated with the "competitive advantages" of the local specialists in the Russian philology that are manifested not in all countries but are currently important for the countries where the Russian language is popular (China, Turkey). For instance, in China the competition among its own Russian teachers is quite high and therefore those invited from Russia are offered to conduct only auxiliary courses ("Speaking and writing practice", "Efficient communication", etc.). In Turkey, the positions in the "Russian schools" are claimed by the immigrants from the Turkic-speaking countries of the former USSR (Azerbaijan, Turkmenistan, Uzbekistan). For obvious reasons, it is rather difficult to compete with them. From time to time more than 20% of respondents felt discomfort related to the direction or subjects of the courses of study.

The fourth group is the factors stipulated by the cultural differences of the parties (a teacher and the students). The analysis of the interview materials shows that one should speak not only about the cultural (cultural and religious) peculiarities of the country of study but about the perception of the profession of a teacher in the certain culture. It can be equally difficult for a female teacher from the psychological point of view to work in a Muslim country where "the profession of a teacher is considered to be male as it is estimated as the too complicated profession for a woman" [20] and in a non-Muslim country in a "masculine oriented group" [21]. In the first case, the difficulties are associated with the traditional preferences of the population when choosing the methods and form of knowledge acquiring, and in the second case, they are associated with the character of the lessons conducted. In the opinion of many social teachers and psychologists, the underestimation of the peculiarities of the gender composition of the students leads to their incomplete acquisition of knowledge as due to the "gender asymmetry" men and women are oriented to the different forms and methods of acquisition ("achievement" of knowledge

[21]. Due to their gender nature the female teachers are far from the "objectivity" and less adaptive to the productive activity, subjected to a higher extent to control and cannot adapt to the changes taking place. As a result, there are possible "hidden conflicts" of the teacher with a group that leads to nervous breaks.

According to the majority of respondents, the ideas of "impossibility for a woman to work in a Muslim country" are too exaggerated: an insignificant amount of teachers (less than 1 %) faced the demonstrations of tactlessness, lack of respect for a teacher as a woman. But teachers faced much more problems due to some other reasons: 1) as a result of the incorrect selection of the teaching material – printed texts, illustrations, audio and video fragments (it was mentioned by 22% of respondents); 2) due to the wrong choice of forms and methods of material presentation (18%); 3) as a result of the wrong selection of the style of communication in teaching (7%). Many problems of this category could be avoided if gender peculiarities of the group were taken into account. Thus it would be more correct to emphasize not the "position of a woman in the Muslim society" but the gender type of the group where the female teacher works.

4 Discussion

Starting the work on the article we were influenced by the well-known stereotypes about the predominance of ethnocultural and religious components in the structure of the gender problems. However, the results of the interview showed that these factors are not the leading ones: a woman "suffers" much more from the non-readiness to the interaction with a foreign cultural environment (of any type) and immature of professional skills. The psychological support of the female teachers of the Russian schools abroad should be carried out to prevent the development of the negative tendencies in human psychology to overcome the difficulties of the personal growth, to eliminate the conflict situations in relationships with students and colleagues. It is reasonable to provide this support in two directions.

The first direction is preventive work with the specified group that helps to form the "image of the country" where the teacher will have to live and the psychological readiness of the teacher to stay in this country.

According to experts in the field of communication, the main means of formation of the "image of the country" are mass media therefore the acquaintance with the materials of mass media of different countries presenting reliable and objective information to the public is quite productive. During such acquaintance, the special attention should be paid to the cultural, civilizational, mental and other significant characteristics of the host country. It is more correct to perform such work within the framework of the general course of the pre-departure training of the teacher that includes the consultations as well as diagnostic measures (questionnaire, situational tests for the solution of practical tasks, etc.)

The second direction is the work to eliminate the competence-based lacunes in the field of psychological knowledge. A teacher can be offered the programs of supplementary psychological and pedagogical training. ("Psychological framework of educational activity in the foreign language", "Professionally significant personal qualities of the language teacher", "Psychological analysis of a lesson of Russian as a Foreign language", etc.) formed on the base of the results of the diagnostic testing and implemented remotely.

5 Conclusion

Revealing of the group of stress factors accompanying the activity of the female teacher of the Russian school abroad and determination of the deficit of professional competences in the field of psychological knowledge allow to describe the structure of psychological support of the teacher including the preventive psychological measures and psychological correction and to form the content of additional professional training programs for a teacher.

References

1. N.V. Bondarenko, D.R. Borodina, L.M. Gokhberg et al., Indikatory obrazovaniya: 2020: statisticheskii sbornik [Educational indicators: 2020: Statistical collection. (RDU HSE, Moscow, 2020)
2. Sh. Rosen, *The Theory of equalizing differences*, in O. Ashenfelter, R. Layard (eds), Handbook of Labor Economics. (Elsevier Science, 1986)
3. P. Dolton, G. Makepeace, Europ Econ Rev, **37**, 1393–1411 (1993)
4. G. Court, S. Morris, B. Reilly, M. Williams, Teachers. Recruitment and the Labour Market. (Institute for Employment Studies, Brighton, 1995)
5. R. Anker, Int Labour Rev, **136(3)**, 315-339 (1997)
6. R. Anker, H. Melkas, A. Korten, Gender-based Occupational Segregation in the 1990s. (ILO, Geneva, 2003)
7. E. Blom, et al., Front Psych, **8**, 552 (2017). https://doi.org/10.3389/fpsyg.2017.00552
8. M. Lehtonen, et al., Psych Bul, **144(4)**, 394–425 (2018). https://doi.org/10.1037/bul0000142
9. M. Antoniou, Ann Rev Ling, **5(1)**, 395-415 (2019). https://doi.org/10.1146/annurev-linguistics-011718-011820
10. M.E. Baskakova, I.V. Soboleva, Quest Edu, **4**, 83-103 (2017). https://doi.org/10.17323/1814-9545-2017-4-83-103
11. M.E. Baskakova, I.V. Soboleva, Population, **21(3)**, 122-135 (2018). https://doi.org/10.26653/1561-7785-2018-21-3-09
12. M.E. Baskakova, I.V. Soboleva, T.V. Chubarova, Bul Inst Econ RAS, **2**, 85-100 (2018). https://doi.org/10.24411/2073-6487-2018-00034
13. G.G. Sillaste, Gender sociology and the Russian reality. (Alfa-N, RDC INFRA-M, Moscow, 2016)
14. G.G. Sillaste, Human South Rus, **6(5)**, 48-62 (2017)
15. G.Z. Efimova, Sci Stud, **7(5)**, 1-12 (2015)
16. TALIS 2018 Results (Volume I): Teachers and School Leaders as Lifelong Learners. (OECD Publishing, Paris, 2019). https://doi.org/10.1787/1d0bc92a-en
17. TALIS 2018 Results (Volume II): Teachers and School Leaders as Valued Professionals. (OECD Publishing, Paris, 2020). https://doi.org/10.1787/19cf08df-en
18. Positive, High-achieving Students? What Schools and Teachers Can Do. (OECD Publishing, Paris, 2021). https://doi.org/10.1787/3b9551db-en
19. E.M. Markova, et al., *Obrazovanie i gorod: Obrazovanie i kachestvo zhizni v gorode* [Education and City: Education and Quality of Living in the City], in The Third

Annual International Symposium. Moscow, 2021, 3013. (2021). https://doi.org/10.1051/shsconf/20219803013

20. I.I. Sokolova, Universum: Bul Herzen Univ, **2**, 3-15 (2010)
21. V.P. Simonov, Bul Moscow Reg State Univ, **1**, 1-13 (2014)

Gender asymmetry peculiarities among lecturers of foreign languages department

Galina Viktorovna Tretyakova[1*], and *Farrukh Begijon* Khudoydodzoda[2]

[1]Financial University under the Government of the Russian Federation, Department of Foreign Languages and Intercultural Communication, Moscow, Russia
[2]Tajik National University, Department of Ontology and Theory of Cognition, Dushanbe, Tajikistan

Abstract. The background of the research: the article deals with the problems of gender asymmetry among teachers of the Department of Foreign Languages and Intercultural Communication of the Faculty of International Economic Relations, which are caused by the socio-economic structure of society, the level of democratization, the development of the spiritual life of society, the peculiarities of the socio-demographic structure. It is also necessary to take into account the experience of the Soviet school, which had a pronounced gender asymmetry, which had led to the feminization of many professions. The gender factor is important when planning the work of universities, determining the areas of training and specialties that are in demand by the market. Under such conditions, gender asymmetry becomes a clear indicator of the educational society. So, the purpose of the study is to conduct a gender analysis of the ongoing changes in the higher education system and to show the professional and social role of women in the modern world.
Keywords: principles of historicism, gender factor, society of teachers, gender strategy

1 Introduction

The concept of "gender" was introduced into scientific circulation in the fifties by American scientists. Now it denotes any mental or behavioral properties that supposedly distinguish men from women and thus is associated with masculinity and femininity. In pedagogy, gender means the socially determined role, identity, and sphere of activity of men and women, depending not on biological sex differences, but on the characteristics of the social organization of a particular society [1]. Gender asymmetry is a phenomenon that reflects the objective regularity of the quantitative distribution of men and women in the social structure of a society and in all the spheres of its life in favor of the male or female community.

* Corresponding author: gvtretyakova@fa.ru

2 Materials and methods

This article develops a set of socio-methodological support for the study of gender asymmetry in the professional society of foreign language teachers, explains the nature of stereotypes, shows their historical variability and social conditionality. Society cannot be fully democratic without the full-scale inclusion of women in the process of education, the development of scientific and methodological thought. Today, Russia is trying to enter the European community and take its rightful place there. Eliminating all the forms of discrimination, including gender-based discrimination, will help it do this. The transition to the information society today – a society of intelligence, knowledge and thinking-creates basic opportunities for self-development and self-realization for each person. The need to adopt a gender strategy is determined by the social problems caused by socio-economic and political transformations. In this regard, an interdisciplinary approach, system analysis, the principles of historicism and continuity, and social determinism were used.

3 Results

Since the 60-70s of the twentieth century, the problem of social inequality between men and women or gender asymmetry in various spheres of human activity, including the sphere of labor and employment, which is inextricably linked with the field of vocational training and education, has taken a central place in gender studies. A huge number of scientific works are devoted to this problem, but it remains relevant in the light of global socio-economic and cultural transformations and the formation of a new strategy aimed at preserving human capital and sustainable development of a society [2]. Among modern researchers, the prevailing opinion is that it is in the sphere of labor and employment that gender asymmetry is most pronounced: for various reasons, women tend to work in less prestigious sectors of the economy, are more often discriminated against at work, and less often occupy high-paid positions and positions associated with making politically significant decisions [3]. Some scientists criticize these provisions, citing the example of the education system, which, in their opinion, has already achieved an equal ratio of men and women in the structure of administrative and teaching staff, including the provisions in terms of salary and career opportunities [4]. Nevertheless, there is no doubt that there is a real problem of professional differentiation based on gender and imbalance in the field of employment in a society. It manifests itself not only in the different levels of the positions prestige which are held by men and women, the amount of their wages, but also in the existence of stereotypical ideas that affect the choice of future specialty and professional orientation within the educational system. Let us turn to the statistics. According to the data of sociological studies conducted in 2015-2019, the level of professions prestige for men was distributed as follows in a descending order: a lawyer, an economist, a programmer, a physician, an engineer, a merchant, a translator, a researcher, etc.; the professions of a psychologist and a teacher have taken the last places in the rating. The professional selection of women was as follows (also in a descending order): an economist, a lawyer, a physician, a translator, a teacher, some creative professions, an engineer, a psychologist, a merchant, etc.; the last place was taken by the profession of a programmer [5].

Modern realities and the total informatization of a society have set new priorities and brought new specialties to the fore. Numerous resources dedicated to the career guidance and the search for future work, orient men to the professions of an IT specialist, banking specialties, the professions of an architect, an engineer, a doctor, a sales specialist, a marketer, an accountant [6]. For women, the list of recommended professions in the first place is as follows: a doctor, a PR specialist, a psychologist and a personal consultant, a marketer, an ecologist, a chemist, a service specialist, an IT specialist (but not in the first positions), an engineer, etc. [7]. Despite some blurring of the boundaries, there are still very significant gender differences in professional orientation and the choice of a future field of work, confirmed, among other things, by professional sociological studies [8]. There is still a clear understanding of the "traditional" male and female professions, as well as the predominance of divergent interests among men (technical and scientific) and women (mainly humanities)[9].

Such an asymmetry can be seen in the composition of teachers of any professional educational institution, and in this regard, the Department of Foreign Languages and Intercultural Communication of the Faculty of International Economic Relations is no exception.

Table 1. Personnel of the Department of Foreign Languages and Intercultural Communication of the Financial University under the Government of the Russian Federation.

Personnel structure	Women	Men
Total	74%	26%
Top managers	98%	2%
Candidates of Science	48%	4%
Associate Professors	39%	2%
Professor	1%	1%

The personnel composition of our Department (Table 1) is represented by the overwhelming majority of the female contingent (74%), including candidates of Science-48%, men-26%, top managers-98% and 2%, respectively, candidates of Science-48% are women, and 4% are men. The asymmetry is obvious, but in favor of women. At the same time, women perfectly cope with their administrative tasks. The Department of Foreign Languages and Intercultural International Communication of the Financial University under the Government of the Russian Federation occupies a leading position in many indicators: in educational activities, science, extracurricular activities. We have a parity in the number of professors. In terms of wages, there is also a full parity.

Is such a distribution considered to be appropriate and even natural, even if it includes many mutually influencing factors for determining professional orientation? Or does it reflect the dependence of a society on established stereotypes that can negatively affect the further development of the field of professional activity? On the one hand, there is no doubt that there are psychophysiological features that manifest themselves in men and women from the first days of their birth [10], and in the future they are displayed in a deeper and brighter way [11]. On the other hand, it is also indisputable that the excessive emphasis on gender differences and the uncompromising opposition of the male and female principles give rise to discrimination, the lack of equal cooperation between the sexes [12]. The gender differences and characteristics dictated by a society become less unambiguous when it comes to the individual characteristics of the development of a particular person, his or her inclinations, life and professional priorities. A person needs to see his or her development prospects and his or her role and place in the social processes that affect his or her future life. Therefore, the process of social adaptation without taking into account the gender factor of self-actualization creates real problems of socialization and career guidance of the individual, an unacceptable narrowing of the boundaries of thinking in a rapidly changing world [13].

Of course, our main mission is to provide educational services in the field of teaching foreign languages. Among the students of the Department of Foreign Languages and Intercultural Communication, a survey of students concerning the professional skills of teachers was conducted according to the following criteria (Table 2):

Table 2. Study of students' opinions on the professional skills of teachers of the Department of Foreign Languages and Intercultural Communication.

Evaluation criteria	Women	Men
AVERAGE (5-point scale)	4.59	4.42
Is able to clearly, easily and consistently present the material	4.57	4.48
Connects the theoretical knowledge with the practical component of the future profession	4.67	4.64
Is able to motivate and engage the student in the learning process	4.81	4.83
From the first lesson in the discipline puts forward clear and understandable requirements for the knowledge assessment at classes, certifications and exams	4.84	4.76
Objective in assessing students' knowledge and their completed work	4.81	4.24
Interested in improving the level of students' knowledge acquiring and their skills in the taught discipline	4.71	4.73
Always friendly and tactful with students	4.92	4.68
Available for the extracurricular communication on educational and scientific issues at consultations or remotely	4.69	4.61

4 Discussion

As a result, it was found that the indicators of male teachers are inferior to the average indicators of female teachers by 0.5%. In addition, during an oral survey of 365 students of the Faculty of International Economic Relations it turned out that 98% of students still do not see the difference between male and female teachers. For them, the priority was such qualities as professionalism, love for the profession, the ability to establish contact with the audience.

Thus, we can conclude that the true equality of rights and freedoms can be achieved only on the basis of recognizing the value of each individual, creating stable conditions for the self-development of the human intelligence, the fullest realization of its creative potential and creative abilities, the full disclosure of essential forces and talents, which will ensure the full contribution of everyone to the development of a society[15]. The transition of many countries to an information society – a society of intelligence, knowledge and thinking-creates basic opportunities for self-development and self-realization for each person. The need to adopt a gender strategy is determined by the social problems caused by socio-economic and political transformations. Along with the opening of new perspectives for the wide application of human abilities, high adaptive capabilities are required from people [14].

5 Conclusion

In order to create favorable conditions for the growth of opportunities for self-realization of the individual, the modern society should take care of rethinking the traditional cultural limitations of the potential of the individual development depending on gender and pay attention to the problem of gender asymmetry, which persists, including in the field of higher education. Gender issues contribute to a deeper understanding of human rights, a modern understanding of the concepts of sustainable democratic development. Mastering the basics of gender knowledge helps to free yourself from the "captivity" of gender stereotypes. Gender education contributes to the growth of self-awareness, active integration into all spheres of a society. As for the Department of Foreign Languages and Intercultural International Communication of the Faculty of International Economic Relations, the pronounced asymmetry of the personnel structure in no way affected the implementation of our main mission – the effective foreign language teaching the future international specialists.

References

1. F.D. Blau, J Econ Lit, **36**, 112-165 (2018)
2. F.D. Blau, P. Brummund, A.Y. Liu, Demography, **50**, 471-492 (2013). https://doi.org/doi:10.1007/s13524-012-0151-7
3. M. Bertrand, E. Kamenica, J. Pan, The Quart J Econ, **130**, 571-614 (2015). https://doi.org/10.1093/qje/qjv001
4. M.Bittman, et al., Amer J Soc, **109**, 186-214 (2013). https://doi.org/10.1086/378341
5. H. Blossfeld, Annual Rev Soc, **35**, 513-530 (2019). https://doi.org/10.1146/annurev-soc-070308-115913
6. H.A. Schwartz, et al., PLoS One, **8(9)**, e73791 (2018). https://doi.org/10.1371/journal.pone.0073791
7. K. Schwab, et al., The Global Gender Gap Report 2018 (World Economic Forum, 2018)
8. M. Hout, Mobility tables (SAGE, Beverly Hills, 2013)

9. N.L. Graf, C.R. Schwartz, Gender Soc, **25**, 101-107 (2011). https://doi.org/10.1177/0891243210390356
10. C. Goldin, M. Shim, J Econ Persp, **18**, 143-160 (2014). https://doi.org/10.1257/0895330041371268
11. F.D. Blau, L.M. Kahn, The Acad Manag Persp, **21**, 7-23 (2017). https://doi.org/10.5465/AMP.2007.24286161
12. K.A. Ratliff, S. Oishi, J Person Soc Psy, **105**, 688-702 (2013). https://doi.org/10.1037/a0033769
13. T. Greenstein, Annual Rev Soc, **35**, 87-105 (2019). https://doi.org/10.1146/annurev-soc-070308-115920
14. S.S. Iyengar, E. Kamenica, I. Simonson, Quart J Econ, **121**, 673-697 (2006). https://doi.org/10.1162/qjec.2006.121.2.673
15. F. Hou, J. Myles, Canad J Soc, **33**, 337–366 (2018) The status of women in the States: 2015. Accessed on: September 24, 2021. [Online]. Available: https://iwpr.org/iwpr-publications/report/the-status-of-women-in-the-states-2015/

Sociocultural peculiarities of gender aspect in Russian education and science

Leyla Buniyaminovna Omarova[1*], *Diana Zurabovna* Muzashvili[2], *Irina Gennadievna* Prokhorova[3], *Natalia Valerievna* Savchenko[4], and *Anna Alexandrovna* Suchilina[5]

[1]Financial University at RF Government, Department of Humanitarian Sciences, Moscow, Russia
[2]Financial University at RF Government, Department of Psychology and Human Resources Development, Moscow, Russia
[3]Financial University at RF Government, Sociology Department, Moscow, Russia
[4]Financial University at RF Government, Department of the English Language and Professional Communication, Moscow, Russia
[5]Financial University at RF Government, Department of Political Science, Moscow, Russia

Abstract. Urgency of the problem is conditioned by the fact that under the effect of globalization processes, digital revolution, and transformation of values the present-day society is changing its attitude towards gender division of labor. Gender problem in traditional communities where, on the one hand, new views on the roles of men and women are being formed, and, on the other hand, traditional attitude towards those roles, is being preserved, is of special interest. The objective of the research is to analyze present-day society for revealing the gender asymmetry in different spheres of science and education in Russia. The Russian society that tends to traditional attitude but is unavoidably a part of innovation society, becomes an interesting object for studying the gender aspect in different professional spheres of education and science. A sociocultural approach forms the methodological base of the research that allows us to analyze the current stage of development of the educational and scientific spheres taking into account social and cultural peculiarities of the society, as well as its gender perspective. The study is based on the works that reveal gender issues in modern society in the sphere of education and science, psychology, culture and philosophy. It also presents the analysis of present-day scientific and educational spheres and gender asymmetry occurred under the effect of some or other social or cultural peculiarities.
Keywords: education, science, professional sphere, gender gap, culture, social environment, traditions, innovations

1 Introduction

Rapid changes in the modern society as a consequence of globalization processes that transform the social platform and values form a new attitude towards a gender aspect in the professional sphere.

* Corresponding author: leylaomarova1@gmail.com

Despite some different feminist movements in the 20th-21st centuries that claimed and defended the women's rights in different spheres of life, many professions remained predominantly men's ones. Moreover, some professional and scientific spheres that emerged with development of digital and informational technologies are immanently assumed to be men's spheres. The percentage of women in sciences related to medical care, in the sphere of social and humanitarian sciences is rather high, but in sciences related to IT, energy, and engineering women are in absolute minority. As to participation of women in large-scale scientific studies, the 'glass ceiling' or 'leaking pipeline' effect is observed. The share of women on the global scale who got bachelor and magister degrees exceeds the share of men and makes 53%, but the share of women with scientific degrees, especially PhD degrees makes only 43%. As to women dealing with scientific practice, their share does not even reach 30% (28%, to be more exact).

Low percent of women involved in fundamental studies and lack of high positions occupied by women in science despite gender equality observed in all the educational spheres during the latest decades, and even gender asymmetry in favour of women in some spheres makes us investigate the causes of this phenomenon.

The gender balance in science and education is affected by many factors that are both of global and local nature. Economic situation and level of scientific and technological development in individual regions are the basic factors. For example, the share of women-scientists in South-East Europe makes 40%, whereas in South African countries it is as low as 30%. Social and cultural traditions are another crucial factor. For example, in Arab countries, despite high economic development, the number of women in science is 12% lower than in European countries, and this number in Japan is three times lower than in Europe.

2 Methods

A method of complex social and cultural analysis applied in the course of the study allows us to reveal factors affecting the low share of women at high positions in science and science management boards. The scope of our research is in studying and searching for some possible solutions as in current science education, on the one hand, women in top positions are in need, on the other hand, there are various social, cultural and psychological barriers to female social mobility.

3 Results

Social studies on the gender aspect in the modern system of education demonstrate a large number of women in the higher education system.

For instance, a famous Harvard astronomer and physicist Edward Charles Pickering employed a group of female scientists who performed boring and unspecialized work. Regardless of the unequal pay and distribution of responsibilities, this work was crucial. Edward Pickering is known to be a greatest scientist who introduced the use of the meridian photometer to measure the magnitude of stars. Whereas female scientists like Annie Jump Cannon, Williamina Fleming, Henrietta Swan Leavitt, Antonia Maury, and Florence Cushman are called Harvard Computers and barely mentioned in the world of science.

This is due to the fact that education nowadays is one of the means of a vertical social lift. It allows a higher level of earnings, but getting a job with such income requires women to demonstrate higher professional level than that of men, i.e., it is more difficult for a woman to get that job as an employer makes them prove their professional competence in a certain sphere. This is not conditioned by cultural or social peculiarities only. Psychologically a woman-professional or a scientist in the majority of cases is doubted by employers.

This explains a gender gap in favour of men at managerial positions in science and education.

A present-day manager or a head, and society as a whole, surely, do not agree to Plato [1] or Machiavelli [2] who makes a woman equal to man's property. But the majority who keeps to an archetype where a woman is a supplement that is needed rather than the main thing, is not ready to hand over the reins of power to a woman in any sphere, to say nothing of science.

The number of women-executives of high educational institutions in Russia, for example, is as low as 10%; among 836 Academicians of the Russian Academy of Sciences there are only 46 women. And this situation persists despite the fact that in 2015 the UNESCO report on science noted that "gender equality (45-55% of women among scientific workers) in many regions is a heritage of the former soviet union that included Middle Asia, Baltic countries, East and South-East Europe" [3].

4 Discussion

The problem of gender equality in science and education is among the most urgent issues of the modern society. It is a concern of specialists on education science, psychology, and sociology. The objective of the modern society and the scientific community is to reveal problems of a gender misbalance and find the best solutions.

Gender sociology in the Russian science is considered in the works by G.G. Sillaste (2016) [4] who analyzes the gender asymmetry in science and education. G.G. Sillaste gives a detailed characteristic of women and men position in the present-day science and education. The works by N.A. Shvedova, I.O. Maltsev, S.S. Balabanov, E.P. Myasin study the gender asymmetry in natural and humanitarian sciences, the problem of a "glass ceiling", issues related to vertical social lift in the scientific and academic environment.

Some foreign researchers analyze and find the solutions to the same problems. Camille S. Johnson, Pamela K. Smith, Chunlei Wang [5], for example, speculate on the low number of women at the world scientific symposia as the main speakers where they are mainly assistants [5]. Cristina Gimenez Elorriaga, Monike Nicolas [6] in their report at the scientific congress Ciencia technologia y Genero, XI Congreso iberoamericano noted that "despite the progress, the programs, laws and policies that were adopted, noted, implemented and evaluated, and were to contribute to equality of men and women in that sphere at the international level" the number of women in science and engineering in such key spheres as STEM (Science Technology Engineering Mathematics) does not remain in favour of women [6] as women's contribution in the majority of cases remains underestimated or not estimated at all.

Marta du Vall, Marta Majorek [7], Polish researchers of gender problems in the modern European society agree to the fact that "stereotypes on the distribution of roles between women and men historically formed in the public conscience were carried over to the realities of the present-day epoch" [7, 8].

American psychologists Lin Bian, Sarah-Jane Leslie, Andrei Cimpian [9] state that girls even in their childhood tend to believe the similar stereotypes about the impossibility to get the educational level in definite sciences. The scientists say that common stereotypes associate high-level intellectual ability (brilliance, genius, etc.) with men more than women.

These stereotypes discourage women's pursuit of many prestigious careers; that is, women are underrepresented in fields whose members cherish brilliance (such as physics and philosophy) [9].

The research studies of Murphy, M. C., Steele, C. M., & Gross, J. J. (Murphy, Steele, Gross, 2007) [10], Leslie, S. J., Cimpian, A., Meyer, M., Freeland, E. (Leslie, Cimpian, Meyer, Freeland, 2015) [11], Guiso, L., Monte, F., Sapienza, P., & Zingales, L. (Guiso, Monte, Sapienza, Zingales, 2008) [8], Ackerman, P. L., & Heggestad, E. D. (Ackerman, Heggestad, 1997) [12], Kanagawa, C., Cross, S. E., Markus, H. R. (Kanagawa, Cross, Markus, 2001) [13], Anna Szlavi (A. Szlavi) [14], Nigora Salieva, Ulugbek Saliev (N. Salieva, U. Saliev) [15] aims at destroying the myth of being bad at maths. Based on many social psychology studies and experiments some scientists consider that the stereotypes in some cultural codes may contribute to the formation of false social identity as belonging to particular sex or gender groups.

5 Conclusion

Despite the fact that in terms of gender the number of women with higher education in the modern society exceeds that of men, the number of women-scientists and heads of large-scale scientific projects on the global scale decreases. After getting a master's degree a woman is lost for science. This is due to some social, cultural and psychological factors.

Because of cultural or social stereotypes even those who start their scientific career, give it up under certain circumstances.

The society declares publicly about the gender equality, although secretly follows traditional archetypes. Therefore, this problem requires a systematic approach with the change of attitude towards women from different standpoints. That is a global challenge and it has to be solved on a global scale. Gender equality is not only a fundamental human right, but a necessary foundation for a peaceful, prosperous and sustainable world. The Sustainable Development Goals (SDGs) and Global Health of the UN, Goal 4 aims to ensure inclusive and equitable quality education and promote lifelong learning opportunities for all. Moreover, Goal 5 is to achieve gender equality and empower all women and girls [16].

We agree with N.A. Shvedova that gender equality is the primary goal and an important tool for development and a better quality of life. Only it may ensure an economic growth.

References

1. Plato, The State (ACT, Moscow, 2020)
2. N. Machiavelli, The Prince (EKSMO, Moscow, 2017)
3. C. Hugher, Is the gender gap narrowing in science and engineering? Accessed on: September 25, 2021. [Online]. Available: https://en.unesco.org/sites/default/files/usr15_is_the_gender_gap_narrowing_in_science_and_engineering_ru.pdf
4. G.G. Sillaste, Gender sociology and Russian reality (Alfa-M, Moscow, 2016)
5. S.J. Camille, K.S. Pamela, W. Chunlei, Person Soc Psy Bul, **43(4)**, 493-507 (2017). https://doi.org/10.1177/0146167216688213
6. C.G. Elorriaga, M. Nicolas, *Impacto de los Premios en Genero, Ciencia y Technologia: del efecto Matilda a la Teoria del Embudo en la creacion del Premio Ada Byron a la*

Mujer Tecnologa de la Universidad de Deusto, in Ciencia technologia y Genero, XI Congreso iberoamericano (2016)
7. M. du Vall, M. Majorek, Aequalitas, **2**(1(2)), 2-17 (2013)
8. L. Guiso, F. Monte, P. Sapienza, L. Zingalcs, Science, **320**, 1164-1165 (2008)
9. L. Bian, S-J. Leslie, A. Cimpian, Science, **355**, 389-391 (2017)
10. M.C. Murphy, C.M. Steele, J.J. Gross, Psy Sci, **1(18)**, 879-885 (2007). DOI: 10.1111/j.1467-9280.2007.01995.x
11. S.J. Leslie, A. Cimpian, M. Meyer, E. Freeland, Science, **347**, 262-265 (2015). https://doi.org/10.1126/science.1261375
12. P.L. Ackerman, E.D. Heggestad, Psy Bul, **121**, 218–245 (1997). https://doi.org/10.1037//0033-2909.121.2.219
13. C. Kanagawa, S.E. Cross, H.R. Markus, Person Soc Psy Bul, **27**, 90-103 (2001). https://doi.org/10.1177/0146167201271008
14. A. Szlavi, J Res Int Edu, **2(2)**, 26-40 (2020). https://doi.org/10.36427/CEJNTREP.2.2.472
15. N. Salieva, U. Saliev, The Amer J Soc Sci Edu Innov, **3**, 292-295 (2021). https://doi.org/10.37547/tajssei/Volume03Issue05-52
16. N.A. Shvedova, Woman Rus Soc, **3/4**, 17-30 (2015)

Female education in post-reform Russia as a social lift

Yulia Viktorovna Bodrova[1], *Aleksey Viktorovich* Vinnik, *Olga Konstantinovna* Ermishkina, *Elena Alekseevna* Makarova, and *Angelina Valerievna* Tsyganova

Tver State University, Department of Social and Cultural Service, Tver, Russia

Abstract. A significant expansion of female education for the representatives of different classes was a progressive phenomenon of the "great reforms" of Emperor Alexander II. Against the background of social changes in Russia, women thus received additional opportunities for their socialization and work. The study of the use of female education as a social lift is of interest from the point of view of the history of social modernization. The study aims to study the experience of using female education as a social lift on the example of the Tver province of the Russian Empire. The study uses a set of methods of female and gender history and the local history approach, which collectively make it possible to focus on the regional specifics of the modernization of the female education system in the context of the individual's everyday practices within the framework of an interdisciplinary field. Published statistical data are involved in the research; unpublished documents from the funds of regional archives are also introduced into scientific circulation. The novelty of the article lies in the recreation of the practices of using female education as a social lift in the second half of the 19th – early 20th centuries using the example of the Tver province. Various levels and forms of female education in Russia in the era of "great reforms" and the period that followed them have been studied. The mechanisms of socialization of Russian women of different classes are revealed, depending on the level of education received.
Keywords: history of education, teaching methods, educational institutions for the local community

1 Introduction

The era of "great reforms" 60s-70s of the 19th century provoked radical socio-economic changes in Russian society, one of the manifestations of which was the reassessment of the concept of female upbringing and education [1, 2]. The public unfolded a wide controversy about the need to introduce a new model of female institutions based on the principles of all-estate, openness, humanistic orientation and public-state partnership [3, 4]. The modernization of female education has led to the development of a network of educational

[1] Corresponding author: Bodrova.YV@tversu.ru

institutions, improving their financial condition by attracting private charitable funds, introducing advanced pedagogical ideas, democratizing the social composition of pupils [5-8]. The availability of education contributed to a gradual change in the social status of women and the formation of new life strategies [9].

2 Materials and methods

Transformation of the position of women in social stratification is one of the categories of scientific analysis of gender history [10, 11]. The gender approach allows us to consider the development of female education as an achievement of a progressive society in the struggle for female equality [12]. The study of the regional specifics of modernization processes in the field of education determines the application of the local history methodology [13-16]. Published statistical documents [17] and unpublished materials of the State Archives of the Tver Region were used as a source base for the study.

3 Results

The post-reform period brought significant social changes to Russian society, during which not only existing class barriers were actively broken down, but also the gender stereotypes. A woman strove for a new social role, as the emancipated person she did not take care of the family and a successful marriage, conversely, she wanted independence, mental and moral improvement. The real way to this was education, equal to that of men.

A key stage in the reform of female education was the approval of the Regulations on Female Schools of the Department of the Ministry of Public Education by Emperor Alexander II in 1858. Similar educational institutions began to be created under the Department of Institutions of the Empress Maria. Due to the support of the public and the attraction of charitable funds, female educational institutions were opened in various regions and differed in their organizational and legal forms, goals, curricula, level of training, and the composition of students.

Tver province was actively involved in the process of reforming the education system. According to the data of the first general census of the population of the Russian Empire in 1897 in terms of literacy of the population, including women, the Tver province occupied one of the first places, which is explained by the proximity to the capital's cultural and educational centers, the presence of qualified and motivated teaching staff, the active position of the provincial government, activities of zemstvo and public organizations. In the province, the number of educational institutions for women, primarily gymnasiums, increased. By the beginning of the 1870s there had been two female gymnasiums in Tver and one in Vyshny Volochyok, moreover, by January 1, 1905 there had been already six female gymnasiums (Bezhetskaya, Vesyegonskaya, Ostashkovskaya, Rzhevskaya, Tverskaya Mariinskaya gymnasium, Tver female gymnasium Rimskaya-Korsakova) [17]. Educational and educational activities in them were organized in accordance with the Regulations on Female Gymnasiums and Gymnasiums of 1870, which emphasized the practical nature of female education, which implied the possibility of applying the abilities of graduates in a wider space than home and family.

Mariinskaya Gymnasium was the first secondary educational institution in the Tver province for girls of all classes and religions. In the post-reform period, the duration of education in the gymnasium increased to 7 years, an 8th pedagogical class was additionally established. Graduates of the 7th grade were issued a certificate that allowed them to work as an elementary school teacher, who graduated from the 8th grade – a certificate of a home teacher and a home tutor. The introduction of an additional class opened up broader prospects for graduates for pedagogical activity. The gymnasium was the most numerous in terms of the number of students. In 1906, the total number of female students was 570 [18].

Girls had the opportunity to receive secondary education in private gymnasiums, one of which was A.A. Rimskaya-Korsakova Tverskaya female gymnasium. Girls of all classes and religions were admitted to it, it was supported by tuition fees and funds from the city budget and zemstvo fees.

Since 1905, a secondary specialized educational institution for women, the Tver City Female Commercial School, has been operating in Tver. It had its own charter and consisted of two classes and five departments. The school accepted girls of all classes and religions at the age of 8, the duration of the study was 7 years. In the first year, 153 people entered the school [19]. Students received general education and the basics of business knowledge. Graduates of the female commercial school had been entitled to enroll in the Pedagogical courses in Tver.

One-year permanent Pedagogical courses for the training of primary school teachers were opened in 1912 in connection with the growing need of the province for school teachers. Those who graduated from secondary educational institutions were admitted to the courses. The first listeners were 40 people, by the beginning of the 1914-1915 academic year their number had reached 78 [20].

It should be noted, that in most cases the daughters of the nobility, clergy and relatively wealthy urban strata studied in gymnasiums and schools. The path to education was much more difficult for the representatives of the lower classes, peasant girls who dreamed of breaking out of poverty and ignorance. It was in Tver that a project was implemented that gave several generations of women a chance for professional self-realization. Tver female teacher's school, founded in 1870 by a prominent zemstvo figure Maksimovich, was a tool for generating and testing many innovations in the organization of an educational institution. First of all, it gave free education to peasant girls, who could not get it anywhere else and in such a volume, since all other secondary educational institutions of Tver were paid. The school teachers tried to equip it with modern teaching aids, new equipment (observatory, laboratory), introduced advanced teaching methods (excursion method, optional classes). Maksimovich believed that students should be taught "smallpox vaccination, familiarize them with the main rules of folk hygiene, practically applicable to rural folk life, with caring for the sick, postpartum women, newborns and, in general, report all medical information available to them by education and applicable to the life of the rural population" [21]. It is no coincidence that the second place in the employment of graduates after teaching activities is occupied by the professions of "paramedic" and "midwife-paramedic", some of the students continued their studies at medical institutes.

The school was distinguished by unique forms of organization of extracurricular activities. Much attention was paid to sports, excursions, aesthetic education, and leisure. Informal, creative traditions in the organization of extracurricular activities have become an example for the pupils in their future life. The school graduated 36-40 female teachers annually; over the 48 years of its existence, it trained 1,200 female teachers [22].

4 Discussion

The study of the activities of female educational institutions in individual provinces allows us to conclude that the availability of education was conditional due to the collection of fees from the pupils. Only due to the attraction of charitable funds, a small number of representatives of the lower classes received the opportunity of free education. The introduction of advanced pedagogical technologies into the educational process depended on public and private support. Despite the fact that education provided women with the opportunity to change their social status, its relevance began to be felt only by the beginning of the twentieth century.

5 Conclusion

Female education in the post-reform period has become a striking social phenomenon. It not only made it possible to solve the problem of training qualified personnel for various social and economic spheres, but also expanded the understanding of the social role of women, allowed her to become a self-sufficient and independent person, provided an opportunity for professional and personal self-realization, and served as a kind of "social lift".

References

1. I.E. Lyskova, Woman Rus Soc, **3-2**, 80-87 (2014)
2. V.Yu. Kapustina, Bul Moscow State Ling Univ. Edu Pedag Sci, **2(766)**, 77-91 (2016)
3. O.S. Buslaeva, Lecturer 21th, **3**, 347-352 (2015)
4. A.V. Spichak, Bylye Gody, **56(2)**, 652-66 (2020). https://doi.org/10.13187/bg.2020.2.652
5. I.V. Kornilova, T.A. Magsumov, R.R. Shakirov, Europ J Contemp Edu, **16(2)**, 217-228 (2016). https://doi.org/10.13187/ejced.2016.16.217
6. E.N. Khabaleva, Woman Rus Soc, **3(84)**, 103-109 (2017). https://doi.org/10.21064/WinRS.2017.3.9
7. M.Yu. Varavva, Edu Stud, **4**, 250-272 (2020). https://doi.org/10.17323/1814-9545-2020-4-250-272
8. N.A. Sodelnikova, Bul Surgut State Univ, **1(70)**, 140-150 (2021). https://doi.org/10.26105/SSPU.2021.1.70.016
9. I.V. Kornilova, T.A. Magsurov, Europ J Contemp Edu, **6(2)**, 352-366 (2017). https://doi.org/10.13187/ejced.2017.2.352
10. M. Makienko, A. Panamaryova, A. Gurban, Procedia – Soc Behav Sci, **166**, 18-23 (2015). https://doi.org/10.1016/j.sbspro.2014.12.476
11. R. Pearse, J.N. Hitchcock, H. Keane, Fem Stud Int Forum, **72**, 109-126 (2019). https://doi.org/10.1016/j.wsif.2018.05.007
12. N.L. Pushkareva, Woman Rus Soc, **2-3**, 32-37 (2002)
13. N.A. Bulygina, V.P. Ermakov, M.S. Bogoslavtseva, Man India, **96(7)**, 2179-2190 (2016)
14. A. Torre, Hist Crítica, **69**, 37-67 (2018). https://doi.org/10.7440/histcrit69.2018.03
15. L.P. Repina, Dial Time, **69**, 5-16 (2019)
16. A. Pons, Hispania Nova, **1** Extraord, 52-80 (2020). https://doi.org/10.20318/hn.2020.5366

17. Uchebnye zavedeniya vedomstva Ministerstva narodnogo prosveshcheniya: Spravochnaya kniga, sostavlennaya po ofitsialnym svedeniyam za 1905 god [Educational institutions of the Department of the Ministry of Public Education: A reference book compiled according to official data for 1905]. (St. Petersburg, 1907)
18. State Archives of the Tver Region. F. 12 – Tver Female Mariinsky Gymnasium, op. 1, c. 157, let. 11
19. State Archives of the Tver Region. F. 842 – Tver City Female Commercial School, op. 1, c. 28, let. 1
20. State Archives of the Tver Region, f. 153 – Permanent Pedagogical Courses of the Tver Provincial Zemstvo, op. 1, c. 114, let. 3
21. A.F. Selivanov, P.P. Maksimovich, osnovatel Tverskoi zhenskoi uchitelskoi shkoly (biograficheskii ocherk s portretom) [P.P. Maksimovich, founder of the Tver women's teacher's school (biographical sketch with a portrait)]. (St. Petersburg, 1901)
22. T.A. Ilyina, Shkola Maksimovicha [Maksimovich's school]. (Tver State University, Tver, 1992)

Labor, gender and governance in terms of establishment of new technological order

Liudmila Aleksandrovna Vasilenko[1] and *Sergey Vladimirovich* Vasilenko

The Russian Presidential Academy of National Economy and Public Administration, Institute of Public Administration and Civil Service, Moscow, Russia

Abstract. The prerequisites for the study are the transition of the world social system to the Technological Orders VI and VII in the context of a digital breakthrough and the hybrid nature of the digital social space, which significantly affects the change in the social structure, the distribution of social roles in society, the formation of the labor market and the formation of a new gender order. The purpose of the study is to analyze the impact of digitalization on the labor market, on the gender balance in the labor market and formulate how the processes of preparing a person for new technological structures should be built. According to the results of the study, the authors note that the formation of social institutions of a new type in a hybrid digital society significantly lags behind the need, digital regulators of social behavior in the labor market have not yet emerged, which does not allow building norms and rules of life in a virtual social space at the legislative level. The cognitive structures of society are immature, adequate incentives and sanctions as a response to the phenomena of crossing the boundaries of the office set framework by social actors have not been formed. There are no frameworks yet, but they can be created by maintaining a balance between the realization of society's needs for development and relying on its ability to maintain the stability of the value system, i.e. property of resistance.
Keywords: digital revolution, artificial intelligence, technological order, public administration, gender order, discrimination

1 Introduction

Technological Order – it is a set of technologies used at a certain historical stage in the development of production. The term was introduced by Academician D. Lvov and S. Glazyev. Modern transition to Technological Order VI and Technological Order VII at the global level, it is similar in scale to the previous technological transitions in the evolution of mankind – to agriculture, industrial production [1]. By 2010, the share of the productive forces of Technological Order V in developed countries was about 60%, Technological Order IV – 20%, and Technological Order VI – about 5%. Technological Order VI is actively advancing almost today. Its main industries are nano- and biotechnology,

[1] Corresponding author: vasilenkola@mail.ru

nanoenergy, molecular, cellular and nuclear technologies, nanobionics, nanotronics; use of stem cells, engineering of living tissues and organs, reconstructive surgery, digital medicine, new types of household appliances, transport, communications. The seventh order grows in the depths of the sixth, gradually forming the shoots of the future. The digital revolution is the most important component of the technical and technological conditions of the new order [2, 3]. Its signs are the complication and hybrid nature of the digital social space, mediated by infrastructure, software and hardware, and the growth in the number of users of the global Internet. By the beginning of 2021, their number amounted to 4.2 billion people; 1.3 million people join social networks every day. Hybridity is expressed in the fact that users spend 3-4 hours a day on the Internet, including interactive viewing of TV broadcasts from tablets, mobile phones, laptops, personal computers, every 12th person owns a smart speaker with a built-in microphone and a voice assistant working on based on artificial intelligence, a third use image recognition tools, 4 out of 10 female users have used an image as a search query [4]. The digital revolution is an important technical and technological condition for new structures, including the industrial Internet of things; virtual and augmented reality; IT solutions in the financial industry and law (blockchain, Smar contract); artificial intelligence (AI) and robotics, neural networks; digital medicine; 3D printing; Big Data, etc. New technological structures significantly change the organization of production and the essence of professional activity in the workplace. Talent management is in demand, which originates from the study "The war for talent" in 2000 [5], GALS technologies, creative industries [6], Agile technologies. In the new technological structures, the social structure, the distribution of social roles in society and the labor market are changing significantly. The breadth and fundamentality of education, the universality of qualifications and the ability to work in interdisciplinary teams in institutionally unformed environments, the ability to solve problems outside the box, and work with limited tools are becoming important.

2 Methods

2.1 Technique

Interaction in a common digital environment of many civilizational communities requires a revision of cultural universals that determine the degree of coherence of relationships [7]. The implementation of digital network platforms of the state expands the network interaction of the state with citizens, business associations and public organizations in all spheres of life. Out-of-state regulators, new local and global parameters of order are being formed. According to the UN approaches [8], it is necessary to create conditions for the full involvement of citizens and their associations in public administration, create network tools for expressing the interests, rights and obligations of citizens and their performance of management functions, and develop network digital institutions. This requires preparing a person for the future in conditions of uncertainty generated by two contradictory tendencies [9]: the desire to preserve the old way of life and the desire of individuals to develop (new, professions, lifestyle), sometimes reaching extreme forms of deviation, called by F. Fukuyama phenomena spontaneous socialization, denoting the elements of a new order through the rejection of some of the traditional values [10].

In these conditions, Galina Sillaste puts forward the idea of a new gender order, understanding it as a social model of new norms, rules, principles of life and work, social and interpersonal interaction of the two largest socio-demographic communities: male and female, which are system-forming communities in the structure population of any state [11]. The ideological foundation is gender equality, in contrast to the model of equality of the status of men and women.

Does digitalization violate the equality of men and women in the labor market and how should the managerial processes of preparing a person for the future be built? It is undeniable that digitalization is changing the labor market. Old professions are becoming a thing of the past, mass layoffs of high and medium-level professionals in obsolete professions, "deprofessionalization", "skill mismatch", and "on-the-fly transformation" are expected. In Russia, 33.9 million people fall into the "skills mismatch" zone, when the professional skills of workers are either insufficient or excessive for the available vacancies. The number of traditionally female jobs is declining, but new opportunities (new specialties) are emerging that can provide employment gaps for women [12]. The problem of discrimination against women in employment in Russia is seen in terms of employers' use of masculine names of professions and descriptions of the proposed position. The problem can be solved by revising by key companies "their mission, social responsibility criteria and the style of posted vacancies with the probable use of verification using artificial intelligence programs" [13]. According to T. Shaulova, digitalization will transform not only the labor market, but also the mentality, and "artificial intelligence" demonstrates "partiality" in connection with the formulation of the problem "using machine learning methods, cognitive bias" at the stage of selecting criteria and inaccurate data into algorithms in period of study, i.e. possible "algorithm bias" is associated with "real world" bias [14]. Artificial Intelligence-as-a-Service (AIaaS) empowers individuals and organizations to access AI on-demand, in either tailored or 'off-the-shelf' forms. However, institutional separation between development, training and deployment can lead to critical opacities, such as obscuring the level of human effort necessary to produce and train AI services. Information about how, where, and for whom AI services have been produced are valuable secrets, which vendors strategically disclose to clients depending on commercial interests" [15].

2.2 Empirical base

1. Analysis of texts Russian media and Russian-language social media, analyzed 168 texts (UNIAN, TASS, Interfax, RBC-Ukraine, IA "Ukrinform", korrespondent.net, "Parliamentary newspaper", IA Reporter, IA Sputnik, zn.ua, IA Ukrainian News, MIA Russia Today, IA Bridge Dnipro, Rossiyskaya Gazeta, Izvestia, Deutsche Welle, gazeta.ua, znak.com, Komsomolskaya Pravda, Kommersant), containing the phrase "public administration" for the period from February 13, 2015 to February 13, 2020, using the Medialogy information system.

2. The problem of discrimination against women in employment in Russia. Poll in Stories [16].

3. Women Business Index (WBI) wave V. Conducted by the Analytical Center NAFI in partnership with Opora Rossii, the Agency for Strategic Initiatives and URALSIB Bank. 600 women who do not carry out entrepreneurial activity were interviewed (the sample is representative in terms of geography, age and type of settlement); 400 women entrepreneurs from eight Federal Districts (the sample of enterprises was built according to the indicators of the size and industry of Rosstat's business [17].

3 Results and Discussion

Features of the modern labor market:

1. An increase in the aging of the population associated with a significant increase in human life expectancy. Russian pensioners will make up 27% of the population by 2022;
2. Problems of employment, impossibility of self-realization, full-fledged life of pensioners and socially vulnerable groups of the population;
3. An increase in the number of divorced women by 6%, the development of "virtual sex", the exaggeration of the rights to register same-sex marriages, a fashion for an extremely high level of material well-being and an unwillingness to combine the responsibilities of a mother and father with building a career, refusal to have children, since keeping children interferes with this;
4. M. Gray and S. Suri showed new forms of exploitation [18] – high-tech companies (Amazon, Google, Microsoft, Uber) exploit ghost work, creating a ghost economy. Freelancing attracts young mothers, early retirement professionals, and recent graduates. They cannot get a job with reliable guarantees of working conditions and are forced for little money to correct large volumes of publications of various formats from prohibited content on sites, etc. They do not have health benefits; they can be dismissed at any time, since there is not enough legislative regulation in this area today.

The topic of freelance work, self-employment, and "nomadic" labor resources in a digital society require legislative registration, since these trends become an objective basis for countering democratic processes and returning part of society to patriarchy. The incompetence of officials in the field of gender relations, the incorrect work of social institutions is dangerous for society. "The average salary of women in 2015 ranged from 68.3% to 82% of the salary of men". The idea of a higher status and more profitable position of men in a digital society leads to an increase in cases of gender reassignment by women, not for medical reasons, but for social reasons [19].

According to the International Economic Forum, from 2015 to 2020, men will lose about 4 million jobs, but get almost 1.4 million new ones, and women will lose 3 million jobs, and get about 550 thousand, the share of new professions by 2022 will grow to 27% [20], which may lead to gender asymmetry regarding professions classified as "male". Freelancing and STEM professions (science, technology, engineering and mathematics) are rapidly developing in 2018. Of the 4,777 employees and 367 employers of Russian organizations, 38% say that there are not enough professional candidates in the labor market; 46% note a shortage of specialists in new professions, and only 16% of respondents believe that there is enough qualified personnel. At the same time, it is noted: the lack of a full-fledged educational structure for the digital economy [21], the low number of women receiving STEM professions; problems of women's entrepreneurship in the field of information technology [22] and problems of perception of the digital world by women of different generations [23].

The authors also note the positive trends in the digital labor market:

1. Robotization and the use of artificial intelligence create a technological basis that allows women to perform former "heavy, male" activities,
2. An increase in the proportion of women in ICT in such professions as operator, software tester, content manager, consultant; it is predicted that gender inequality will be smoothed out in some industries, for example, in dairy farming and the dairy industry [24];
3. An increase in the number of freelancers in the field of online services and cryptocurrencies (digitizers, webmasters, miners, etc.);
4. Cloud-watchers who are able to improve themselves, work in different modes, make extraordinary decisions in non-standard situations and act with a limited set of tools are in demand in high-tech industries.

In Russia, self-organizing processes of activity of women entrepreneurs are observed. For example, the WBI-5 Entrepreneurial Activity Index introduced an index of business readiness for the digital economy among women entrepreneurs (WBI). It amounted to 45 out of 100 items. 12% of women entrepreneurs sent their employees to IT training, 79% would like to have a mentor, 32% noted positive dynamics in their personal / family life (hobbies, sports, interests, travel). They see the problem in the insufficient representation of their business on the Internet – 40% do not have a website, 47% do not have pages in social networks, 27% maintain paperwork. 67% of non-entrepreneurial women consider starting a business, 64% would like to learn from well-known entrepreneurs.

In these conditions, the weakness of the institutions of socialization in a digital society is noted. The participation of civic organizations, the publicity of public administration in the context of digitalization is little covered in modern media. So, the only one of 168 articles in the regional press in five years (2016-2020 raises the problems of introducing an intelligent management system in public administration through the introduction of KPIs in rural administrations, expressing doubts: is it possible to increase the efficiency of public administration by technocratic methods, and what place should citizens in the public administration system [25]. It is necessary to monitor the gender structure of employment and the management of these processes, deprofessionalization processes. It is in demand to search for new models and drivers of development in the formation of appropriate qualifications, assessment and certification of specialists of a new format.

At the same time, the property of resistance is revealed in the virtual environment, the ability of society to preserve the value, worldview system in the format of the return of a part of the population to patriarchal relations. In the context of digitalization, the state as a platform, there is an effect of the primacy of the state and the secondary nature of the structures and organizations of civil society, the institutions of the state primarily do what contributes to the effectiveness of its own administrative structures. The issue of creating a civil society as a platform is out of the question.

4 Conclusion

In September 2018, at the Second Eurasian Female Forum [26], V. Putin emphasized the vital role of women in the modern world, noted the need to solve the problem of gender inequality, remove existing restrictions, create the necessary conditions for professional activity and doing business, provide ample opportunities for girls to receive education, but at the same time preserve the value of motherhood and a full-fledged family. Here, not only comfortable and comfortable conditions are important, ensuring the guarantee of legal and social protection so that a woman can feel independent and protected, but also the ability of all state and non-state actors to quickly and constructively master the incoming information, conditioning the organizational integration of the emerging complexity and ensuring the stability of the social system. In this regard, it is important to form a new concept of public administration, which should be based on the actor-network concept and ethics of digitalization, analysis of social networks, relational sociology, the concept of figuration and fractal-evolutionary concept, the concept of configuring the relational network of stakeholders [27].

The research is supported partially by Russian Fond for Basic Research, grant No. 20-011-00694 "Public Administration as Configuring of Relational Networks in the Public Space of a Digital Society".

References

1. V.V. Glushchenko, Int J Sci Adv, **2(4)**, 554-563 (2021)
2. C.E. Shannon, W. Weaver, The mathematical theory of communication. (University of Illinois Press, Urbana, 1963)
3. M. Hilbert, L. Priscila, Science, **332(6025)**, 60-65 (2011). https://doi.org/10.1126/science.1200970
4. Digital 2021: Global Overview Report. Accessed on: October 4, 2021. [Online]. Available: https://datareportal.com/reports/digital-2021-global-overview-report
5. Ed. Michaels, H. Handfield-Jones, B. Axelro, The War for Talent. (Massachusetts Harvard Business School Press, Boston, 2001)
6. Kreativnyye industrii – sektor novoy ekonomiki [Creative industries – a sector of the new economy]. Accessed on: October 4, 2021. [Online]. Available: http://uldelo.ru/2018/05/03/b-kreativnye-industrii--b-br-sektor-novoi-ekonomiki
7. L.A. Vasilenko, *Tsifrovoe soobshchestvo i rasprostranenie novogo gendernogo poryadka* [Digital community and dissemination of the new gender order], in G. Sillaste (ed), Gender resources and the formation of a new gender order in the 21st century, 122-126. (Moscow, 2020)
8. Reconceptualizing Governance. (Management Development and Governance Division, Bureau for Policy and Programme Support, UNDP, New York, 1997)
9. L.A. Vasilenko, Soc Manag, **4(3)**, 45-56 (2018). https://doi.org/10.18413/2408-9338-2018-4-3-0-4
10. F. Fukuyama, Trust: The Social Virtue and the Path to Prosperity. (Free Press, New York, 1995)
11. G.G. Sillaste, *Novyi gendernyi poryadok kak sotsialnyi fenomen epokhi globalizatsii i ego rossiiskaya traektoriya* [New gender order as a social phenomenon of the age of globalization and its russian trajectory], in G. Sillaste (ed), Gender resources and the formation of a new gender order in the 21st century, 18-29. (Moscow, 2020)
12. N.V. Nikiforova, Sci Notes Rus State Soc Univ, **19(4(157))**, 69-76 (2020). https://doi.org/10.17922/2071-5323-2020-19-4-69-76
13. S.A. Anisimova, *Problema diskriminatsii zhenshchin pri trudoustroistve v Rossii* [The problem of discrimination against women in employment in Russia], in Collection of articles of the International Scientific and Practical Conference: in 2 parts, 2021, Penza, January 23, 2021. (Science and Education, Penza, 2021)
14. T.V. Shaulova, Int Rel Dial Cult, **7**, 42-54 (2019)
15. G. Newlands, Big Data Soc, (2021). https://doi.org/10.1177/20539517211016026
16. @sofya_kings. Pools. Accessed on: October 4, 2021. [Online]. Available: https://www.instagram.com/s/aGlnaGxpZ2h0OjE3ODU1NTg0NDkzMzk0MzQw?story_media_id=2477917158413655647_55447732&utm_medium=copy_link
17. Komitet po razvitiyu zhenskogo predprinimatelstva "Opora Rossii" [Committee for the Development of Women's Entrepreneurship "Support of Russia"]. Accessed on: October 4, 2021. [Online]. Available: https://www.womanopora.ru
18. M. L.Gray, S. Suri, Ghost Work: How to Stop Silicon Valley from Building a New Global Underclass. (Eamon Dolan Books, Boston, 2019)

19. V.G. Ushakova, *Tsifrovoe i (informatsionnoe) obshchestvo: gendernye aspekty* [Digital and (information) society: gender aspects], in Gender Dimension of the Digital Economy: From Strategy to Action (2018-2030). Materials of the All-Russian conference with international participation. (2018)
20. The Future of Jobs Report 2018. Accessed on: October 4, 2021. [Online]. Available: https://www3.weforum.org/docs/WEF_Future_of_Jobs_2018.pdf
21. O.O. Korobova, *Zhenskoe Predprinimatelstvo: Tsifrovye Biznes-modeli* [Female Entrepreneurship: Digital Business Models], in Citizens and Citizens in the Political, Economic and Cultural Processes of Russian Urbanization in the XIV-XXI Centuries, **2**, 18–21. Proc. XI Int. Scientific. conf. RAIZHI and IEA RAS, October 4-7, 2018, Nizhny Novgorod. (Moscow, 2018)
22. A. Larsson, Y. Viitaoja, *Identifying the digital gender divide*, in A. Larsson, R. Teigland (eds), The Digital Transformation of Labor: digest. (Routledge, 2019)
23. L.A. Gromova, L.S. Kiseleva, Woman Rus Soc, **3(88)**, 85-96 (2018)
24. M.N. Dudin, E.V. Levina, A.N. Anishchenko, Soc Labour Res, **38(1)**, 73-84 (2020). https://doi.org/10.34022/2658-3712-2020-38-1-73-84
25. Vrio gubernatora Lipetskoi oblasti planiruet vnedrit KPI v administratsiyakh selskikh raionov [Acting Governor of Lipetsk Region plans to introduce KPI in rural administrations]. Accessed on: October 4, 2021. [Online]. Available: https://www.kommersant.ru/doc/3764386
26. The Second Eurasian Women's Forum plenary session, Women for Global Security and Sustainable Development. Accessed on: October 4, 2021. [Online]. Available: http://council.gov.ru/en/events/news/96167/
27. V.V. Zotov, L.A. Vasilenko, Manag Consul, **5**, 98-109 (2021). https://doi.org/10.22394/1726-1139-2021-5-98-109

Labour activities of disabled people in the Russian Federation: features of legal regulation

Ruslan Madamindzhanovich Khakimov[1*], *Igor Leonidovich* Rudenko[2], *Galina Vasilievna* Martyanova[3], *Alla Ignatovna* Kairova[4], and *Natalya Valerievna* Cherkashin[3]

[1]Moscow State University for the Humanities and Economics, Faculty of Law, Moscow, Russia
[2]Moscow State University of Humanities and Economics, Faculty of Psychology and Pedagogy, Moscow, Russia
[3]Moscow State University for the Humanities and Economics, Department of Civil Law and Procedure, Moscow, Russia
[4]Moscow State University for the Humanities and Economics, Department of Theory and History of State and Law, Moscow, Russia

Abstract. The purpose of the research is to provide a legal analysis of the labour activities of persons with disabilities in the Russian Federation. The authors consistently address employment issues of the persons with disabilities; also based on a review of Russian judicial practices, they formulate recommendations for the harmonisation of this vital area of labour relations. The research used the methods of system-oriented analysis, content analysis and comparative analysis. Based on the scientific analysis, the features and prospects of the development of legal regulation of employment of persons with disabilities were identified, and the comparative analysis method to identify areas of practice in the application of Russian federal and regional laws in Russia. At the beginning of the article, a historical excursion to the creation of the institution of labour relations with the involvement of persons with disabilities is provided; it starts from the second half of the 18th century until the end of the Soviet era, in the 1990s of the 20th century. Further, based on an analysis of the main legal instruments regulating the specified area of relations, the authors consider issues of the legislative consolidation of the procedure for quoting jobs for persons with disabilities, and also analyse the legal forms of protecting the labour rights of persons with disabilities and preventing discrimination on the basis of disability. Special focus is put on exercising control over the fulfilment by employers of the legal requirements concerning the provision of disabled people with jobs according to the established quota, as well as compliance with the related requirements established by sanitary rules. In Russia, the realisation of the right to work of disabled people is regulated not only at the federal but also regional level. Top priority is given to the regulation of labour relations involving disabled people and creation of the most favourable conditions for them in the constituent entities of the Russian Federation. In conclusion, the authors provide a list of proposals and recommendations for improving the

* Corresponding author: rmkhakimov.ru@gmail.com

legal regulation of labour relations in the Russian Federation involving disabled people.
Keywords: right to work, legal regulation, rights of persons with disabilities, labour relations, job quotas, social obligations, human rights, disability discrimination, prohibition of discrimination

1 Introduction

The Russian Federation is a legal social state in which social problems are a major focus of interest. Certainly, one of them is the employment of disabled people.

The relevance of the research is due to a number of interrelated social and legal factors and, first of all, the high social problem of disability. All countries of the world and every group of society have people with disabilities. The number of them is significant and it keeps growing. The number of disabled people in Russia as of 01 April 2021 is 6,132,491 among women (57%) and 4,688,234 among men (43%) of the total population of the Russian Federation; and taking into account their family members, the topic of disability directly affects a quarter of the total population [1].

Despite the government's efforts, social policy towards people with disabilities is not sufficient; these efforts have a largely delayed effect, which increases the social stratification of Russians. In this context, a primary objective of the government is to carry out integration with less social difficulties for people with disabilities, for whom the level of social and labour adaptation is significantly reduced. Currently, it seems necessary to develop and pursue new state policy priorities focused on people with disabilities as far as labour relations are concerned. Such a policy should promote the principles of equal opportunities, focus bringing their life to normal and prohibiting discrimination.

By analysing the scientific literature on the issues in question, their narrow focus should be noted. Among the few analysed theses of legal, pedagogical and sociological focus, it is worth noting the scientific research of Gusev [2], Mozdokova [3], Tkachenko [4].

During the research, the regulatory documents of the Russian Federation, regional legislation and historical legalisations of the Russian Empire were widely used and analysed.

2 Materials and methods

The purpose of the research is to analyse the employment of people with disabilities in Russia and its legal regulation, identify legislative gaps in the implementation of the rights of people with disabilities in the world of work and promotion of the theoretical and practical mechanisms.

As part of the research, the following tasks were solved. The main stages of the history of regulation of labour relations of persons with disabilities were systematised; the key trends in the creation and legal regulation of the policy of equal opportunities in the field of employment in Russia and the regions were traced; based on the study of the regulatory framework and law enforcement practices, some problems of employment of people with disabilities in modern society and ways to solve them were identified.

Modern developments undoubtedly rely on the foundations created by the predecessors of modern scientists and researchers. By applying the historical research method, the development of labour legislation and its "echoes" in the modern regulation are considered. At the same time, methods of comparison and analysis are applied when studying the application of the existing law in regions and the Russian Federation as a whole. Also, a scientific and applied approach to study the conceptual apparatus in the field of regulation of labour relations for a certain category, namely disabled people. By applying these methods, the relationship between modern scientific achievements and legislative regulation of the employment of persons with disabilities is addressed.

3 Results

The contemporary Russian legislation concerning the employment of persons with disabilities is not fully implemented. In practice, some social barriers are encountered in the way of effective employment of persons with disabilities.

The analysis of the employment of people with disabilities in Russia can be clearly seen on the example of the statistical data presented in the tables below.

Table 1. Information about working disabled people [5].

	2014	2015	2016	2017	2018	2019	2020	2021
People with disabilities, total, thsd	2,407	2,473	2,543	2,012	1,644	1,655	1,571	1,517
Group I, thsd	82	82	82	69	49	53	40	41
of which lifelong disabled persons, thsd	3	3	3	6	7	7	7	7
Group II, thsd	913	923	939	704	557	556	525	472
of which lifelong disabled persons, thsd	23	23	24	49	58	60	63	63
Group III, thsd	1,409	1,466	1,520	1,235	1,038	1,046	1,006	1,004
of which lifelong disabled persons, thsd	58	69	83	110	139	146	153	156
Working disabled people in the total number of disabled people, %	19.1	19.8	20.5	16.9	14.3	14.7	14.0	13.9

When reviewing the situation over time, it is noteworthy that until 2016 in Russia there was an annual increase in the number of employed people with disabilities, but since 2017, the official statistics of the Russian Federation began to show a decrease in the number of working people with disabilities. This is due to the entry into force of the federal law dated 29 December 2015 No. 385-FZ "Suspension of certain provisions of regulatory instruments of the Russian Federation, amendments to certain regulatory instruments of the Russian Federation and the specifics of increased insurance pension, a flat-rate payment to the insurance pension and social pensions" providing for non-indexation of pensions for working pensioners [6].

In the dynamics of employment of persons with disabilities until 2017, there were multidirectional trends depending on the group of disabilities. According to I.V. Kalashnikova and A.A. Trinadtsatko, the increase in the proportion of employed persons with disabilities of Group 1 was due to a gradual decrease in their total number, approximately 50,000-70,000 per year [7].

In 2017, the Russian Federation Government approved an action plan to increase the employment rate of persons with disabilities in 2017-2020, which provided for an increase in the efficiency of employment services and the creation of conditions for expanding employment opportunities for persons with disabilities, which significantly increased their employment rates.

The presented statistics make it possible for us to conclude that the employment rates of the disabled people are largely determined by the government's policy in this area, as well as by the general socio-economic situation in the country. This circumstance necessitates the adoption of effective statutory instruments that change the institutional conditions for the employment of persons with disabilities, which should be based on an assessment of the effectiveness of measures taken in the future.

4 Discussion

The peculiarities of employment of persons with disabilities have attracted the attention of scientists from overseas countries for a long time. In particular, certain aspects of the employment of people with disabilities were studied by scientists from Poland [8], Kazakhstan [9], Hungary [10], Moldova [11], Republic of Belarus [12] and other countries.

Issues of people with disabilities in the area of labour relations were the subject of regulation by the Russian state like in previous eras. The state took measures to create conditions for the employment of certain categories of persons with disabilities [13]. At first, disabled people were recruited to work in dedicated institutions created by the state, for example, by the decree of Empress Catherine II "Creation of public charity orders as part of provincial institutions" in each province of the Russian Empire, special bodies were set up – orders of public charity – in charge of the organisation and activities "workhouses" locally. In the second half of the 19th century, the so-called "houses of industriousness" were organized. In the "houses of industriousness" people with disabilities were provided with the necessary training followed by paid work.

Subsequently, employment became a full-fledged priority of the public policy and implied a series of medical, economic and organisational mechanisms for the realisation by disabled people of the right to work and protection of their health. The creation of the Soviet state significantly changed the country's political agenda. Able-bodied citizens of both sexes (from 16 to 55 years old) were required to get a job and work. In 1918, the workers' and peasants' government issued the Council of People's Commissars of the RSFSR [14]. This document guaranteed disabled citizens a pension that ensured their existence, in an amount that directly depended on the degree of disability. It is notable how precisely the early Soviet state defined disability – it was measured literally as a percentage. The state medical expert examination was responsible for that measurement. Only those persons, who were not able to work due to illness (congenital or acquired) or injury, were entitled to full exemption from labour service [15].

The labour activities of disabled people in the early 1930s became the focus of scientific analysis. In 1932, research institutes that focused on involving people with disabilities in the labour force were established. In addition, it was in the 1930s when special workshops appear for certain categories of disabled people on the basis of nosologies (pulmonary tuberculosis, diseases of the nervous system, visually impaired, etc.). The following important stage in promoting the realisation of the rights of persons with disabilities to work was associated with the adoption of the Constitution of the USSR in 1936. The 1936 Constitution recognised the right to work and pension for all citizens with disabilities irrespective of the type of disability and the cause of it [16]. Employment of people with disabilities in the post-war period primarily included their preliminary preparation for work, from supplying the necessary prostheses to short-term training.

During and after the Great Patriotic War, the number of people with disabilities in the USSR increased sharply as a result of the disablement of servicemen who were wounded during hostilities. Responsibility for the realisation of the right to work of persons who received the status of people with disabilities was entrusted to the People's Commissars by the decree of the Council of People's Commissars No. 640 "Employment of people with disabilities of the Patriotic War" [17]. In accordance with that document, workers with disabilities received both wages and pensions irrespective of the size of their previous income which significantly improved their financial situation.

The legislation on the regulation of labour relations with the participation of persons with disabilities was further developed in the 1971 Labour Code [18] which increased labour guarantees for persons with disabilities. This document brought to the legislative level the employer's obligation to hire people with disabilities and establish a special conditions specifically for them, for example, part-time work and other preferential working conditions in accordance with medical recommendations.

Contemporary Russian legislation, first of all, the Constitution prohibits discrimination at work [19]. In accordance with Article 3 of the Labour Code of the Russian Federation [20], Russian citizens may not have any advantages in labour relations including when applying for a job, including gender, age, other characteristics, if they are not related to the professional aptitude of an employee. Part 2 of the same article explains that it is not discrimination to establish any exceptions or preferences if they are related to the specifics of this type of labour activities and (or) are conditioned by the state's increased concern for persons in need of special legal and social protection.

With regard to persons with disabilities, who have restrictions on the performance of labour activities, the state guarantees them job quotas, that is, the employer's obligation to enter into an employment agreement with a certain number of applicants with disabilities, subject to the rule that it is possible to hire workers in excess of the quota, and less is unacceptable.

In accordance with Articles 20, 21 of the federal law "Social Protection of Persons with Disabilities in the Russian Federation" for organisations, irrespective of organisational and legal forms and forms of ownership, the number of employees of which is more than 100 people, the legislation of the constituent entity of the Russian Federation establishes a quota for hiring disabled people – at least 2 and no more than 4 per cent of the average number of employees [21]. As for employers with the number of employees from 35 to 100 people, the constituent entities of the Russian Federation are not obliged but can set a quota for hiring people with disabilities in the amount of no more than 3% of the average number of employees. For example, in St. Petersburg, for organisations with more than 100 employees, a quota of 2.5% of the average number for employment of people with disabilities is established [22]. The effectiveness of the quota method in most regions of Russia is extremely low. Due to the lack of rigid control, employers do not pay the stipulated fines for evading quotas. Enterprises prefer to independently select employees from among vulnerable groups of the population for existing vacancies without using the services of employment centres. As a result, vulnerable individuals registered with the employment service remain unemployed [23].

The law "Employment of the Population in the Russian Federation" [24] was amended by the federal law No. 11-FZ dated 23 February 2013 [25] in accordance with which employers are obliged to provide monthly employment agencies with information on the availability of vacancies (positions), on the fulfilment of the quota for the employment of persons with disabilities, as well as on created or allocated workplaces for the employment of people with disabilities in accordance with a certain quota, inform the employment agencies about local regulations containing information about those workplaces.

For violation of the rights of persons with disabilities in the field of employment, namely for the employer's failure to fulfil the obligation to create or allocate jobs for the employment of persons with disabilities on the basis of the established quota, for the employer's refusal to hire persons with disabilities on account of the established quota by Article 5.42 of the Code of Administrative Offences of the Russian Federation for officials, administrative liability in the form of a fine from 5,000 to 10,000 roubles [26]. Moreover, the federal law of 23 February 2013, No. 11-FZ, the same liability was extended to officials of the employment agency of the Russian Federation for the unjustified refusal to register a person with disabilities as unemployed.

The working conditions for people with disabilities should meet individual rehabilitation or habilitation programmes. In addition, for people with disabilities of Groups I and II, reduced working arrangements are introduced – no more than 35 hours per week, subject to the preservation of full wages (Article 92 of the Labour Code of the Russian Federation). This rule does not seem fair enough, since in accordance with the Order of the Russian Ministry of Labour dated 27 August 2019 No. 585n "Classifications and criteria used in the implementation of medical and social examination of citizens by federal state institutions of medical and social examination" [27], Disability Group I can be characterised by a violation of human health with a persistent and significantly pronounced disorder of body functions caused by diseases, consequences of injuries or defects leading to a restriction of one of the following categories of life activities or their combination (ability to self-service; ability to move; to orientation; to communication; to control their behaviour; to learning and work activities).

A 35-hour working week should rather be established for people with disabilities of Groups II and III, and for Group I, the reduction in the number of working hours should be more significant (for example, up to 30 hours per week).

In accordance with Article 23 of the law "Social Protection of Persons with Disabilities in the Russian Federation", Articles 96, 99 and 113 of the Labour Code of the Russian Federation, persons with disabilities can be recruited to work at night, overtime work, to work on weekends and non-working holidays only if the following conditions are met:

It should be noted that in the area of job quotas for persons with disabilities, there are, in our opinion, some legal gaps that need to be addressed, namely:

- work should not be contraindicated for them in accordance with the medical assessment report;
- it is allowed to carry out such work only with their written consent;
- persons with disabilities should be informed in writing of their right to refuse the specified work.

Otherwise, such work should be considered as compulsory (Article 4 of the Labour Code of the Russian Federation), and for such a violation, the employer shall be brought to administrative responsibility under Article 5.27 of the Code of Administrative Offences of the Russian Federation.

The health status of working people with disabilities requires a longer time to recover physical, intellectual and mental costs; therefore, people with disabilities are provided with extended annual basic paid leave of 30 calendar days. In accordance with Article 128 of the Labour Code of the Russian Federation, persons with disabilities, upon their written application, are granted unpaid leave for up to 60 calendar days a year.

The health status of working people with disabilities requires a longer time to recover physical, intellectual and mental costs, therefore people with disabilities are provided with extended annual basic paid leave of 30 calendar days. In accordance with Article 128 of the Labour Code of the Russian Federation, disabled persons, upon their written application, are granted unpaid leave for up to 60 calendar days a year.

- the federal legislation does not contain the concept of a quota, which causes difficulties in its calculation. The 1995 law does not disclose the concept of "quota". If we turn to the Recommendations on 1995 Quotas, which have now lost, then they understood the quota as the minimum number of jobs for citizens in particular need of social protection and experiencing difficulties in finding a job, which was determined as a percentage of the average number of employees of enterprises, institutions and organisations.

This definition of a quota is not entirely correct, since it does not allow calculating the number of jobs allocated as a quota if a fractional value is formed, i.e. federal law does not set rules for rounding fractional parts of numbers.

For example, with the maximum quota for the employment of people with disabilities at 4% and the minimum number of employees of the organisation, at which the quota is introduced, i.e. 101 people, as determined in the 1995 Law, the number of jobs will be 4.04. It is obvious that it is impossible to employ a person with disabilities at 0.04 of a workplace;

- it should be noted that at present the level of quotas for people with disabilities has decreased: previously, the quota for hiring people with disabilities was set for entities exceeding 30 employees, and the percentage of quota jobs should have been at least 3 per cent.

If we compare the level of quotas in Russia and abroad, then it is much lower in Russia. For example, in Germany, every employer who employs more than ten employees is obliged to provide 6% of jobs for people with disabilities. In France, in enterprises with the same number of employees, 3% of the staff should be workers with disabilities and 10% war disabled veterans [28];

- in the 1995 Law, when establishing a quota, the statistical categories "number of employees" and "average number" of employees, on the basis of which the quota is established, are not used correctly. The number of employees is calculated as of a specific date. In fact, it is determined daily, including weekends and holidays.

Given the dynamism of labour resources in a market economy, the number of employees can, for a short period of time, for example, a month, repeatedly exceed 100 people and be less than this value, which makes it difficult to fulfil the quota, makes this category unsuitable for quotas and creates an opportunity for employers to evade the quota.

The category "average number of employees" is less dynamic, since it is determined by summing the number of employees on the payroll for each calendar day of the month and dividing the amount received by the number of those days. Therefore, it is more preferable;

- the federal law does not establish any features (requirements) in respect of jobs allocated on account of the quota.

The 1995 law does not take into account that quotas are carried out in relation to citizens who have a limited ability to work (they cannot be involved in work with harmful (hazardous) working conditions). For some citizens, it is required to create special jobs allowing employers to allocate either the lowest-paid jobs or jobs that are not suitable for the work of the corresponding category of citizens.

Therefore, when establishing a quota, it is necessary to use such criteria for the differentiation of labour as the industry affiliation of the employer, the nature of working conditions in the related production, the availability of specially equipped workplaces, etc. [29].

Thus, in the Russian Federation as a whole and in its regions separately, many regulatory documents aimed at exercising the right to work by persons with disabilities were adopted but not all of them are observed. Punishments for violation of the rights of persons with disabilities and the legal provisions in the area are too mild, and the powers of the regulatory authorities are too limited.

5 Conclusion

The state and society should be interested in adapting this social group so that people with disabilities can freely work in the profession that they consider most suitable for themselves. Employers should not remain indifferent to those people's problems. Enterprises should be supplied with custom equipment for people with disabilities, so that they feel like full-fledged people capable of working, so that they feel on an equal footing with healthy people. It is necessary to take into account the fact that it is economically more profitable for society and the state to have an employee with a disability – a taxpayer than a non-working disabled person – a dependent. All relevant services and departments, as well as the developed and existing legal provisions concerning the rights of persons with disabilities to work are aimed at this.

Priority should be given to the following measures to improve the employment of persons with disabilities:

- measures to design and adapt workplaces and work premises so that they are accessible to persons with various forms of disabilities;
- measures of support in the use of new technologies, development and production of auxiliary devices, fixtures, equipment and measures to facilitate the access of persons with disabilities to such means and equipment, so that they can get and keep a job;
- the state should organise and support companies that contribute to overcoming negative attitudes and prejudices towards employees with disabilities;
- the state as an employer should create favourable conditions for the employment of persons with disabilities in the public sector;
- measures to maximise the involvement of persons with disabilities in the development of training programmes and employment programmes in the private and non-governmental sectors should be taken;
- the state and organisations, employers should cooperate with organisations of persons with disabilities in all efforts taken to create opportunities for training including the establishment of flexible working hours, part-time work and job splitting.

To eliminate the problems identified in the Russian Federation, the following solutions are proposed: to conduct an annual monitoring of employment at workplaces equipped for people with disabilities and determine the level of need for people with disabilities in employment; it is necessary to develop centres of assistance in the employment of students and graduates with disabilities, at public or municipal organisations of vocational education; to strengthen the interaction of educational authorities and employment agencies with medical and social assessment institutions as well as social welfare bodies in order to implement an individual programme for the rehabilitation and habilitation of a person with disabilities; to ensure that tax incentives can be provided for employers with employees with disabilities; to toughen employers' liability for refusing to hire a disabled person within the established quota; make the above changes to the existing Russian law, specify the provisions concerning job quotas for people with disabilities, in the Labour Code of the Russian Federation; combining miscellaneous provisions on labour relations with the involvement of persons with disabilities, add a separate chapter "Peculiarities of regulation of labour of persons with disabilities".

References

1. Chislennost invalidov [Number of people with disabilities]. Accessed on: October 17, 2021. [Online]. Available: https://sfri.ru/analitika/chislennost
2. A.Yu. Gusev, Sudebnaya zashchita prava rossiiskikh grazhdan na sotsialnoe obespechenie [Judicial protection of the right of Russian citizens to social security] Abstract of a PhD thesis in Legal Science (The Institute of State and Law of the Russian Academy of Sciences, Moscow, 2017)
3. Yu.S. Mozdokova, Kommunikativno-deyatelnostnyi podkhod k integratsii invalidov: sotsialno-kulturnyi aspekt [Communicative-activity approach to the integration of disabled people: social and cultural aspect], Abstract of a PhD thesis in Pedagogic Sciences (Moscow State University of Culture and Arts, Moscow, 2006). Accessed on: October 17, 2021. [Online]. Available: https://static.freereferats.ru/_avtoreferats/01003319584.pdf
4. V.S. Tkachenko, Integratsiya v rossiiskom obshchestve lyudei s invalidnostyu [Integration of people with disabilities in the Russian society] Abstract of a PhD thesis in Sociology (North-Caucasus State Technical University, Stavropol, 2007. Accessed on: October 17, 2021. [Online]. Available: https://www.dissercat.com/content/integratsiya-v-rossiiskom-obshchestve-lyudei-s-invalidnostyu
5. Polozhenie invalidov [The situation with the disabled people]. Accessed on: October 17, 2021. [Online]. Available: https://rosstat.gov.ru/folder/13964
6. A.A. Voropaev, Young Res, **17(307)**, 341-346 (2020)
7. I.V. Kalashnikova, A.A. Trinadsadko, Newsl Pacific State Univ, **1(44)**, 141–154 (2017)
8. M. Garbat, Econ Trends, **1**, 17-25 (2017)
9. M.M. Yessirkepova, R. M. Anarova, Newl Gumilyov Euras Nation Univ. Ser: Law, **3(128)**, 127-136 (2019). https://doi.org/10.32523/2616-6844-2019-128-3-127-136
10. E.O. Olabisi, Polish J Sci, **26-2(26)**, 3-6 (2020)
11. E. Aculai, L. Saghin, Econ Soc: Theor Sci J, **1**, 31-42 (2020). https://doi.org/10.36004/nier.es.2020.1-03
12. I.A. Komotskaya, Labour Soc Law, **4(32)**, 39-42 (2019)
13. N.V. Cherkashina, A.V. Ognev, Int Res J, **5-3(107)**, 109-114 (2021)
14. Dekret SNK RSFSR ot 31.10.1918 "Polozhenie o sotsialnom obespechenii trudyashchikhsya" [Decree of the Council of People's Commissars of the RSFSR of 31.10.1918 "Regulations on social security of workers"]. Accessed on: October 17, 2021. [Online]. Available: http://www.consultant.ru/cons/cgi/online.cgi?req=doc&base=ESU&n=4478#NBWHJfS4id1oTqs02
15. Sobranie uzakonenii i rasporyazhenii rabochego i krestyanskogo pravitelstva. Statya 905. Kodeks zakonov o trude [Collection of laws and regulations of the workers' and Peasants' government. Article 905. Labor Code]. Accessed on: October 17, 2021. [Online]. Available: https://vcpi.nlb.by/static/pdf/kodexy-BSSR/st_905_kodeks_zakonov_o_trude.pdf

16. Konstitutsiya (Osnovnoi zakon) Soyuza Sovetskikh Sotsialisticheskikh Respublik (utverzhdena postanovleniem Chrezvychainogo VIII Syezda Sovetov Soyuza Sovetskikh Sotsialisticheskikh Respublik ot 5 dekabrya 1936 g.) [Constitution (Basic Law) of the Union of Soviet Socialist Republics (approved by the resolution of the Extraordinary VIII Congress of Soviets of the Union of Soviet Socialist Republics of December 5, 1936)]. Accessed on: October 17, 2021. [Online]. Available: https://constitution.garant.ru/history/ussr-rsfsr/1936/red_1936/3958676/
17. Postanovlenie ot 6 maya 1942 g. No. 640 O trudovom ustroistve invalidov otechestvennoi voiny [Resolution of May 6, 1942 no. 640 On the labor arrangement of the disabled of the patriotic war] Accessed on: October 17, 2021. [Online]. Available: https://www.alppp.ru/law/trud-i-zanjatost-naselenija/trud/151/postanovlenie-snk-sssr-ot-06-05-1942--640.html
18. "Kodeks zakonov o trude Rossiiskoi Federatsii" (utv. VS RSFSR 09.12.1971) (red. ot 10.07.2001, s izm. ot 24.01.2002) ["The Labor Code of the Russian Federation" (approved by the Supreme Council of the RSFSR on 09.12.1971) (ed. of 10.07.2001, with amendments. from 24.01.2002)] Accessed on: October 17, 2021. [Online]. Available: http://www.consultant.ru/document/cons_doc_LAW_1035/
19. "Konstitutsiya Rossiiskoi Federatsii" (prinyata vsenarodnym golosovaniem 12.12.1993 s izmeneniyami, odobrennymi v khode obshcherossiiskogo golosovaniya 01.07.2020) ["Constitution of the Russian Federation" (adopted by popular vote on 12.12.1993 with amendments approved during the all-Russian vote on 01.07.2020)]. Accessed on: October 17, 2021. [Online]. Available: http://www.consultant.ru/document/cons_doc_LAW_28399/
20. "Trudovoi kodeks Rossiiskoi Federatsii" ot 30.12.2001 No. 197-FZ (red. ot 28.06.2021, s izm. ot 06.10.2021) (s izm. i dop., vstup. v silu s 01.09.2021) ["Labor Code of the Russian Federation" dated 30.12.2001 No. 197-FZ (ed. dated 28.06.2021, with amendments. from 06.10.2021) (with amendments and additions, intro. effective from 01.09.2021)]. Accessed on: October 17, 2021. [Online]. Available: http://www.consultant.ru/document/cons_doc_LAW_34683/
21. Federalnyi zakon "O sotsialnoi zashchite invalidov v Rossiiskoi Federatsii" ot 24.11.1995 No. 181-FZ (poslednyaya redaktsiya) [Federal Law "On Social Protection of Disabled Persons in the Russian Federation" dated 24.11.1995 No. 181-FZ (latest edition)]. Accessed on: October 17, 2021. [Online]. Available: http://www.consultant.ru/document/cons_doc_LAW_8559/
22. O kvotirovanii rabochikh mest dlya trudoustroistva invalidov v Sankt-Peterburge (s izmeneniyami na 19 fevralya 2020 goda) [On the quota of jobs for the employment of disabled people in St. Petersburg (as amended on February 19, 2020)]. Accessed on: October 17, 2021. [Online]. Available: http://docs.cntd.ru/document/8377307
23. N.V. Gashenina, Regionology, **3(60)**, 222-227 (2007)
24. Zakon RF "O zanyatosti naseleniya v Rossiiskoi Federatsii" ot 19.04.1991 No. 1032-1 (poslednyaya redaktsiya) [The Law of the Russian Federation "On employment of the population in the Russian Federation" dated 19.04.1991 No. 1032-1 (latest edition)]. Accessed on: October 17, 2021. [Online]. Available: http://www.consultant.ru/document/cons_doc_LAW_60/
25. Federalnyi zakon "O vnesenii izmenenii v otdelnye zakonodatelnye akty Rossiiskoi Federatsii po voprosu kvotirovaniya rabochikh mest dlya invalidov" ot 23.02.2013 No. 11-FZ (poslednyaya redaktsiya) [Federal Law "On Amendments to Certain Legislative

Acts of the Russian Federation on the Issue of job Quotas for disabled people" dated 02/23/2013 No. 11-FZ (latest edition)]. Accessed on: October 17, 2021. [Online]. Available: http://www.consultant.ru/document/cons_doc_LAW_142512/

26. "Kodeks Rossiiskoi Federatsii ob administrativnykh pravonarusheniyakh" ot 30.12.2001 No. 195-FZ (red. ot 01.07.2021) (s izm. i dop., vstup. v silu s 01.10.2021) ["The Code of the Russian Federation on Administrative Offenses" of 30.12.2001 No. 195-FZ (ed. of 01.07.2021) (with amendments and additions, intro. effective from 01.10.2021)]. Accessed on: October 17, 2021. [Online]. Available: http://www.consultant.ru/document/cons_doc_LAW_34661/
27. Prikaz Ministerstva truda i sotsialnogo razvitiya RF ot 27 avgusta 2019 g. № 585n "O klassifikatsiyakh i kriteriyakh, ispolzuemykh pri osushchestvlenii mediko-sotsialnoi ekspertizy grazhdan federalnymi gosudarstvennymi uchrezhdeniyami mediko-sotsialnoi ekspertizy" [Order of the Ministry of Labor and Social Development of the Russian Federation No. 585n dated August 27, 2019 "On classifications and criteria used in the implementation of medical and social expertise of citizens by Federal state institutions of medical and social expertise"]. Accessed on: October 17, 2021. [Online]. Available: https://www.garant.ru/products/ipo/prime/doc/72921006/
28. I.Ya. Kiselev, Sravnitelnoe i Mezhdunarodnoe trudovoe pravo [Comparative and International Labour Law] (PRIOR, Moscow, 1999)
29. Yu.B. Korsanenkova, Labor Law Rus Abroad, **2**, 23-27 (2010)

Professional self-realization of disabled persons in an urban environment

Dmitry Sergeevich Raidugin[1*], *Elmira Kamalovna* Naberushkina[1], *Ekaterina Vladimirovna* Voevodina[1], *Yevgeny Olegovich* Kubyakin[1], and *Vladimir Valeryevich* Plotnikov[2]

[1]Financial University under the Government of the Russian Federation, Department of Sociology, Moscow, Russia
[2]Krasnodar University of the Ministry of Internal Affairs of Russia, Department of Philosophy and Sociology, Krasnodar, Russia

Abstract. The article is a review of the part of the research conducted in the framework of the grant of the Russian Foundation for Basic Research "Urban well-being of people with disabilities". The immediate subject of the article is professional self-realization as a factor of urban well-being of a person with a disability. Well-being is understood as a subjectively constructed phenomenon, that is, through the prism of people with disabilities' own perceptions of the sufficiency of conditions created in the urban environment for their professional self-realization and, as a consequence, for well-being in general. Professional self-realization has become one of the six factor groups that directly or indirectly influence the urban well-being of citizens with disabilities. Along with professional self-realization, the factors of urban well-being in the framework of the study were: trust in the government and involvement in management processes; the level of implementation of basic values; psychological comfort; accessibility of the spatial environment; and finally, the quality of social services. The five factor groups listed above are not discussed in the article, but the reader should have an idea of the context in which professional self-realization is studied. The study included two sub-samples, reflecting the specifics of a megalopolis (Moscow) and a regional center (Saratov).

Keywords: urban environment, disability, well-being of a person with a disability, urban well-being, professional self-realization

1 Introduction

Every person, at least once in his life, has asked himself the question of how much his everyday life corresponds to his own expectations. And hardly anyone has ever been able to answer this question in the affirmative, that is, to state that in each of the areas that are important to him, everyday life meets all his desires without exception, from "comfortable

* Corresponding author: raydmis@mail.ru

living conditions" to having people in his social environment who are exclusively "pleasant" to him.

If we imagine that everyday life includes a number of conditions that make it difficult, due to the presence of a disability, then in addition to typical problems, issues related to the environment are actualized.

There are many environmental factors on which the happiness of each of us depends. In this work, one of the most essential conditions for a person with a disability will be investigated: accessibility of urban space. Space should not be understood in a narrow sense, that is, as a set of buildings, structures, roads, and other material conditions, in which a person finds himself when he leaves the walls of his home. We interpret urban space with its manifold architectural, power and social ways of differentiation and distancing as a form reflecting social interactions in society. Spatial forms arise from social forms, urban architecture is shaped by social attitudes, and the city's physical barriers that "immobilize" disabled people reflect social interactions between disabled and non-disabled people [1]. In other words, together with the disability factor, the lack of adaptability of urban space not only contributes to the strengthening of physical barriers, but also stimulates the emergence of new circumstances that hinder the well-being of people with disabilities.

In this paper, well-being is thought of as a social reflection of happiness [2]. The basis for the restriction is the addition of such a property as the social conditionality of happiness, transferring the latter from the categorical plane to the conceptual plane and, as a consequence, from the disciplinary framework of philosophy to the sociological research field.

Confirmation of the validity of this approach can be found in many publications from universally recognized dictionaries to scientific articles. Thus, the Encyclopedic Dictionary of F.A. Brockhaus and I.A. Efron identifies the two concepts: "Well-being (Greek ευ'δαιμονία) is the same as happiness, always subjectively understood by the attainment of all the goods (see this sl.) to which one aspires" [3]. This is also characteristic of a number of other encyclopedic sources. As far as scientific literature is concerned, the concept of well-being, as already noted, has found its reflection in many publications, the most methodologically referential of which we will take as a basis.

S.V. Yaremchuk defines well-being as "a person's satisfaction with the main spheres of life and life in general" [4]. Her study identifies indicators of well-being according to six main "modules" previously proposed by R.M. Shamionov [5]: 1) financial situation; 2) personal self-determination (assessment of the degree of self-realization); 3) social self-determination; 4) personal well-being (assessment of own character and individual characteristics); 5) professional self-determination and 6) physical and psychological health. Foreign colleagues Tom Rath and Jim Harter identify five main domains of well-being: 1) occupational well-being; 2) physical well-being; 3) social well-being; 4) financial well-being; and 5) environmental well-being [6], noting that to achieve happiness, success must be achieved in all five domains and failure in at least one domain cannot be compensated by well-being in the others. Another group of foreign authors led by I. Giovannini identifies six factors of well-being: 1) physical and mental health; 2) freedom and self-determination; 3) a good job; 4) satisfactory interpersonal relationships; 5) material well-being; 6) knowledge and understanding of the world [7].

Of the most pressing problems related to the one considered in this article, it is necessary to highlight: the impact of urban transformations on the well-being of citizens [8]; the relationship between income and subjective well-being [9]; the relationship between the quality of life and life crises [10]; social support and well-being in the student environment [11]; the impact of work on life satisfaction [12]; various aspects of well-being in the era of a pandemic [13-15]. It should be noted separately that we deliberately avoided studying the problem of the pandemic due to the fact that the data collection was carried out before the Covid-19 pandemic became a global phenomenon.

As we can see, most of the presented authors directly or indirectly include professional self-realization among the factors of well-being. This state of affairs allows us to formulate the thesis that the conditions of the urban environment, created for professional self-realization of citizens with disabilities, directly affect their well-being in general.

2 Materials and methods

As part of the implementation of the "Urban Well-Being of Persons with Disabilities" grant, a field survey was conducted in March-June 2019 to collect data on the well-being of people with disabilities in the urban environment. The quota sample of the questionnaire survey included two sub-samples in Moscow and Saratov. The total sample size was 497 people, which allows achieving a confidence interval of 4.4%. Accordingly, the margin of error is 2.2%.

The main quotas were distributed as follows:

- by gender (men – 48.5%, women – 51.5%);

- by disability group (group 1 – 26.6%; group 2 – 35.4%; group 3 – 27.4%; other categories – 10.7%).

In addition to these grounds for quotas, the age composition, specifics of the underlying disease, and occupation were also taken into account. A more complete description of the research methodology is presented in the previously published article "Methodology and Methodology for Researching Opportunities for Professional Self-Actualization of People with Disabilities in the Context of the Problem of Urban Well-Being" [16].

To describe and analyze the data, mainly methods of descriptive statistics were used, which allowed to obtain the following statistically significant results:

- univariate frequency distributions for each of the studied indicators (will be described below in the relevant section of the article);

- description of modal groups;

- description of rank indicators;

- identification of an index of urban well-being for each indicator in two sub-samples: Moscow and Saratov;

- identification of the integral index of urban well-being of disability, as well as the integral index of well-being in the field of professional self-actualization, which allows comparing the results in general with the results of the factor group under study.

The purpose of this paper is to describe the level of satisfaction with professional self-realization opportunities of citizens with disabilities in the context of the problem of urban well-being.

3 Results

As stated earlier, the factor group "well-being in the sphere of professional fulfillment" was measured in our study through eight indicators, which include: standard of living; number of vacancies; level of offered salaries; opportunities for career development; availability of current professional connections; assessment of potential professional connections; diversity of employers; and finally, assessment of the level of trust in employers.

Accordingly, the results were described for each of these eight indicators in two sub-samples – in Moscow and Saratov.

The well-being index was calculated according to the formula:

$$\frac{0\sum r1+0.25\sum r2+0.5\sum r3+0.75\sum r4+1\sum r5}{R} \quad (1)$$

where $r1$ – respondents who chose a score of "1" (metric value "0");

$r2$ – respondents who chose the mark "2" (metric value "0.25")

$r3$ – respondents who chose a score of "3" (metric value "0.5")

$r4$ – respondents who chose a score of "4" (metric value "0.75")

$r5$ – respondents who chose the rating "5" (metric value "1");

R – total number of respondents who chose one of the non-zero values.

Accordingly, the index is distributed over five values: 0-0.2 – low well-being; 0.21-0.4 – below average; 0.41-0.6 – average; 0.61-0.8 – above average; 0.81-1 – high well-being.

Below we will consider each of the indicators separately.

Indicator 1. Standard of living

The overall level of well-being according to the indicator "standard of living" is average (0.58).

The difference between the Moscow (0.7 – above average) and Saratov (0.4 – below average) sub-samples is 0.3.

Modal groups:

- Total (34.8%) – in the value of "4";
- in the Moscow subsample (50%) – in the value of "4";
- in the Saratov subsample (30.8%) in the value of "1".

Ranking scores:

- Overall, 21.6% of respondents are dissatisfied with their standard of living, 26.2% are satisfied, and 51.1% are satisfied;
- in Moscow 6.6% of the respondents are dissatisfied with the standard of living, 24.2% – satisfied, and 68.9% – satisfied;
- in Saratov, 44.6% of those polled were dissatisfied with their standard of living, 29.2% were satisfied, and 23.6% were satisfied.

Indicator 2. Number of vacancies

The overall level of well-being according to the indicator "number of vacancies" is average (0.44).

The difference between the Moscow (0.49 – average) and Saratov (0.35 – below average) subsamples is 0.14.

Modal groups:

- general (22.7%) – in the value of "2";
- in the Moscow subsample (27.8%) – in the value of "3";
- in the Saratov subsample (28.7.8%) in the value "1".

Ranking scores:

- overall 44.7% of respondents are dissatisfied with the number of vacancies, 22.1% are satisfied, and 33.2% are satisfied;
- in Moscow 37.5% of respondents are dissatisfied with the number of vacancies, 27.8% – satisfied, 34.7% – satisfied;
- In Saratov, 65.9 percent of the respondents are dissatisfied with the number of vacancies, 13.3 percent are satisfied, and 20.8 percent are satisfied.

Indicator 3. Wage level

The overall level of well-being according to the indicator "level of salaries" is below average (0.38).

The difference between the Moscow (0.49 – average) and Saratov (0.21 – below average) subsamples is 0.28.

Modal groups:

- General (29.8%) – in the value of "1";
- in the Moscow subsample (31.8%) – in the value of "3";
- in the Saratov subsample (48.7%) in the value "1".

Ranking scores:

- overall 48.2% of respondents are dissatisfied with wages, 26.2% are satisfied, and 25.6% are satisfied;
- in Moscow 34% of respondents are dissatisfied with the level of salaries, 31.8% – satisfied, 34.2% – satisfied;
- in Saratov 69.7% of the respondents are dissatisfied with the level of wages, 17.4% – satisfied, 8.2% – satisfied. Characteristically, "5" was not chosen by any of the Saratov respondents.

Indicator 4. Opportunities for career development

The overall level of well-being according to the indicator "career development opportunities" is average (0.47).

The difference between the Moscow (0.58 – average) and Saratov (0.29 – below average) subsamples is 0.29.

Modal groups:

- Total (23.7%) – in the value of "3";
- in the Moscow subsample (27.8%) – in the value of "4";
- in the Saratov subsample (33.3%) in the value of "1".

Ranking scores:

- overall, 39.6% of respondents are dissatisfied with career development opportunities, 23.7% are satisfied, and 36.7% are satisfied;
- In Moscow, 24.6% of respondents are dissatisfied with career development opportunities, 25.8% – satisfied, and 49.6% – satisfied;
- In Saratov, 62.6% of the respondents are dissatisfied with the career development opportunities, 20.5% – satisfied, and 16.9% – satisfied.

Indicator 5. Current Professional Connections

The overall level of well-being according to the indicator "current professional connections" is average (0.45).

The difference between the Moscow (0.47 – average) and Saratov (0.42 – average) subsamples is 0.05.

Modal groups:

- Total (29.4%) – in the value of "3";
- in the Moscow subsample (34.1%) – in the value of "3";
- in the Saratov subsample (27.7%) in the value "1".

Ranking scores:

- overall 37.4% of respondents are dissatisfied with the number of actual professional connections, 29.4% are satisfied, and 33.2% are satisfied;
- in Moscow 34.6% of respondents are dissatisfied with the number of current professional relations, 34.1% – satisfied, and 31.3% – satisfied;
- In Saratov, 41.5% of the respondents are dissatisfied with the number of current professional contacts, 22.1% are satisfied, and 36.4% are satisfied.

Indicator 6. Potential professional connections

The overall level of well-being according to the indicator "potential to acquire professional connections" is average (0.44).

The difference between the Moscow (0.49 – average) and Saratov (0.35 – below average) sub-samples is 0.14.

Modal groups:

- Total (29.8%) – in the value of "3";
- in the Moscow subsample (34.1%) – in the value of "3";
- in the Saratov subsample (35.4%) in the value "1".

Ranking scores:

- overall, 38.7% of respondents are dissatisfied with professional networking opportunities, 29.8% are satisfied, and 31.5% are satisfied;
- In Moscow, 30.7% of those polled were dissatisfied with the opportunity to acquire professional connections, 34.1% were satisfied, and 35.2% were satisfied;
- In Saratov 50.8% of those polled were dissatisfied with the opportunity to acquire professional connections, 23.1% were satisfied, and 26.1% were satisfied.

Indicator 7. Diversity of the circle of employers

The overall level of well-being according to the indicator "diversity of the circle of employers" is average (0.45).

The difference between the Moscow (0.5 – average) and Saratov (0.39 – below average) subsamples is 0.11.

Modal groups:

- Total (25.6%) – in the value of "3";
- in the Moscow subsample (31.1%) – in the value of "3";
- in the Saratov subsample (28.2%) in the value "2".

Ranking scores:

- overall 41.4% of respondents are dissatisfied with the diversity of employers, 25.6% are satisfied, and 33.3% are satisfied;
- in Moscow 32.8% of the respondents are dissatisfied with the diversity of the circle of employers, 31.1% – satisfied, and 36.1% – satisfied;
- in Saratov 54.4% of the respondents are not satisfied with the diversity of employers, 16.9% are satisfied, and 28.7% are satisfied.

Indicator 8. Confidence in employers

The overall level of well-being according to the indicator "confidence in employers" is average (0.51).

The difference between the Moscow (0.52 – average) and Saratov (0.49 – average) subsamples is 0.02.

Modal groups:

- Total (27.6%) – in the value of "3";
- in the Moscow subsample (35.1%) – in the value of "3";
- in the Saratov sub-sample (20.5%) Modal groups were distributed in two polar values at once: "1" and "5".

Ranking scores:

- overall, 31.1% of respondents do not trust employers, 27.6% show an average level of trust, and 41.3% are above average or high;
- In Moscow, 25.5% of those polled did not trust their employers, 35.1% showed an average level of trust, and 39.4% an above average or high level of trust;
- in Saratov 39.5% of the interrogated do not trust their employers, 15.9% have an average level of trust, and 44.6% have an above average or high level of trust.

Conclusion on the group of indicators "opportunities for professional self-realization"

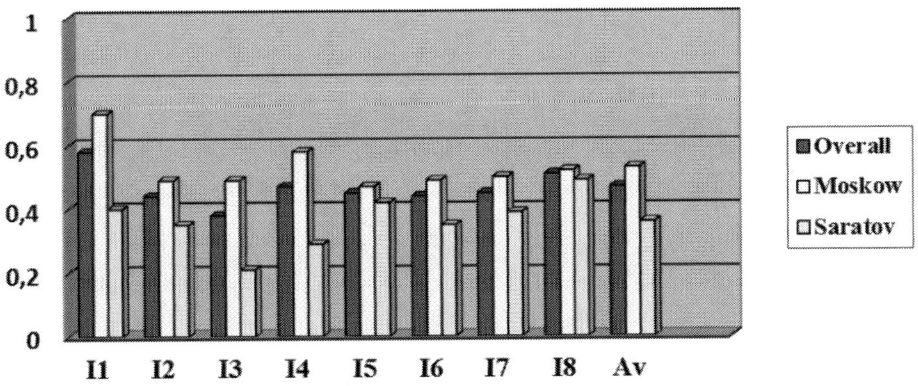

Fig. 1: Comparison of indicators of well-being in the sphere of professional self-actualization.
Source: compiled by the authors.

4 Discussion

The overall well-being index for the group of indicators "opportunities for professional self-actualization" is 0.47. The difference between the Moscow (0.53) and Saratov (0.36) sub-samples is 0.17. In our opinion, this difference is very significant and reflects the real difference in the possibilities of professional self-realization in the capital and the regional center, which is Saratov. If the trend recorded in Saratov is confirmed in other regional centers, then it is fair to assume that with a more expanded study with all-Russian geography and the corresponding sample, reflecting the real distribution of population between the capital and regional centers, the index of well-being in the sphere of professional self-realization will be lower than the obtained (0.47), approaching the negative zone of 0.4 or lower.

Overall sample. There are no indices in the "positive" zone (0.61 and above). In the "negative" zone (0.4 and below) is the index of satisfaction with the level of wages (0.38). The most favorable situation – 0.58 – is observed in assessing the standard of living, the least favorable – 0.38 – in assessing the level of wages. The difference between the highest and lowest indices is 0.2. This result suggests the presence of environmental barriers that accompany the professional self-realization of citizens with disabilities.

Moscow sub-sample. In the "positive" zone there is an index of satisfaction with the standard of living (0.7). There are no indices in the "negative" zone. The most prosperous situation – 0.7 – is observed in the assessment of the standard of living, the least prosperous – 0.47 – in the assessment of actual professional connections. The difference between the highest and lowest indices is 0.23. Despite the presence of one of the indices in the positive zone, we are forced to make the disturbing conclusion that even in the capital the index of well-being in the sphere of professional self-realization is mostly in the average values. However, even if we talk about the value of 0.7 in the assessment of living standards in Moscow, it does not add optimism for city residents with disabilities: this variable does not reflect the respondent's assessment of living standards, but only reflects his opinion about living standards of city residents in general. In other words, the question about the standard of living in the city referred the respondent not to the assessment of their own standard of living, but to the assessment of the standard of living of others.

Saratov sub-sample. There are no indices in the "positive" zone. In a negative zone there are indices of satisfaction with living standards (0.4), number of vacancies (0.35), salary level (0.21), possibilities for career development (0.29), potential professional connections (0.39) and variety of employers (0.35). The most favorable situation – 0.49 – is observed in assessing confidence in employers, the least favorable – 0.21 – in assessing the level of wages. The difference between the highest and lowest indices is 0.28. As can be seen from these summary statistics, six out of eight indices of the factor group "professional self-actualization" are in the negative zone. As previously stated, the significant difference between the integral indices in the two subsamples reflects the real difference between the level of well-being in Moscow and the regional center. Moreover, one cannot ignore the fact that the assessment of wages in Saratov is close to critical (0.21). The desire of regional residents to move to the capital is understandable. However, as we noted above, even in the capital they can expect only an average level of well-being in the sphere of professional fulfillment.

5 Conclusion

As the results of the questionnaire survey showed, the level of well-being of citizens with disabilities in the field of professional self-realization is located in the average values (0.47), which fully correlate with the overall index of well-being in all six indicator groups, which according to the results of the study was 0.52. Professional self-actualization, being an inherent attribute of urban well-being in general, has a direct impact on the satisfaction of city residents with disabilities with their lives.

As a practical conclusion of this study, it should be suggested to the authorities and employers to pay attention to the problem areas, which include the level of wages, accessibility of jobs, career opportunities, and some others.

In the future, a similar study should be conducted on an all-Russian sample and regular measurements should be taken to assess the effectiveness of federal and regional programs aimed at improving the well-being of people with disabilities, as well as the dynamics of change in the situation.

References

1. E.K. Naberushkina, Disabled People in the Big City: Problems of Social Citizenship (Variant, Moscow, 2012)
2. D.S. Raidugin, E.V. Voevodina, Man. Soc. Incl, **34**, 23-28 (2018)
3. Entsiklopedicheskii slovar F.A. Brokgauza i I.A. Efrona. Blagopoluchie [Encyclopedic dictionary of F. A. Brockhaus and I. A. Efron. Wellbeing]. Accessed on: September 26, 2021. [Online]. Available: https://dic.academic.ru/dic.nsf/brokgauz_efron/13583/
4. S.V. Yaremchuk, Psych J, **34**, 89 (2013)
5. R.M. Shamionov, Psychology of the subjective well-being of personality (Publishing house of Saratov University, Saratov, 2004)
6. T. Rath, J. Harter, Wellbeing: The Five Essential Elements (Gallup Press, New York, 2010)
7. L.I. Galiakhmetova, Bul Bashkir Univ, **20**, 1115 (2015)
8. Y.-C. Chiang, et al., J Urban Plan Devel, **147(4)** (2021). https://doi.org/10.1061/(ASCE)UP.1943-5444.0000740

9. T. Muhammad, S. Srivastava, T.V. Sekher, BMC Psych, **9**, 82 (2021). https://doi.org/10.1186/s40359-021-00588-5
10. X. Li, H. Chen, Z. Zhu, BMC Pub Health, **21**, 774 (2021). https://doi.org/10.1186/s12889-021-10861-2
11. M. Yildirim, F.Ç. Tanrıverdi, J Pos Sch Psy, **5(2)**, 127–135 (2021). https://doi.org/10.47602/JPSP.V5I2.229
12. M.D. Merino, J. Privado, R. Durán, Anales de Psica, **37(3)** (2021). https://doi.org/10.6018/analesps.430801
13. R. Clair, M. Gordon, M. Kroon, C. Reilly, Human Soc Sci Comm, **8**, 28 (2021). https://doi.org/10.1057/s41599-021-00710-3
14. M. Juchnowicz, H. Kinowska, Information, **12(8)**, 293 (2021). https://doi.org/10.3390/info12080293
15. V. Van Acker, L. Ho, C. Mulley, Transp Res Part A: Pol Prac, **151**, 214-227 (2021). https://doi.org/10.1016/j.tra.2021.07.010
16. D.S. Raidugin, E.K. Naberushkina, S.P. Gorbunov, Human, Soc-Econ Soc Sci, **12**, 110-116 (2019)

Employment problems of disabled persons: stereotypes and employer missing

Elmira Kamalovna Naberushkina, *Dmitry Sergeevich* Raidugin[1], *Ekaterina Vladimirovna* Voevodina, and *Valeria Evgenievna* Ikhneva

Financial University under the Government of the Russian Federation, Department of Sociology, Moscow, Russia

Abstract. The article presents the results of an empirical study aimed at examining social stereotypes and established attitudes that hinder the employment of people with disabilities. To identify stereotypical ideas about the employment of people with disabilities, at the beginning of 2021, we undertook a study of two subgroups: employers (1) and people with disabilities (2). The paper presents a comparative analysis of the responses of representatives of the two groups, describes deep-rooted stereotypes, identified similarities and differences in existing attitudes about the barriers to employment of people with disabilities. The results of the study showed that people with disabilities and employers have stereotypes that either promote or, on the contrary, hinder the employment of a person with a disability. Negative stereotypes include: the idea of people with disabilities as inconvenient workers, dependents. On the other hand, employers have a fairly firmly formed setting of social responsibility and the obligation to hire people with disabilities, to promote their labor adaptation. In people with disabilities, although there is an attitude towards expectation of paternalism, the attitude towards self-development and social activity dominates. Turning to research within the framework of a qualitative methodology clarified the position of employers and showed the most vulnerable areas of employment of people with disabilities, namely, the unattractiveness of certain categories as workers and the absence of significant incentives for the employer to induce to hire people with disabilities.

Keywords: labor market, employment, people with disabilities, stereotypes, social attitudes, employers

1 Introduction

One of the most important aspects of personal socialization is professional self-determination and employment. This issue is particularly acute for people with disabilities. According to some studies, people with disabilities are, on average, less happy and satisfied with life than those without disabilities [1]. Employment for people with

[1] Corresponding author: raydinis@mail.ru

disabilities is not only about gaining financial independence, but also about gaining a new social status, the normalization of life, and the expansion of social connections. Under the conditions of current state social policy in Russia, efforts are being made to form a system to promote employment of people with disabilities. However, certain barriers arise on the way of implementing these measures. In this article, we focus on the consideration of social stereotypes and prevailing attitudes that hinder the employment of people with disabilities. A preliminary analysis of scientific publications has shown that sociological research on this issue is insufficient, and the labor market problems brought up to date by the pandemic require additional study. Studies devoted to the analysis of disability issues range from consideration of the historical aspects of social protection of people with disabilities [2-5] to the issues of modern practice of inclusion of people with disabilities and discussions of social citizenship in the context of modern urbanization [6-15].

The American sociologist W. Lippmann introduced the concept of "stereotype" into scientific usage. He understood a stereotype as a "pattern of perception, filtering, interpretation and information accepted" in a historical community when recognizing and recognizing the world around, based on previous social experience" [16]. It should be noted that in Lipman's view, the stereotype acts as a rather negative phenomenon. Stereotypical representations, at different time, were considered in the works of G. Allport, T. Adorno, B. Davis, T. Shibutani and other authors, who worked in the framework of social behaviorism, phenomenology, symbolic interactionism, social constructivism theories. Later, scientists began to consider the stereotype as a certain condition for the formation of a picture of the human world, and stigma as a social label constructing a situation where an individual is considered incapable of full social life. I. Goffman in his work "Stigma: Notes on the Management of Corrupt Identity" managed to show in detail the structure and mechanisms of stereotypes [17]. O.N. Shestopalova suggests classifying social stereotypes on the following grounds: orientation, degree of stereotype adequacy, subject and object of stereotyping, sphere of manifestation, degree of changeability, nature of comprehension of reality and interaction with it, functional meaning and content [18]. True and false, positive and negative stereotypes are distinguished according to their orientation. Positive stereotypes allow quickly orienting oneself in the social space, giving the ability to generalize and synthesize. Negative stereotypes simplify approaches to assessment of various phenomena, social groups and individuals. According to the subject of formation can be distinguished: spontaneous and conscious stereotypes. Depending on the object distinguish individual, group and mass stereotypes. Spheres of manifestation of stereotypes allow to speak about stereotypes of thinking (images and evaluations) and stereotypes of behavior.

At least three main fields of construction of stereotypes regarding disability and the labor potential of people with disabilities can be distinguished:

1. Political, which includes the model of disability that dominates state social policy and current legislation in the area of employment and social protection of disabled people. The position of employers is significantly influenced by requirements to comply with regulations on equipping workplaces for disabled persons, creating an accessible environment at enterprises, and enforcing the norms of the labor code. In the field of Russian social policy along with approval and acceptance of ideas of integration of people with disabilities a narrow focus on costs and benefits from the policy of creation of accessible employment for disabled people dominates. A secondary issue remains the quality and range of existing measures of social inclusion in labor activity.

2. The economic field includes questions of the social well-being of the disabled, the risks associated with state control of enterprises, the possible desire of enterprises to attract

cheap labor, the focus on profit and cost reduction. Economic factors that influence the formation of stereotypical perceptions among employers include opinions about the social well-being and human potential of people with disabilities, and employers' ideas about the resources that can be built into the pay of an employee with a disability.

3. The socio-cultural field is characterized by historical traditions characteristic of a given nation and state, the level of education and upbringing of the population, and the discourse of disability in the mass media. Sociocultural factors of formation of stereotypes are caused by the ideas formed in a society about disability as a phenomenon, about a place and a role of the person with disability in society. L.A. Torlopova notes that until the 16th century, people with bodily and other injuries were completely excluded from the circle of attention and care due to their uselessness and inconsistency with the aesthetic ideals of bodily beauty. As a phenomenon, disability could only be present in the minds of certain professional groups, primarily doctors [19]. In Russia, the attitude toward the disabled, on the one hand, has always been associated with "pauperism", treating them as God's people, and on the other, turning them into a kind of closed caste, very limited in their rights and opportunities. During Soviet times, despite the development of state social protection, the problem of disability was silenced, and social assistance practices were associated with exclusion and institutional social segregation. At the present stage, especially after the ratification of the UN Convention on the Rights of Persons with Disabilities, the situation is changing for the better, but practices of implementing measures of social protection and integration of people with disabilities require permanent and detailed consideration.

2 Materials and methods

To identify existing social attitudes and stereotypical perceptions of employment of people with disabilities in early 2021, we undertook a study of two subgroups: employers (1) and people with disabilities (2). A total of 160 respondents were interviewed in equal shares. We relied on the methods accepted in the world empirical practice, relevant studies of issues of social justice [20]. The sample of employers included people authorized to make personnel decisions. The sample of respondents with disabilities included both employed people and job seekers.

The study allowed for a comparative analysis of the responses of the two groups, to identify ingrained stereotypes, and to identify similarities and differences in the current attitudes about barriers to employment for people with disabilities. The research was conducted by means of questionnaires and interviews.

3 Results

To develop possible strategies to transform stereotypical perceptions of employment of people with disabilities in the mass consciousness, it is possible to understand the main motives and social attitudes of both employers and employees with disabilities. To this end, we conducted a survey among job seekers and employees with disabilities living in Moscow, as well as interviewing a number of employers in Moscow. Among respondents, 21.3% belonged to the age group 18 to 30 years old, 49.2% were 31 to 45 years old, 18% were aged 46 to 60 years old, 9.8% were 61 to 65 years old, and finally 1.6% of respondents were over 65 years old. In terms of gender composition, 43.3% of respondents were male and 56.7% were female.

In general, all respondents viewed the employment of people with disabilities as a rational and socially significant process. When comparing answers to the question about the necessity to solve the problem of employment of the disabled people it became clear that 50% of employers and 61% of disabled people believe that the disabled are an additional labor potential, effective use of which will positively influence the development of the economy. 24.8% of respondents authorized to make personnel decisions and 27.5% of respondents with disabilities believe that "solving the problems of the disabled: it is a moral obligation facing society", and according to 6.5% of employers and 7.5% of the disabled "this problem should be solved for reasons of mercy and humanity". About nine percent of employers and less than four percent of respondents with disabilities believe that the problem of disability employment is farfetched and not relevant at all.

Next, we compared opinions on the responsibility to address the issue of employment for people with disabilities. Half of the employers and 55 percent of respondents with disabilities placed it on the state, while 6.5 percent of employers and 5 percent of respondents with disabilities believed that such responsibility should be placed on educational institutions. At the same time, a quarter of respondents among employers answered that the responsibility for solving the problem of employment of disabled people should be placed on enterprises and employers, and 12.5% of respondents with disabilities think so. The results suggest that some employers have internalized the attitudes of socially responsible business, according to which the employer must solve socially significant tasks, which include the employment of people with disabilities. Finally, 15.2% of employers and 21.25% of citizens with disabilities believe that such responsibility lies with disabled people themselves, arguing that it is the drowning people's responsibility to save them and nothing should be done for the disabled without their participation. The rest indicated their own variants of the answer, the general sense of which is that the responsibility should be distributed equally among all participants in the process of employment of disabled people.

When discussing the obstacles to the employment of people with disabilities the positions of disabled people and employers in some cases coincided, but significant discrepancies were also revealed. 17.3% of employers and 11.25% of respondents with disabilities considered an insufficient level of qualifications received by disabled people to be the main factor impeding the employment of disabled people. Employers in a context of reflections about the development of new professions and demands of modern business designated a problem of inconsistency of personnel training programs in Russian universities with modern requirements of the labor market. Traditional education is recognized as poorly functioning to develop dynamic skills in students. Many employers believe that most schools and universities teach outdated models, while advanced 21st-century skills are required today.

For most employers, pragmatic considerations are the primary motivation for deciding to hire people with disabilities. A quarter of employers consider the legally prescribed necessity and responsibility to equip a workplace for a disabled person to be the main reason why they do not want to hire disabled people, while 12.5% of respondents in the second sub-sample think so. The difference in the answers is due to the fact that it is the employer who is responsible for creating special working conditions for an employee with a disability, and it is the employer who may be subject to sanctions if these conditions are not created. The imperfection of Russian labor legislation as an obstacle was mentioned by 13% of employers and 18.75% of disabled people.

21.7% of employers and 21.25% of respondents with disabilities consider bureaucratic obstacles to employment (certification of specialized jobs, execution of individual programs of rehabilitation and habilitation of disabled people, etc.). The inability of people with disabilities to join a work team was mentioned by only 2% of respondents among employers and 1.25% of respondents with disabilities. The incompetence of employers deciding on the employment of disabled people was mentioned by 25% of respondents with disabilities. The refusal can be sounded in connection with the incompetence of the employer concerning the features of this or that nosology and misinterpretation of labor recommendations fixed in IPRA, sometimes employers are not sufficiently aware of modern opportunities of disabled employees, which, for example, can include possession of a personal computer by the disabled person with a visual impairment.

Often employers question the potential abilities of a person with a disability to quickly adapt to the work environment and cope independently with job duties. The vast majority of employers and respondents with disabilities believe that employees with disabilities need additional coaching and mentoring. The fact that a person should figure things out for themselves, regardless of disability, was stated by 17.4% of employers and 15% of people with disabilities. The rest believe that this issue should be solved on a case-by-case basis, based on the level of complexity of the work process and the severity of the illness of the employee with a disability.

The answers to the question whether an employer should treat an employee with a disability differently from other employees when setting tasks and assessing the results of their work were interestingly distributed. 2.1% of employers and 8.75% of respondents with disabilities answered affirmatively. A strictly negative answer was given by 54.3% of employers and 31.25% of disabled respondents. The significant difference in responses suggests a stereotypical attitude of expectation of paternalism and leniency among some disabled people, while employers believe that they are not obliged to attach a "babysitter to an employee with a disability". The majority of employers believe it is wrong to pay more attention to an employee with a disability than to others, but at the same time just under half of employers are willing to be attentive to an employee with a disability and facilitate his or her full adaptation at the workplace.

Respondents of both subgroups, noting that the main responsible for employment of disabled people is the state, argued about effective measures to support employment of disabled people among which, first of all, they named:

- creation of new and stimulation of existing specialized enterprises using labor of disabled people (34.8% of employers and 33.75% of disabled people)
- tougher sanctions against employers who do not fulfill the law "On Quota of Workplaces for Disabled Persons" (6.5% of employers and 11.25% of disabled people)
- reservation of work places within individual industries most accessible for workers with disabilities (13% of employers 11.25% of the disabled)
- improvement of measures aimed at rehabilitation and habilitation of disabled people (23.9% of employers and 18.75% of the disabled)
- change of the information policy of the state aimed at creating a positive image of a person with disability in the mass consciousness (15.2% of respondents having the right to make personnel decisions and 18.75% of respondents with disabilities).

A rather large percentage of respondents in both the first and the second groups consider it necessary to create and stimulate already existing specialized enterprises for the disabled, as well as advocate the reservation of jobs for the disabled in certain industries. This may indicate that the open labor market in our country is not fully accessible for workers with disabilities, as well as the problems associated with the competitiveness of the disabled and attitudes to social exclusion of this category of citizens.

It turned out that 33.3% of respondents with disabilities are employed and they are completely satisfied with their current status, 28.3% are employed but would like to find another job, 26.7% are unemployed and are in search of a job and finally 11.7% of respondents are unemployed and they are completely satisfied with their status.

Slightly more than a quarter of respondents with disabilities said that they could do any job on an equal basis with others, 41.7% explained that they could do any job on an equal basis with others, but in specially created conditions, 18.3% believed that they could do any job, but for a longer time, and 13.3% of respondents explained that only light work was available to them. It is possible to pay attention to a rather large percentage of disabled people (41.7%) who think they can work in specially created conditions and disabled people who think that a disabled person who is employed needs additional mentoring or supervision only at the first stage (58.7%).

Just 27.9% of respondents were interested in full-time employment, and this result was roughly the same as the percentage of respondents who thought they could do any job on an equal basis with others. For 39.3% of the respondents the most optimal is full-time employment with flexible schedule, which also correlates with the percentage of respondents willing to work in specially created conditions.13.1% of the respondents are only interested in part-time employment, and the rest would like to have part-time employment with flexible schedule and willing to do any job, but for a longer time.

It was extremely important for us to catch the main motive that motivates a person with a disability to look for a job. The answers were distributed as follows: for the overwhelming majority of respondents (60.7%) work is a way to increase their welfare, for 16.4% work is a way to improve their social status, for 6.6% work is an opportunity to fill free time, for the rest work is an opportunity to expand their social circle.

To get a desired job, 60.7% of respondents are ready to undergo professional retraining, but only at the expense of the state or the employer, 23.2% are ready to invest their personal funds into such training, and 16.1% of respondents believe that their level of qualification is sufficient and they do not need professional training. It is interesting that those who are satisfied with their professional status and employed are ready to invest their own money into professional training. Only 13.3% of respondents are ready to change their place of residence for the sake of employment, 26.7% of respondents are ready to change their place of residence if an employer decides on their housing problem, and the majority of respondents (60%) answered this question rigidly in the negative.

The last question addressed to the respondents with disabilities was intended to clarify the optimal ways of job search in their opinion. The answers were distributed as follows: 34.4% of the respondents consider that the easiest way to find a job is to use the services of the employment center, 3.3% consider that the best way to find a job is to use the services of recruiting agencies, 26.2% of respondents consider that the best way to find a job is to place their resume on the Internet, 31.1% of respondents think that the only thing that can help a disabled person to find a job is personal contacts. The rest believe that all the proposed ways should be used in the aggregate. The result shows, on the one hand, rather high evaluation of the work of the employment centers by the respondents, but the high percentage of the respondents who do not believe in the possibility to get a job without using personal connections, indicates the need to further improve the work to promote the employment of the disabled.

The interviews with Moscow employers were aimed at clarifying the information obtained during the quantitative surveys. Thus, as a result of the interviews, the issue of possible conflicts in the workplace, the source of which may be the fact of disability of one of the employees, was raised in a slightly different light. The question was asked: "Should the head of the unit where a person with a disability is employed notify the other employees in advance? " The most typical response was formulated this way:

"I think it's more likely yes than no. As an employer, I am very concerned about the psychological climate in the team. If I say that disability can't be a cause of conflict in a team, I'm lying. It can. Because not all people are tolerant and willing, sometimes these conflicts do happen. Just recently I was talking to a relative and they had this conflict at work. A boy with a disability works there. In order to get paid a salary they had to make 30 applications a day, and he, for example, makes only three, but gets the same salary. And accordingly from ignorance, not understanding the capabilities of the person, some such people began to resent this fact. I think that the task of preventing and resolving such conflicts should lie with line managers".

There is another opinion:

"If an invalid gave a kindness, perhaps I would say. Maybe a person does not want to be considered weak because he or she is disabled. That is, I would clarify this question at the interview: should I tell the staff about it, or will you do it yourself. I would go that way".

Thus, when considering a disabled person as a possible candidate for a particular vacancy, many employers, subconsciously, are afraid that an employee with a disability may become a possible source of conflict and an unfavorable psychological climate in the team. When discussing why some employers refuse to hire people with disabilities, despite their skill level, informants talked about fears and possible risks.

"Probably fear. People with disabilities, they're different, too. Maybe they're people of some other sort of mindset".

"I think it's not readiness, fear, or rather not fear, but lack of experience with this category of citizens. Some unfortunate experience of the employer may be a factor: that is, a citizen with disabilities worked and the experience was unsuccessful. Then the existing stereotype works on other people with disabilities".

"There is a fear of managers that a person with a disability and the associated right will begin to perform his or her duties worse and cover up for it. That's why my employees, department heads are reluctant to hire people with disabilities. I have positive experience of cooperation with such people, they don't have it, there is a resistance: on the one hand managers are afraid, on the other hand, if there is a person who creates a negative attitude towards the disabled, this attitude works on everyone".

Thus, an obstacle to the employment of disabled people is the employer's previous negative experience of interaction with disabled people. Often people with disabilities with a rent attitude "I am owed everything" when interacting with an employer, create a stereotype of a disabled person as a prone to scandals and a disabled person as a source of additional problems".

A slightly different view, based on the results of the interviews, is the additional attention that the employer must pay to the employee with a disability. According to the results of the questionnaire survey, most employers were inclined to believe that an employee with a disability is the same as everyone else and should not be given more attention than others. In the interviews, employers were not as unambiguous. When asked "should the head of the structural subdivision where a disabled person works pay more attention to such an employee than to others", they answered:

"more attention is not necessary, but to be more attentive... Try to be tactful. Understand what a person's limitations are".

"You don't need to single out this person, but as a manager I would take an interest. I wouldn't show special visible attention, but for myself, I would keep this person under control".

Thus, the employer's approach to an employee with a disability will inevitably be somewhat different than to other members of the workforce. It does not matter what the employer's motives are for keeping the employee under special control. It is important for any disabled person to have an understanding that this control will inevitably occur.

Speaking of factors that may contribute to a positive solution to the employment of disabled people, employers speak primarily of pragmatic interests. In discussing the motivation for their decision, informants answered:

"Self-interest: some subsidies or the Ministry of Labor gives some preferential treatment. Well, a purely selfish motive. We are economic agents, first of all, think about some pragmatic things".

"If a person sees, first of all, a professional, then there are no obstacles. One of the factors should be an administrative resource, when the employer bears a certain responsibility for the failure to comply with the law on quotas".

4 Discussion

According to the results of earlier studies [21], we concluded that the situation in Russian society has changed for the better with the understanding of disability issues, the ability to form an accessible infrastructure of social services and build social communications of a non-discriminatory order. However, the situation with employment of disabled people has changed to a lesser degree, tensions in the employment of people with disabilities remain, and the facts of exclusion of disabled people in the labor market are hidden by legitimate formulations of refusals to hire them. Today it is obvious that the situation with employment of disabled people has passed a negative point of bifurcation and there is a tendency of formation of loyal attitudes concerning employment of disabled people. Previously, employers denied their responsibility to employ the disabled, saying that it was the responsibility of the state, which should create specialized enterprises, and tried to circumvent the law on quotas. Today we note an increase in social responsibility of business, people with disabilities are perceived as full labor potential, and among the cohort of people with disabilities dependency attitudes have significantly decreased. However, as before, we note that the sociological reading of the problems of disability, served the growing recognition by society of the rights of people with disabilities, reformatting the discourse of social protection, but the situation with employment of people with disabilities remains one of the most tense, especially not in such large cities as Moscow.

5 Conclusion

To summarize, it should be noted that in the minds of both the disabled person and the employer, there are a number of stereotypes that can either promote or, on the contrary, hinder the successful employment of a person with a disability. Negative stereotypes can include: the stereotype of the disabled person as a scandalous person in the mind of the employer, or the dependent stereotype in the mind of the disabled person. On the other hand, we can talk about the recent formation of some positive attitudes among both disabled people and employers. First and foremost, they should include "Everything, first of all, is in my hands" in the mind of the disabled person, and the attitude of social responsibility in the mind of the employer.

Considering possible strategies for transforming negative attitudes, it is necessary, first and foremost, to develop measures for displacing negative stereotypes. Thus, the image of a disabled dependant should be gradually displaced by the image of a socially active disabled person, striving to be independent. The mass media can be of particular help in this regard.

Considering that a significant percentage of respondents consider the media to be the main source of information about people with disabilities, the development of social journalism today seems extremely relevant. It is necessary to work with people with disabilities themselves; success stories of individuals should be cultivated. Cooperation between employment centers and public organizations of people with disabilities can be very useful.

State control in resolving the issue of the employment of the disabled remains key; it is the state that has been named as the main subject responsible for resolving the issue of employment. Strict control over the execution of the law "On Quota of Jobs for Disabled Persons" is necessary, but the employer must also be interested in hiring an employee with a disability. The application of measures of control and encouragement must be proportional. It is important to develop a sense of responsibility for one's decisions, both for the employer and the disabled person. Only an understanding by both parties of their responsibility, both to themselves and to society, will make the process of inclusion of people with disabilities in the employment system as effective as possible.

References

1. E.Nastina, A. Almakaeva, Values As Determinants of Social Capital: Regional Perspective (September 30, 2020). Higher School of Economics Research Paper No. WP BRP 95/SOC/2020, Available at SSRN: https://ssrn.com/abstract=3702389 or http://dx.doi.org/10.2139/ssrn.3702389
2. P. Hongling, *Study on the Effect of Grassroots Community Services on Disabled Elderly Care System*, in Proceedings of the 2016 International Conference on Economy, Management and Education Technology (2016). https://doi.org/10.2991/icemet-16.2016.356
3. L.K. McCrary, *Hannah Arendt and Disability: Natality and the Right to Inhabit the World*, in B. Arneil, N.J. Hirschmann (eds), Disability and Political Theory (Cambridge University Press, Cambridge, 2016)
4. A.S. Kovalev, The J Soc Pol Stud, **12(1)**, 77-94 (2014)
5. V.N. Yarskaya, The J Soc Pol Stud, **1(1)**, 11-28 (2010)
6. T. Loreman, Ch. Forlin, Measuring Inclusive Education. International Perspectives on Inclusive Education, vol. 3 (Emerald Group Publishing Limited, Bingley, 2014)
7. H. Dempster, K. Hargrave, Understanding of social attitudes towards refugees and migrants. Working Paper 512 (The Chatham House, London, 2017)
8. P.Romanov, (2013). Industrial Paternalism in a System of Enterprise Welfare. The Journal of Social Policy Studies, **3(3),** 287-304. Retrieved from https://jsps.hse.ru/article/view/3745
9. M. Yasin, The J Soc Pol Stud, **17(4)**, 601-614 (2019). https://doi.org/10.17323/727-0634-2019-17-4-601-614
10. T.A. Vlasova, The J Soc Pol Stud, **17(3)**, 407-422 (2019). https://doi.org/10.17323/727-0634-2019-17-3-407-422
11. N. Astakhova, N. Bolshakov, The J Soc Pol Stud, **15(1)**, 51-66 (2017). https://doi.org/10.17323/1727-0634-2017-15-1-51-66
12. A. Goriainova, The J Soc Pol Stud, **17(4)**, 555-570 (2019). https://doi.org/10.17323/727-0634-2019-17-4-555-570

13. V.N Yarskaya-Smirnova, (2018). Attitudes among the Professional Groups of the Welfare State towards Migrants. The Journal of Social Policy Studies, 16(4), 691-700. https://doi.org/10.17323/727-0634-2018-16-4-691-700
14. I.M. Lid, P.K. Solvang, Europ J Disab Res, **10**, 181-194 (2016). https://doi.org/10.1016/j.alter.2015.11.003
15. S.H. Zdrahat, Sci Theor Bul "Grani", **20(4)**, 10-16 (2017)
16. W. Lippmann, Obshchestvennoe mnenie [Public Opinion] (Institute of the Public Opinion Foundation, Moscow, 2004)
17. E. Goffman, Notes on the Management of Spoiled Identity Chapter 5. Deviations and deviation / translation. A. Maktas // Sociological Forum **(3–4).** (2000)
18. O.N. Shestopalova, Proc Ural State Univ, **51(3)**, 106-110 (2007)
19. L.A. Torlopova, Bul Tomsk State Univ, Philos, Soc, Pol Sci, **37**, 92-102 (2017)
20. C. Sabbagh, M. Schmitt (eds), Handbook of Social Justice Theory and Research (Springer, New York, 2016)
21. E.K. Naberushkina, The J Soc Pol Res, **15(2)**, 333-340 (2017)

New paradigm of the Russian labor market in the context of the formation of an inclusive economy and a new gender order

Galina Georgievna Sillaste[1]

Financial University under the Government of the Russian Federation, Department of Sociology, Faculty of Social Sciences and Mass Communications, Moscow, Russia

Abstract. The author considers the profound changes that have occurred in the Russian labor market during the transition from a transitive economy to an inclusive one and the expanding influence of the emerging gender order, the theoretical concept of which is the theory of gender neutrality. A new paradigm for the development of the Russian labor market is proposed, which meets the radically changed economic and social conditions for the development of socio-gender labor resources in the conditions of digitalization. According to the author, a new multi-vector structure has formed in the Russian economic space, consisting of four closely interrelated markets: labor, employment, professions, and educational services. The labor market and the employment market have become the epicenter of the development of social inclusion and an inclusive economy, which have presented qualitatively different requirements for socio-gender labor resources. The complex interrelations between the four segments of the new structuration are analyzed as deep structural changes and tectonic shifts in the formation of socio-gender labor resources, the implementation of their social functions, and the reformatting of social mechanisms for managing ongoing processes. The inertial view of the labor market as a single social space does not correspond to the realities of its modern development. The article reveals the essence of the new social structuration in the context of the theory of Giddens, its architectonics, logistics of development, dialectics, and social multidimensionality. The trends in the development of a new social structuration are evaluated through the standpoint of the formation of a new gender order and the theory of gender neutrality.
Keywords: socio-gender and labor resources, world gender order, inclusive economy, four-segment structure, social space, labor market, employment market, profession market, educational services market, gender neutrality

1 Introduction

[1] Corresponding author: galinasillaste@yandex.ru

The development of the economy and types of labor activity is closely interrelated with the changes in the field of labor employment, the distribution of labor resources, including socio-gender resources. This process is currently determined by two factors: a developing inclusive economy and a new gender order developing in the world. The goal of an inclusive economy is to ensure equal access for people to economic independence and unhindered interaction between social groups. The driving force of an inclusive economy is the development of a socially-oriented approach in two directions: the first is to ensure economic growth by involving the poor population in economic activity; the second is to create resources, primarily human, labor, socio-gender, for the functioning of an inclusive economy.

There have been profound structural changes and movements of labor resources over the past 30 years that have changed the traditional perception of the labor market as a single social whole for Russians. Earlier, in the sphere of labor, the state was the determining regulator, the monopolistic institution responsible for the formation of the labor sphere, and the institutions of education, labor, production, distribution and consumption, law and management were a single well-coordinated mechanism. In a transitive, and even more so in an inclusive economy, the social interests of the population and those who are preparing to enter the labor market do not always coincide with the interests of the institutions of the state, economy, and production. Moreover, education as a social institution has greatly changed its social functions, becoming one of the markets for intellectual services. This market, which is directly connected with the digitalization of the economy, is the generator and driving force of a new four-segment structure. At the same time, structuration is considered by the author (in accordance with the theory of Giddens) as a process of organization and reproduction in the system of social relations, accompanying the social interaction of actors (subjects). A new system of rules and the availability of resources generate a new social system of labor relations and the distribution of labor resources.

The architectonics of the new social structuration, within which four interconnected markets function, nevertheless preserve the dialectics and consistency of the social process, production and reproduction of basic labor resources. In fact, the institution of education remains the recreator of new labor socio-gender resources. This institution has changed its social functions, opened its social and educational space to market mechanisms of profit and commercialization, and become a leading producer of the educational services market. Along with it, the budgetary basis for the admission of students to universities is preserved as proof of the implementation of the constitutional right of an individual to affordable education.

The theory of gender neutrality, which is the leading conceptual basis of the new gender order, assumes the removal of restrictive barriers from the sign of gender not only for obtaining any kind of education but also for any profession at the request of men and women. Thus, the new gender order contributes to the expansion of gender equality in the new structuring of the market space.

To analyze the complex and contradictory, hidden from the external view, social processes accompanying the development of labor markets, employment, professions, and educational services, the author uses a multi-paradigm approach. This approach synthesizes a set of theories, concepts, and paradigms in the field of economics and labor sociology: Changli [1], Slezinger [2] (in the field of social economics), Oleynik [3] – institutional economics, and also the paradigms of "new economic sociology" by Svedberg and Granovetter – a concept of the author's approach to the analysis of new structuration develops. A number of particular sociological theories complement the indicated methodological principles of analysis: the sociology of labor (Yadov, Fetisov, Toshchenko,

Aitov), the sociology of markets (Polanyi, Braudel – theories of the public and private market), Jevans (theory of political economy), risk theory (Beck – *Risk Society. Towards a New Modernity*), Prigozhin, Kravchenko, Krasikov [4], the theory of gender sociology and the new world gender order (Sillaste [5]), the theory of structuration (Giddens) and the theory of social change (Shtompka). In the analysis of the gender specifics of the workforce, the emphasis is on the theory of gender economics (Jacobsen, Kalabikhina), theories of gender discrimination and gender segregation.

2 Materials and methods

Various methods of analytical search were used for studying the functioning of the new structuration: the results of a public opinion poll, regional sociological studies, analysis of gender statistics at the country level, methods of standardized interviews and included observation, and others. These methods were used as a tool for studying socio-gender labor resources, labor activity, and labor force of a developing inclusive economy.

The study proceeds from the simplest postulate of the structuration theory, which states that resources arise only as a result of human activity and are changed and maintained only by people. Thus, a specific labor resource type arises – socio-gender resources. To analyze their development and influence on the social tectonics of the labor market, one should turn to the gender structure of society and the new norms of the gender order. The universal norm of the new gender order is the legal and social norm for overcoming segregation and social discrimination, which can ensure the observance of gender equality in society.

The achievement of gender equality is closely related to the gender structure of the population both in society as a whole and in the economy specifically. If the global gender landscape of the world's population structure is dominated by men – 51%, then in Russia there is a steady predominance of women (by 11 million) over the number of men. Russia is traditionally the most feminized country in the world, which affects the system of gender interaction: from the family to the markets of labor, employment, professions, and educational services. The feminization of society has a strong influence on the formation of modern social and gender policy of the state and is fully reflected in the distribution of socio-gender labor resources in all areas of employment.

Gender labor resources, according to the author's definition, are a type of social resources of male and female gender communities, which in the course of labor activity are transformed into an effective mechanism for the development of the national economy, including a mechanism for the development of a new structure of the markets of labor, employment, professions, and educational services. As a type of resources, gender resources are considered as a means, value, reserves of opportunities, and a source of income. They belong to the group of renewable, depletable, and unchangeable labor resources. These resources are measurable, organized, and managed in the process of achieving social and economic goals.

Gender resources are made up of able-bodied men and women and can be attributed to the most universal resources that can act as a kind of "universal equivalent" and a measure of social development along with information, time, and money. At the same time, gender resources are fundamentally different from other types of economic resources, since they are a factor of social production and development, in which human capital is accumulated. Gender labor resources are directly related to the development of the gender economy, the subject of which is the gender factor of economic development and its impact on the economy. Based on the social characteristics of gender resources, intellectual capital is developing in the economy as the basis of its new industries, and it is based on the intensive use of knowledge (knowledge-based industries). Gender labor resources can be considered unique and universal resources, without which the creation of public goods is impossible, and the effective use of these resources can be considered an axiom of reasonable state management.

3 Results

Further, four market types in the new social structuration are considered. For the first, the labor market is not adequate to the employment market. These are markets with different social objects, differing in the degree of their mobility, labor activity, and types of labor activity. The employment market has the highest mobility today. The social object of the employment market is the able-bodied population and also persons older and younger than the working age, who are able to work. Among them, 76.1% are women, 82% are men [1]. There may be migrants among the employed, unemployed, job seekers and those who have found a job, men and women working under an employment contract, including those who perform work for remuneration on full or part-time terms, and those who have other paid work (service), including seasonal, temporary work, with the exception of public works. A certain part of the employed are teenagers who can earn money (from the age of 14), the economically active population, retired, but ready to work or earn extra money, receiving a certain income. According to the form of employment, it can be temporary, discrete, long-term, short-term, full or partial, primary or secondary, legal or shadow.

The labor market is a sphere of supply and demand for labor that can perform various types of labor activities. The volume of the labor market is the entire working-age population (82.2 million people), 49% of whom are women. In 2017, the labor force of the Russian labor market was 76.1 million people, whose gender structure is asymmetric in favor of men (38.2 million people against 35 million women). At the same time, if in 1991, 91% of enterprises in the Russian economy were state-owned, then in 2019, 4% of employees worked at state-owned enterprises, and 87% worked in privately-owned companies. Also, women prefer to work in the public sector of the economy, considering it socially more protected, stable, and gender-equal. The educational potential of Russian women in the labor market is higher than that of men. Important changes in the characteristics of the labor market were made by the pension reform of 2019, which changed the gender "rocker" of retirement, raising the age limit of male pensioners to 65, and women to 60 years.

The market of professions is a sphere of supply and demand for various types of specialties and professions that are in demand by the economy at different stages of its development. The volume of the market of professions is reproduced by the institution of education, supported by the change in the request of employers and the dramatically expanded alternatives to choosing employment in the digital economy. The market of professions in terms of its volume and variety of professional activities directly depends on the historical conditions and the level of technical development of society. If in the 17th century, there were only 200 professions in Russia, and they were mostly male, then the international standard of classifications of professions in 2019 recorded 9,993 professions, of which 7,000 are sold in the Russian market of professions. In megacities, the market for professions is about 5,000, and in medium-sized cities, it reaches 3,000 specialties. The theory of gender neutrality in the context of the formation of a new gender order has a positive impact on the market of professions, since in recent years it has contributed to the removal of discriminatory barriers for women from 1,356 professions in which Russian women can now work. There are only 100 specialties left in the barrier zone that have gender restrictions on their accessibility for women.

Sociologically, professions can be classified into 6 groups of demand: dying, obsolete intellectual professions, obsolete working professions, popular professions, required professions, and mass ones. The modern market of professions is characterized by their accelerated renewal.

As for the development of the educational services market and the institution of education as a generator of the professions market, due to the "Education without Borders" strategy, the educational services market has expanded the traditional age-limiting framework of students of socio-gender communities to age infinity. The introduction of "Silver Education" programs, education beyond the age significantly prolonged the creative and intellectual capabilities of men and women. The issue of gender equality in the educational services market is also being formatted in a new way, based on their new norms of the new gender order. However, this is the subject of further analysis.

4 Discussion

The problem of the development of a new social structuration instead of the once unified labor space can be considered as a subject of scientific discussion for economists and sociologists. However, it is obvious that the transition to an inclusive economy, the inclusion of new groups in the labor process (and these are not only groups of disabled people and people with disabilities) make corrections to the traditional ideas about a certain single "labor market". This process results in tectonic shifts within the old model of the labor market, multidimensional expansion of employment opportunities, and employment of representatives of male and female socio-gender communities under the influence of digital technologies and new norms of the gender order.

5 Conclusion

1. In Russia, a qualitatively new social structure has been formed and it is developing in the conditions of an inclusive economy and the formation of a new global gender order, consisting of 4 markets: labor, employment, professions, and educational services. Each of them is distinguished by existing social facilities, their scale, prospects, and social mechanisms of employment and internal development.

2. The study of the new social structuration is carried out from the point of view that improving the quality of socio-gender resources of economic development will make it possible to better predict, plan, and redistribute their flows along the trajectories of four dependent markets. This will make it possible to integrate the related demands of economic and social policy, to increase the effectiveness of the social mechanism for regulating the reproduction of socio-gender labor resources of a new quality that meets the implementation of the sustainable development strategy-2030.

References

1. I.I. Changli, Rabota [Work] (TSP, Moscow, 2002)
2. G.E. Slezinger, Sotsialnaya ekonomika [Social Economy] (Publishing house "Business and Service", Moscow, 2001)

3. A.N. Oleynik, Institutsionalnaya ekonomika [Institutional Economics] (INFRA-M, Moscow, 2000)
4. S.A. Kravchenko, S.A. Krasikov, Sotsiologiya riska: Poliparadigmalnyi podkhod [Sociology of Risk: A Polyparadigm Approach] (Ankil, Moscow, 2004)
5. G.G. Sillaste, Woman Rus Soc, **2**, 3-16 (2019). https://doi.org/10.21064/WinRS.2019.2.1

Digital transformation of management resources for a comfortable tax environment

Valeriy Ivanovich Bratcev and *Anastasiya Mikhailovna* Simachkova[1]

Plekhanov Russian University of Economics, Department for Accounting and Taxation, Moscow, Russia

Abstract. With the use of digital technologies, a person's daily life, industrial relations, economic processes and education are changing, as well as new requirements for communications, computing power, information systems and services arise. The article reflects the main results of the digital transformation of tax authorities, contributing to the creation of a comfortable tax environment in the context of the general digitalization of the economy. The study aims to determine the causal relationship between the management decisions made by the tax authorities and the dynamics of obligatory payments to the budgets of the budgetary system of the Russian Federation, as well as to formulate proposals for improving the advanced developments of tax authorities aimed at creating a comfortable tax environment, reflecting the principles of the "economy of trust". The solution of the tasks set in the work was carried out on the basis of the application of general scientific research methods within the framework of comparative, logical and statistical analysis, etc. Among key findings of the research, we should mention the substantiation of the transformation of the management resources of the tax authorities under the influence of the development of digital technologies, as well as the development of practical recommendations for improving the advanced developments of the tax authorities (tax monitoring system, special tax regime for self-employed citizens, single tax payment system) in the framework of creating a comfortable tax environment and building trusting relationships with taxpayers.
Keywords: digital transformation, tax authorities, tax administration, economy of trust

1 Introduction

Digital transformation affects all social and economic spheres, including posing new challenges for the tax authorities: the need to introduce advanced information technologies in a short time, the importance of qualitative changes in the organizational structure and human resources, the imperfection of the current legislation in the context of the accelerated pace of digitalization development, etc. In addition, along with the development

[1] Corresponding author: anastasiya.simachkova@inbox.ru

of information technologies, the needs of taxpayers to receive services provided by the Federal Tax Service of Russia in a fast, affordable and high-quality way are growing [1].

In such conditions, like business, tax authorities need to use new modern approaches to data flow analytics, as well as maintain constant communication with taxpayers in order to implement high-quality tax administration and create a comfortable tax environment [2].

In modern scientific works, the digital transformation of tax authorities is often reduced exclusively to the service component:

- fast and contactless work with conscientious taxpayers is considered as a priority area of work of tax authorities [3];
- among the indicators of "digital maturity" of tax authorities, the key role is assigned to the development of multitasking services for electronic interaction with taxpayers [4];
- the development of electronic services is highlighted as a determining factor affecting the rate of tax authorities [5].

At the same time, this is not enough for creating a comfortable tax environment. An integrated approach is needed when considering the impact of digitalization of the economy on the transformation of the management resources of tax authorities [6]. The digital transformation of tax authorities is not only the development and implementation of new technologies, but also organizational changes, the emergence of new competencies, infrastructure renewal, etc. [7].

So, nowadays, the Federal Tax Service of Russia has increased administrative capabilities through the digitalization of its activities:

- the quality of tax control is increasing, including through the introduction of modern software products that allow taking into account new tax risks arising from the general digitalization of the economy (ASK "NDS-2", AIS "Tax-3", good traceability system, online checkout system, etc.);
- the efficiency of interaction with taxpayers is increasing, including due to the introduction of digital (electronic) services (today, more than 60 different electronic services are presented on the official website of the Federal Tax Service of Russia), as well as a qualitative change in the approach to interaction with taxpayers (special attention is paid to feedback);
- human resources are being strengthened by improving recruitment opportunities based on remote technologies (personnel projects of the Federal Tax Service of Russia "Make a career yourself!", "PROmotion", etc.), which contributes to the creation of effective project teams capable of proposing new ideas aimed at modernizing tax authorities and improving tax administration;
- the quality of tax administration is improved through the use of a scientifically grounded system of employee motivation.

2 Materials and Methods

2.1 Descriptive Design

The digital transformation of the activities of tax authorities has made it possible to fundamentally change the approach to tax administration of certain categories of taxpayers, creating a comfortable and not burdensome administrative environment for both taxpayers and tax authorities [8]. The introduction of a tax monitoring system and a new special tax regime for self-employed citizens (tax on professional income) is the most striking example of such a transformation.

The tax monitoring system in 2020 at the OECD International Forum on Tax Administration was named the best practice for building effective digital interaction between the state and taxpayers. It replaces traditional forms of tax control with online interaction based on remote access of tax authorities to the accounting and tax reporting of taxpayers. At the end of 2020, tax monitoring covered 30% of tax revenues [9].

A special tax regime for the self-employed was also named among the world's best practices. It is based on contactless and non-declaration taxation due to technical solutions built into the information system "My Tax".

It should be mentioned that the number of taxpayers wishing to become participants is growing on an ongoing basis, including due to transparency, convenience and ease of interaction with tax authorities.

For example, over the five years of tax monitoring, the number of participants has increased 30 times. In 2021, 209 taxpayers representing 15 industries (114 new and 95 existing members) participate in tax monitoring. Today, about 93% of participants provide remote access to accounting information or to the corresponding data marts, which allows tax authorities to fully use the capabilities of remote control, significantly reducing the volume of document reclamation [10].

In addition, the development of digital technologies has made it possible to expand the list of potential participants in tax monitoring by changing the relevant criteria.

As for the special tax regime for the self-employed, in general, since the beginning of 2019, the number of participants has increased to 2 million. It should be mentioned that, in accordance with the explanations of the Ministry of Finance of the Russian Federation, it is possible to register as a self-employed person from the age of 14. As of 23.03.2021, 25 thousand minors (1.3%) were registered as self-employed.

During the period of the said tax regime, the self-employed registered 327 billion rubles in income, which led to the payment of tax in the amount of 8 billion rubles. At the same time, 85% of registered taxpayers had not previously declared income from business activities, and 40% had no official income at all in the year before registration [11].

2.2 Exploratory Design

Another striking example of creating a comfortable tax environment in the context of digital transformation is providing taxpayers with the opportunity to use the mechanism of a single tax payment (STP). It allows you to transfer obligatory payments in one payment order without specifying the type of payment, the term for its payment, belonging to the budget, which significantly reduces the risk of making a mistake, as well as the formation of debt. The tax authority independently, on the basis of available information, set off the transferred funds against the obligations of the payer.

The STP mechanism for citizens was introduced in 2019 to pay property taxes, and from 2020 the ability to pay personal income tax was added in this way. The number of individuals using this mechanism is gradually growing: in less than 10 months of 2020, over 110 thousand citizens took advantage of the possibilities of a single tax payment, transferring 2.7 billion rubles. This is more than five times the corresponding figure for the whole of 2019 [12].

Today, it is planned to extend the possibility of using the STP also to individual entrepreneurs and legal entities. It is expected that the State Duma of the Federal Assembly of the Russian Federation will consider draft law No 1141868-7 in the second reading. If the relevant draft law is adopted, organizations and individual entrepreneurs will be able to pay all obligatory payments also with one payment order. First, the amount of the STP will be used to pay off the existing debt, then on the account of forthcoming payments with the earliest due date, and in their absence – on account of the debt for payment of penalties, interest and fines.

The adoption of such changes will significantly reduce the time for processing, settlement documents, and will also prevent taxpayers from creating tax debt.

At the same time, it is rather difficult to motivate the commercial sector to use STP, given the peculiarities of this system. It is highly undesirable for a business to withdraw funds for the purpose of early fulfillment of tax obligations, since these funds could be used, for example, to expand production or invest, which could potentially lead to an increase in profits. In this case, for business, STP is a kind of budget advance. In addition, the STP must be paid on the first of all available due dates for the payment of taxes and insurance contributions, which occur during the entire reporting period.

3 Results

The use of modern digital technologies allows tax authorities to significantly simplify and make transparent interaction with taxpayers [13]. Technologies of "contactless" work with taxpayers in real time are the foundation of the tax administration of the future.

Digital transformation efforts are delivering strong results. According to the Federal Tax Service of Russia, the consolidated budget of the Russian Federation in 2020 received 21,013.9 billion rubles (8.2% less than in 2019, which is due to restrictions caused by the COVID-19 pandemic), in 2019 received 22,737.1 billion rubles (6.6% more than in 2018) and in 2018 21 328.5 billion rubles were received (23.0% more than in 2017). A similar dynamics is observed in relation to revenues to the federal budget and to the consolidated budgets of the constituent entities of the Russian Federation. Structure of revenues to the consolidated budget of the Russian Federation in 2018-2020 is presented in Fig. 1.

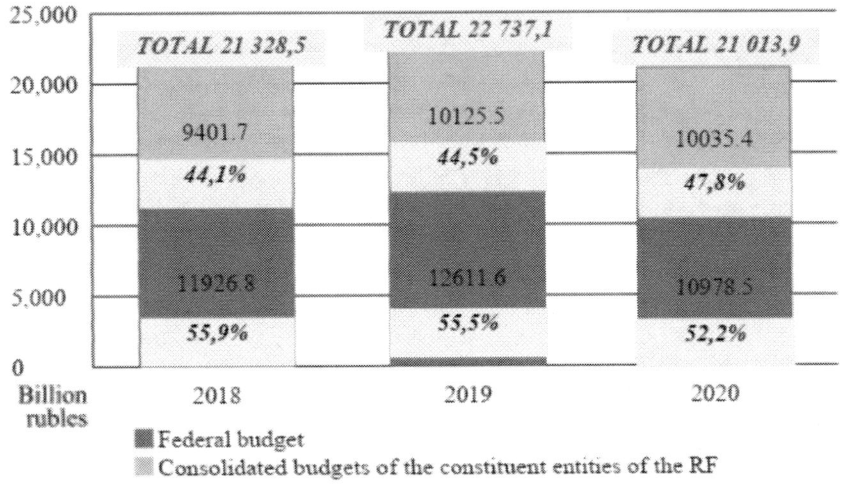

Fig. 1. Analysis of revenues to the consolidated budget of the Russian Federation in 2018-2020.
Source: Compiled by the authors based on statistics from the Federal Tax Service of Russia [14].

These results were achieved precisely due to the synergistic effect from the introduction of modern digital technologies in tax authorities and, as a result, the digital transformation of the management resources of the Federal Tax Service of Russia.

The introduction of innovative technologies into the work allowed the tax authorities to conceptually change the paradigm of control work, systematically reduce the administrative burden on taxpayers and increase the efficiency of interaction with them as part of the formation of a comfortable tax environment based on the principles of customer-centricity in the context of the digitalization of the economy [15].

The transition to working with big data, attention to the interaction with taxpayers, increased transparency of activities allowed the tax authorities of the Russian Federation to become like the leading digital platforms.

4 Discussion

At the same time, it should be mentioned that, according to a number of experts, digitalization of tax administration not only contributes to the creation of a comfortable tax environment, but also becomes one of the factors of tax risk for entrepreneurs [16]. For example, one of the key tasks of tax authorities in the framework of digital transformation is to ensure reliable information protection and security of tax administration systems [17].

Moreover, even the advanced developments of the tax authorities need to be refined, both from the technical side and from the side of legal regulation. For example, the considered achievements of digital transformation of tax authorities today have a number of significant disadvantages that affect the use of these systems by taxpayers.

1. Small business cannot become a member of tax monitoring (as of June 10, 2021, the number of small and micro enterprises exceeds the number of medium-sized enterprises by 5,836 thousand units, or more than 330 times) [18]. Of course, the connection to the tax monitoring system for this category of taxpayers due to the significant cost of the required software is quite problematic. At the same time, the creation by the tax authorities of a mechanism for obtaining a motivated opinion about the specific life situation of the taxpayer (for example, about the completion of a specific transaction) based on the analysis of the submitted materials (documents, information) using modern digital technologies, including artificial intelligence technologies, will allow taxpayers to avoid errors in the calculation of obligatory payments, will increase their financial literacy, and will also allow the tax authorities to analyze the issues that pose the greatest difficulties for taxpayers and opportunely respond to the corresponding challenges.

2. According to a study of the Higher School of Economics (NRU HSE), there are 8 million self-employed citizens in Russia in 2021, or about 11% of the total employed population (the average estimate of Rosstat, NRU HSE, World Bank, OECD and Russian Presidential Academy of National Economy and Public Administration is 7.2 million self-employed) [19]. At the same time, as noted earlier, in 2021 the number of registered self-employed is only about 2 million people. Thus, today the tax authorities need to take additional measures to stimulate the registration of citizens as self-employed, including taking into account the possibilities of modern digital technologies. For example, one of the most effective ways is the integration of the information system "My Tax" with digital platforms that unite service providers and their clients (Profi.ru, "Yandex.Services", etc.).

3. Partial refund of funds transferred as a STP is not provided. Only the remainder after the STP is set off towards the payment of obligatory payments after the expiration of the payment due date is subject to return. The introduction of such a mechanism will allow the business, if necessary, to return a part of the funds transferred as the STP. Due to modern

digital technologies, it will be possible to set a limit for such a partial refund (for example, no more than the amount of the next obligatory payment). Thus, the use of the STP system will not be as burdensome for the commercial sector, which will lead to an increase in the use of such a mechanism, which in turn can potentially affect the reduction of tax debt.

At the same time, the implementation of proposals developed by the authors on the basis of theoretical research and the established practice of introducing digital technologies will make it possible to fully realize the potential of digital transformation of the management resources of tax authorities within the framework of creating a comfortable tax environment. Despite the presence of certain difficulties and risks, the introduction of modern digital technologies in tax administration significantly increases its quality and efficiency, including making the process of paying taxes understandable and convenient for taxpayers [20].

5 Conclusion

The introduction of modern digital technologies allows tax authorities to simultaneously improve all areas of their activities: to increase the efficiency of tax control, the quality of interaction with taxpayers, as well as other ministries and departments, to develop international cooperation, etc.

The digital transformation of the management resources of the tax authorities in the future can lead to an improvement in the quality of budget planning, the stability of obligatory payments to the budget system, despite the influence of external factors, as well as an increase in the transparency of the economy and the formation of a comfortable tax environment [21].

The accelerated pace of development of the digital economy has largely influenced the transformation of the management resources of the tax authorities. Today, it is no longer just the development and implementation of modern digital technologies, it is also the renewal of infrastructure, organizational changes, rethinking of the existing approaches to tax control and interaction with taxpayers [22].

At the same time, in order to create an adaptive digital platform, tax authorities need to give priority to not only the introduction of new mechanisms (information systems, digital services, etc.) that will significantly simplify the process of tax administration (make it more transparent and invisible to the taxpayer, reduce tax risks, etc.), improve the efficiency of interaction between tax authorities and taxpayers. It is also necessary to improve on an ongoing basis (refine) existing solutions that have demonstrated their effectiveness, including tax monitoring systems, a special tax regime for the self-employed, the STP system, etc.

So, when studying the impact of digitalization of the economy on the transformation of management resources within the framework of creating a comfortable tax environment, it is necessary to apply an integrated approach, taking into account the changes covering all activities of tax authorities.

References

1. S. Huseynli, J Curr Debates Soc Sci, **4**, 33-47 (2021). https://doi.org/10.37154/cudes.2021.30

2. A. Mustafa, *Digital service taxes as the fiscal result of digital transformation*. In U. Hacioglu, T. Aksoy (eds), Financial ecosystem and strategy in the digital era. 355-378. (Springer, Istanbul, Turkey, 2021). https://doi.org/10.1007/978-3-030-72624-9_15
3. T.V. Muravleva, D.D. Orlovskaya, Sci Soc, **3(35)**, 31-34 (2019)
4. L.N. Galieva, *Organization of work with taxpayers in the context of digital transformation of the economy*, in Improving tax administration, Materials of the fourth scientific and practical conference, December 06, Ufa (2019)
5. T.Yu. Solodimova, E.G. Grineva, Econ Manag: Curr Trends, **5**, 154-157 (2019). https://doi.org/10.31483/r-33410
6. T. Krieger, Int J Innov Tech Econ, **2(34)**, 1-12 (2021). https://doi.org/10.31435/rsglobal_ijite/30062021/7543
7. D. Bentley, eJournal of Tax Research, **18(2)**, 353-381 (2020)
8. M. Kačaljak, Bratislava Law Rev, **4**, 21-30 (2020). https://doi.org/10.46282/blr.2020.4.2.209
9. OECD. Tax Administration 3.0: The digital transformation of tax administration. Paris, France: OECD Publishing (2020). Accessed on: September 24, 2021. [Online]. Available: https://www.oecd.org/tax/forum-on-tax-administration/publications-and-products/tax-administration-3-0-the-digital-transformation-of-tax-administration.pdf
10. 209 companies will participate in tax monitoring from 2021 (2020). Retrieved from https://www.nalog.gov.ru/rn77/news/activities_fts/10222192/
11. Two million self-employed registered in Russia. (2021). Accessed on: September 24, 2021. [Online]. Available: https://www.nalog.gov.ru/rn77/news/activities_fts/10747467/
12. I. Zubkov, R. Markelov, Edinyi nalogovyi platezh rasprostranyat na biznes [A single tax payment will be extended to businesses] (2021). Accessed on: September 24, 2021. [Online]. Available: https://rg.ru/2021/04/01/edinyj-nalogovyj-platezh-rasprostraniat-na-iurlic-i-ip.html
13. C. Williams, E-tax: electronic tax filing (Regional Cooperation Council, Sarajevo, 2021)
14. Tax analytics (2021). Accessed on: September 24, 2021. [Online]. Available: https://analytic.nalog.ru/portal/index.ru-RU.htm
15. P. Bechko, N. Lysa, V. Bechko, S. Ptashnik, Col Works Uman National Univ Horticulture, **2**, 143-153 (2021). https://doi.org/10.31395/2415-8240-2021-98-2-143-153
16. S.L. Moiseenko, N.P. Malysheva, Vestnik Altai Acad Econ Law, **3**, 194-200 (2021). https://doi.org/10.17513/vaael.1630
17. E.A. Kirova, A.V. Kozhebatkina, Vestnik Univ, **9**, 94-99 (2020). https://doi.org/10.26425/1816-4277-2020-9-94-99
18. Unified register of small and medium-sized businesses (2021). Accessed on: September 24, 2021. [Online]. Available: https://rmsp.nalog.ru/
19. Report of the National Research University Higher School of Economics "The market of services for self-employed citizens" (2021). Accessed on: September 24, 2021. [Online]. Available: https://www.hse.ru/mirror/pubs/share/442787036.pdf

20. B. Firmansah, N. Rahayu, J Manaj Pelayanan Pub, **4**, 51-69 (2020). https://doi.org/10.24198/jmpp.v4i1.28005
21. N. Respati, Scientax, **1(2)**, 109-130 (2020). https://doi.org/10.52869/st.v1i2.38
22. K. Cipek, I. Ljutić, Oditor, **7(1)**, 37-69 (2021). https://doi.org/10.5937/oditor2101037c

Is Coronavirus a new business reality facilitator?

Yulia Vitalievna Kim[1]

Financial University under the Government of the Russian Federation, Department of English Language and Professional Communication, Moscow, Russia

Abstract. This article considers the spread of the SARS-CoV-2 virus, known as the COVID-19 pandemic, to have far-reaching consequences in terms of employment, production of goods and services worldwide. The article aims to research the dynamics of changes in attitudes towards the disease resulting in reshaping labor market strategies worldwide which include forming stay-at-home economy, digitization of business operations and emerging start-ups to facilitate delivering and consuming goods and services. A few research designs are used in this article with quantitative and case studies being major methods. The main results of this research allow one to associate coronavirus disease with a new business reality facilitator in the global economy rather than a mere risk factor. The topicality and urgency of this article are based on the novelty of the pandemic itself having started less than 2 years ago. This research may be helpful for upcoming fundamental studies in the field of 21st-century global emergencies.
Keywords: pandemic, coronavirus, digital economy, risk factor, economic development

1 Introduction

Ironically, Coronavirus, despite its biological nature, is now considered as a purely external economic risk factor. For the last few decades, the world has faced several virus outbreaks, but none of them ever acquired the global scale and impact that SARS-CoV-2 has. As of June 27, 2021, the cumulative number of infections and death due to COVID-19 worldwide was more than 180 million and 3.9 million respectively [1]. It is obvious that little time has passed since the first wave of contamination occurred in China towards the end of 2019 so that one cannot expect scientists to have much data and insights to fully analyze this issue as a global matter with far-reaching consequences in the following years or even decades.

Nevertheless, much has been said and written regarding the impact of coronavirus on global economies worldwide. The current paper is primarily based on articles in business magazines and scientific journals, rather than on the more fundamental researches that for the most part are still to be published. While examining all the information and speculations concerning this pandemic, one will inevitably notice that it falls into two categories, the

[1] Corresponding author: YVKim@fa.ru

first being a descriptive analysis (mainly encountered in scientific journals), whereas the second represents an action research design (business press). For example, E. Smirnov and E. Karelina in their article *Assessment of Paradoxes and Risks of the Development of the Global Economy in the Context of the Coronavirus Crisis* aim to project how national economies recover from the initial shock caused by the disease outbreak [2]. While we respect the profound work undertaken by these prominent Russian scientists, we cannot ignore the fact that their article, though it provides a deep understanding of the processes that have emerged and the measures that have been taken, treats this pandemic as a mere economic crisis rather than a global shift towards a new economic agenda that is here to stay. In this respect we are more inclined to support the position of another Russian scientist, A, Nikiforova, expressed in her article *Labor Market in the Russian Federation in Terms of the Coronavirus Pandemic: The Impact of the Virus on Employment, Production of Goods and Services.* While Nikiforova's article restricts the scope of her research to the Russian economy, her findings can be still applicable to understanding global processes as the author links the pandemic with the "beginning of a new, modern society" [3]. The author also states that this disease "will lead to the development of many spheres of public life" [3]. A similar, yet more elaborated, attempt to frame the upcoming economic changes in the realms of the Coronavirus pandemic is shown in A. Tsatsulin's article *Digitalization of the Population as a Determinant of the Virtual and Real Labor Market during the Pandemic.* The author provides a deep analysis of the current sociopolitical situation in both Russia and the USA to identify digitalization as one of the ways to survive through crisis [4].

This present research aims to fill in the gap in the way Coronavirus is being primarily treated as an emergency which is bound to cease, since the approach of this article is based on the notion that SARS-CoV-2 is a facilitator towards 21st-century development of the economic factors paradigm. With that being stated, it is also essential to define the key issues of the research. Can Coronavirus be treated as a development factor? Is Coronavirus an occasional matter or an imminent feature of the economic development of the next decades? These questions are of primary significance for our research. In this article we will (1) provide materials and methods used to gather and analyze data; (2) make an effort to answer the key questions mentioned above; (3) interpret the results obtained in order to situate them within the frame of contemporary researches on the same subject; (4) outline the practical importance of our findings.

2 Materials and methods

While our paper is aiming to identify the contribution of the SARS-CoV-2 pandemic towards forming new economic realities, it is necessary to outline the research steps required. First of all, quantitative data is gathered to shape our understanding of how values and output were changed and relocated during the pandemic. For this, we are using secondary data collected by business journalists and economy websites. Since this is descriptive data, gathering observations does not mean any intervention and is unable to determine "cause and effect" relationships. To link the event (pandemic) to its outcome (new economic reality), a "cause and effect" method is applied. We believe that this approach is the most favorable in order to establish the relationship between various processes and provide a projection in the future as it is a standard methodology in our field which does not require justification. Since quantitative method is considered to provide valid generalizable results, it is also important to include/exclude certain criteria to choose our variables by using a sampling method which includes both the most affected and improved economic spheres in the last 18 months.

3 Results

Since our goal is to consider economy formation during the Coronavirus outbreak as a global trend, our study focuses on this process worldwide. Many experts and sources comment on this matter as a significant decrease in economic development. The World Bank, being one of the major international financial institutions, has estimated the proportion of national economies facing recession as the largest in the last 150 years (Fig. 1). According to World Bank data, in 2021 more than 90 % of states are experiencing a downturn and an undoing of the previous years of prosperity and development, with one-third having already been suffering from losses from 2002 to 2018.

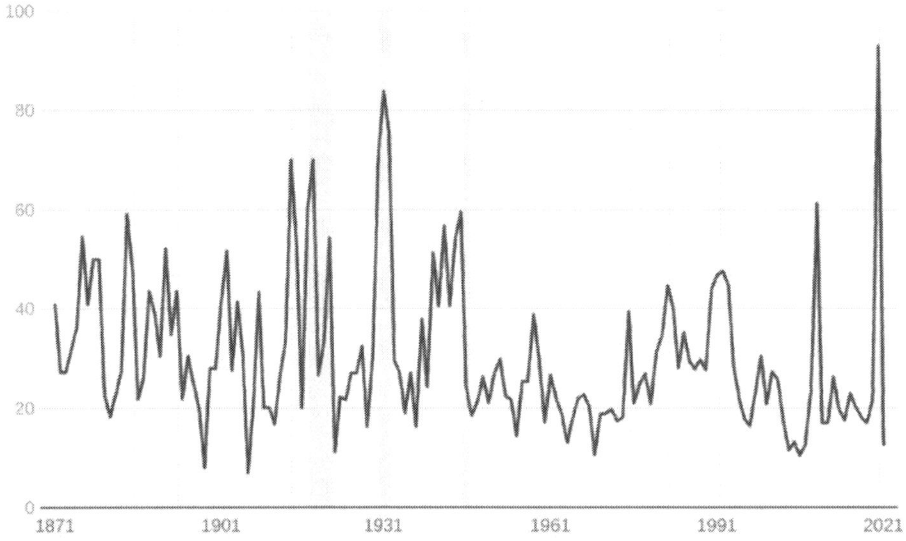

The proportion of economies with an annual contraction in per capita GDP. Shaded areas refer to global recessions. Data for 2020-21 are forecasts.
Source: World Bank

Fig. 1. Share of economies in recession, 1871-2021 [5].

The World Bank also states that the decline in the social economy sphere that was mostly caused by COVID-19 worldwide has "triggered unprecedented collapse in oil demand and a crash in oil prices" [5]. According to Dr. Fatih Birol, Executive Director at IEA (International Energy Agency), "the outbreak of Coronavirus disaster is somewhat distressing the broad range of global energy markets — comprising all types of fossil fuel like renewable, gas and coal, nonetheless its effect on crude oil markets is mainly harsh as it has stopped movement of people and merchandises from moving around, thus causing a severe setback for demand of crude oil" [6]. However, it is also important to note that the economies of developing countries, being mainly dependent on oil prices, have already been viewed as in crisis long before the COVID-19 pandemic outbreak which makes us believe that the notion of Coronavirus being at fault for their declines is not completely justified. In summer 2020, OECD claimed that "prior to the crisis, financing had already fallen short of the spending needs to achieve the SDGs by 2030, and fiscal space was limited by rising public debt levels and servicing costs in emerging and developing economies" [7].

While it is essential to note the overall downturn in national economies, there is one process that is observed in all countries throughout the pandemic – start-ups playing a core role in both developed and developing states. It is known that start-ups are considered to be innovative drivers of economic development and able to offer work places that never existed before. "Young firms indeed account for about 20% of employment but create almost half of new jobs on average across OECD countries, and innovation by young firms significantly contributes to aggregate productivity growth, accounting for half of it in the United States" [8]. Even though the Coronavirus pandemic hindered many innovative firms from continuing to grow, their role in the economic development during the last 18 months has been revolutionary in terms of forging a new business reality.

By 30 March 2020, the early beginning of the pandemic in Europe, it had been estimated that more than 100 start-ups were launched in EU in order to provide social assistance to people left in predicaments [9]. Those include health-tech, mental health and wellbeing, food delivery, fintech, software, transport, remote working, travel, hospitality, events, prop-tech, ed-tech and consumer relations. With this list of different products and services it is clear that customers have gone for online channels since the pandemic started so the businesses have had to respond respectively. According to McKinsey&Company research [10], the rate at which businesses managed to digitize their services and products exceeds the previous years' speed by almost 30% worldwide with this acceleration being equal to 7 years development in precrisis circumstances (Fig. 2). Even the Asia-Pacific region, which is mostly developing economies, has increased its digitization by more than 20% and adopted the proportion of digitized performance that accounts for more than a decade ahead (Fig. 2).

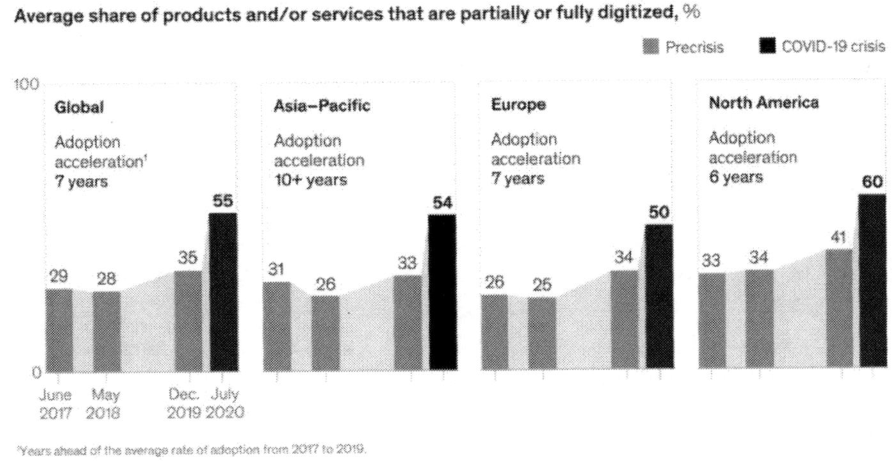

Fig 2. Speedup in acquiring digital or digitally enhanced offerings worldwide [10].

4 Discussion

While it is obvious that the fastest adoption of digitalized products and services is observed in the businesses which are digital in nature, many companies chose to provide online channels refocusing their performance rather than investing in product development. JPMorgan, AT&T, Google, Amazon, Nike, Facebook are hustling to virtualize various operations with Coronavirus being the catalyst for this business transformation [11]. Amazon alone, since mid-March 2020, has hired 175,000 workers to meet surging demand for online shopping [12]. Nevertheless, a paradigm shift towards digitization and servitization of the economy was already underway prior the pandemic [13]. Yet, many European experts believe that the coronavirus will indeed accelerate the pace of the digital transformation and eventually strengthen digital economy [14]. It is also important to note that those huge leaps that are seen in Fig. 2 were made under huge pressure and may have been facilitated by a decrease in offline portfolios in the meanwhile. Besides, it is also crucial to underline that pre-pandemic rate of very high capacity networks availability in EU did not cover even the half of the households with 44% in 2019 [15]. However, seeing how traditional offline channels are failing to deliver at a previous rate and scale, new economic realities might need to get used to the Coronavirus threat as a constant factor in the market field which would inevitably transform present economic reality towards larger integration of digital tools.

5 Conclusion

The impact of the Coronavirus on the global economy has appeared as a twofold hit. While it is too early to make any predictions regarding the longevity of the state of emergency we still live in, one can be sure that the COVID-19 pandemic is yet to determine the future of business realities. The last 18 months have shown that the disease is not only a risk factor but also a facilitator in forming a new business reality in which digital products and services will be the main drivers of the economy.

References

1. Coronavirus Resource Center, John Hopkins University (2021). Accessed on: September 25, 2021. [Online]. Available: https://coronavirus.jhu.edu/map.html
2. E.A. Karelina, E.N. Smirnov, Innov Invest, **2**, 33-38 (2021)
3. A.A. Nikiforova, Econ Bus, **12-2(70)**, 181-183 (2020)
4. A.N. Tsatsulin, SPb State Polytech Univ J. Econ, **13(4)**, 19-35 (2020)
5. The World Bank (2020). Accessed on: September 25, 2021. [Online]. Available: https://www.worldbank.org/en/news/feature/2020/06/08/the-global-economic-outlook-during-the-covid-19-pandemic-a-changed-world
6. IEA (2020). Accessed on: September 25, 2021. [Online]. Available: https://www.iea.org/authors/dr-fatih-birol
7. OECD (2020). Accessed on: September 25, 2021. [Online]. Available: https://www.oecd.org/coronavirus/policy-responses/the-impact-of-the-coronavirus-covid-19-crisis-on-development-finance-9de00b3b
8. OECD (2020). Accessed on: September 25, 2021. [Online]. Available: https://www.oecd.org/coronavirus/policy-responses/start-ups-in-the-time-of-covid-19-facing-the-challenges-seizing-the-opportunities-87219267
9. A. Lewin, C. Bilboe, How Startups are supporting Europe's Coronavirus Efforts (2020). Accessed on: September 25, 2021. [Online]. Available: https://sifted.eu/articles/startup-initiatives-coronavirus/

10. McKinsey&Company (2020). Accessed on: September 25, 2021. [Online]. Available: https://www.mckinsey.com/business-functions/strategy-and-corporate-finance/our-insights/how-covid-19-has-pushed-companies-over-the-technology-tipping-point-and-transformed-business-forever
11. H. McGowan, How the Coronavirus Pandemic Is Accelerating The Future Of Work (2020). Accessed on: September 25, 2021. [Online]. Available: https://www.forbes.com/sites/heathermcgowan/2020/03/23/the-coronavirus-pandemic-accelerates-the-future-of-work-and-provides-opportunity/?sh=69fa9a7b317f
12. Z. Ingilizian, Is Staying In The New Going Out? How The COVID-19 Pandemic Is Fuelling The Stay-At-Home Economy (2020). Accessed on: September 25, 2021. [Online]. Available: https://www.weforum.org/agenda/2020/05/covid19-coronavirus-digital-economy-consumption-ecommerce-stay-at-home-online-education-streaming/
13. BDO (2020). Accessed on: September 25, 2021. [Online]. Available: https://www.bdo.com/insights/business-financial-advisory/strategy,-technology-transformation/covid-19-is-accelerating-the-rise-of-the-digital-e
14. DMEXCO (2021). Accessed on: September 25, 2021. [Online]. Available: https://dmexco.com/press/corona-pandemic-will-strengthen-digital-economy/
15. European Commission (2020). Accessed on: September 25, 2021. [Online]. Available: https://ec.europa.eu/commission/presscorner/detail/en/qanda_20_1022

Fraud in the Russian labor market: a systematic review

Anna Vergazova[1]

Financial University under the Government of the Russian Federation, Department of Sociology and Mass Communications, Moscow, Russia

Abstract. This article synthesizes the current knowledge available on the fraud practices in the Russian labor market by conducting a systematic literature review among various databases. The focus of this review is to categorize and conceptualize the diversified types of fraud in the labor market and understand the challenges such practices present. The study is based entirely on the available secondary data collected from Russian and non-Russian academic databases for the last 5 years (2017–2021) in line with the guidelines into best practice. The aim is to promote social security by combining the state-of-the-art body of knowledge of the Russian labor market with respect to all its agents in terms of deviant sociology measures. The findings of this review show current gaps both in academic and legislative practices for approaching fraud in the labor market and present the challenges to overcome for avoiding further negative effects on the economy and society.
Keywords: fraud, labor market, deviant behaviour, systematic literature review, workers' rights

1 Introduction

The current state of the global economy, part of which the labor market is, could be characterized as unstable. This is also applicable for the Russian market, which due to the pandemic, foreign sanctions, fluctuations of the ruble, and other factors is in a high state of vulnerability.

According to the recent studies [1] conducted by the Ministry of Labor of the Russian Federation, the forecast for the Russian labor market before the global pandemic included the following trends: (i) decline of unemployment (ii); rise of the working-age population (iii); decline of the employed population (iv), increase in migrant workers. The pandemic has changed the dynamics by heavily affecting various sectors of the economy, as well as by the fact that the Russian Federation has closed its borders to migration for preventing the rise of infection rates.

With the increasing dynamics of social processes, there is also a parallel trend, namely, the rise of deviant behavior practices in society, both negative and positive. In that sense, citizens become less safe in their own state as there is no protection from the emerging

[1] Corresponding author: atvergazova@fa.ru

deviant practices on the corporate and individual levels. Therefore, the need for accumulating knowledge of what those behaviors are, how they are practiced, what are their social effects and how the damage can be prevented is rising [2].

Moreover, the market is also affected by modern processes. With the rise of digitalization, development of the digital economy [3] and new forms of employment [4], the market and social policies are yet to adapt to the new reality and form mechanisms for protecting its agents.

Having said the above, the labor market is threatened by deviant behaviors, one of which is fraud. This study defines fraud according to Article 159 of the Criminal Code of the Russian Federation (*theft of another's property or acquisition of the right to another's property by deception or abuse of trust*) [5]. Fraud could be conducted by all the agents of the labor market – employers, employees, job-seekers, or job agencies. This review focuses on fraud that has workers as victims.

Literature suggests that the reason fraud occurs is not the complete absence of law, but politics that are more protective of stakeholders rather than workers along with the fact there are not so many possibilities for workers to protect themselves. The reason for that is mainly the absence of such knowledge [6].

This makes studying fraud as deviant behavior in the labor market a vital concern due to its evolving and rising character. Therefore, the study seeks to (i) categorize fraud in the Russian labor market; (ii) highlight current challenges; (iii) outline potential negative effects of practicing such behaviors. This is going to be achieved via a systematic literature review.

2 Materials and methods

A systematic review of the literature was conducted by adapting the Preferred Reporting Items for Systematic Reviews and Meta-Analysis (PRISMA) protocol [7]. The review included academic literature in Russian and English that was focused on fraud in the Russian labor market.

Using various databases ensures that the study achieves the objective of collecting the entire available body of knowledge. The string used for the search was "fraud labor market" (in Russ.). For the search in English, the word "Russia" was also included in the string. Between all the words, the "AND" operator was used to ensure all of the key terms for this study were included in the text of the articles and for excluding articles non-related to the topic in focus. The search was also limited by articles published in the timeframe between 2017 and 2021.

A total of 108 articles were included in the final corpus of analysis. Table 1 displays the identification and selection protocol.

Table 1. Identification and selection protocol

Database	Total	Taken into analysis	2017	2018	2019	2020	2021
eLibrary	12	7	0	3	2	2	0
EBSCO	6	1	0	0	1	0	0
JSTOR	20	1	0	0	0	0	1
ScienceDirect	68	0	0	0	0	0	0
Scopus	2	0	0	0	0	0	0

Web of Science	0	0	0	0	0	0	0
Academic Reference	0	0	0	0	0	0	0

3 Results

Types of deviant behaviors. Deviant behaviors in the Russian labor market started developing in the 1990s but yet they have not been established as a subject of study in the academic literature [8]. Therefore, fraud mainly comes from the fact that usually the relationship between the employer and the employee is developing based mostly on the word of trust with no back-up documentation that ensures the rights of each agent. Kiryukhina et al. [9] identify the following types of deviant behavior in the labor market from the side of the employers or employing agents (fraud being highlighted):

- Empty vacancies – available job announcements that are not monitored anymore by the employer, meaning that the applicants never receive any feedback;
- Theft of the employee's funds – the most common way to do so happens by assuring the employee they are a good fit for the placement but asking to pay for some educational or other materials as the last step of the selection process. As a result, the employer gets either directly the money or the employee's credit card information;
- Personal data theft – as a result, the employee becomes a potential victim of spam, fishing, or financial fraud;
- Using potential employees as free labor – the employer asks the potential candidates to perform a task and after receiving the results stops any further communication;
- Recruiting the employee in a pyramid or other scheme – the employee is asked to invest a certain amount of money for becoming part of some pyramided scheme or another scheme that is supposed to lead to profit from the invested money which in reality never takes place;
- Asking the candidate to pay for a vacancies database – such behaviors usually are implemented by agencies that promise to send the candidate a database of vacancies that fit their profile. The outcome could be either theft of money when the applicant pays for a service but does not receive any further communication, or when the applicant receives a database that includes vacancies available in open-sourced resources.

Other types of deviant behavior may be related to schemes for not paying salary to the worker by not following the law or completely dismissing workers' rights. As a result, when an investigation is initiated the worker has to prove the fact a relationship with the employer existed and fraud actions indeed took place. Moreover, types of fraud in the labor market become more and more complicated, which makes it harder to catch such behavior and put it under investigation properly [10].

Effects of deviant behaviors. Apart from regular employment types, workers may also belong to the category of the precariat. Being included in such a category could be either voluntarily or for reasons beyond control. The precariat consists of: (i) temporary workers; (ii) part-time workers with seasonal or occasional payments; (iii) unemployed; (iv) freelancers; (v) trainees and students. Some authors include in that group also employees of borrowed labor (against the law since 2016) and migrants.

The precariat does not enjoy the social policies applicable for contracted workers which makes them more vulnerable to acts of fraud. Moreover, the group is growing as the reasons for becoming part of it are the same as the reasons for unemployment. Dudina and Golovanova [11] suggest that in the future, precarization would grow in size. Currently, this group does not have internal instruments for aid, neither their relationship with the employers is regulated to a full extend. It is expected that if this factor does not change, fraud in that group will rise as its size grows.

Academic literature mentions the vulnerable state of the precariat, namely the migrants that have to navigate the labor market being forced to operate by the rules of the shadow economy [12]. Such informal employment is estimated to constitute between one-fifth and one-third of all employment. Given this number, it is highly unlikely that there are ways to measure it precisely, as well as to assess its current and future effects. Authors mention among the reasons why this is happening the complicated legislative process to legalize such typed of employments, especially among migrants to Russia.

The effects for the employees that have been subject to fraud are the following: (i) loss of money may put the household financially to a critical level which can also ruin family life; (ii) time that could be used productively in the economy is wasted on dealing with fraud; (iii) the person's ability to work decreases, while mental illnesses such as depression, irritability, and psychopathic disorders may appear; (iv) the trust in the labor market is undermined, which complicates further job search [13].

Some authors also state that due to the extensive Internet coverage, entering the labor market is easier than ever before. However, some employers choose to hire workers using traditional methods [14] for avoiding complexities. Apart from that, some employers choose not to invest enough for having a comprehensive system for their job placements online which also makes them more vulnerable to fraud [15].

As stated above, there is a risk that the financial situation of the victims of fraud could worsen, and their trust in the labor market will decrease. This affects the economy, while workers are left alone to deal with their problems and there is a chance that they will remain either unemployed or become less economically active.

4 Discussion

Protection from fraud remains now mainly a problem solved by its own victims. Russian workers have to take measures themsleves to ensure they do not become a subject of fraud and there is little the state or the employer can do to ensure their safety, especially if the workers belong to certain more vulnerable categories such as the precariat [11].

However, certain measures were suggested in the literature such as involving the police to a bigger extend in the investigation of fraud cases [13] and working on awareness in the digital environment [15]. Apart from that, some authors state that one of the reasons for fraud in the labor market is that workers are not familiar with the law and their own rights [8].

The legislation system in Russia that has yet to adapt to the new realities is posing challenges for the market that has developed its own shadow part with a parallel order [12]. Informal employment becomes a space that is not institutionalized and, therefore, all types of deviant behaviors could be easily adopted there.

As not much knowledge is accumulated on what exactly fraud is in the all-Russia labor market, the authors tend to develop ideas that are coming close to each other yet do not provide a full picture of how it can be eliminated and fought. Having said that, there is no current assessment of how big the actual level of fraud is neither how much it does affect the society and behaviors of the agents of the labor market. By providing reviews on a regional level or by mentioning fraud in the context of the developing digital economy, there is only a common agreement that it should be expected that fraud levels will rise if no measures are taken though there is no common agreement on how they should be implemented.

5 Conclusion

This review highlighted the vulnerable state of the Russian labor market. This vulnerability is caused by the unstable economy which as an effect brings unemployment rates higher and contributes to the practice of deviant behaviors. The development of shadow markets, the complexity of the law, and lack of awareness from the side of workers combined with the effects of digitalization and challenges posed by the aftermath of the coronavirus crisis may bring in the future even more instability. Adaptation to the new realities, the inclusion of emerging employment types in institutions, and building trustful relationships among the agents of the market are tasks that have to be first solved on a state level.

Despite the fact the subject of the review is understudied, there is a common agreement that if measures are not taken, fraud in the labor market may start affecting other sectors of the economy, as well as have negative social effects such as lack of social security. In the context of growing uncertainty, deviant behaviors are expected to only grow further meaning that fraud and other deviances have to be assessed, analyzed, and slowly eliminated by a complex strategy including educating the agents of the market and proposing a new legislative framework that can be adapted by employers.

References

1. I.G. Andreeva, Edu Law, **4**, 387-394 (2020). https://doi.org/10.24411/2076-1503-2020-00247
2. Ya.I. Gilinskii, Soc Stud, **4**, 72-78 (1991)
3. D.G. Tkachenko, Sports: Econ, Law, Manag, **3**, 29-31 (2020)
4. D.V. Golovanova, O.M. Dudina, Creative Econ, **13(6)**, 1253-1268 (2019)
5. Statya 159 UK RF. Moshennichestvo (deistvuyushchaya redaktsiya) [Article 159 of the Criminal Code of the Russian Federation. Fraud (current edition)]. Accessed on: October 4, 2021. [Online]. Available: https://www.zakonrf.info/uk/159/
6. K. Greenfield, Yale LJ, **107**, 715 (1997)
7. D. Moher, A. Liberati, J. Tetzlaff, D.G. Altman, Brit Med J, **339**, b2535 (2009). https://doi.org/10.1136/bmj.b2535
8. A.S. Kupreeva, A.R. Gazizov, *Fraud as a problem in the labor market*, in World Science: Problems and Innovations. Proceedings of the 18th International Scientific and Practical Conference. In 2 Parts, 56-58 (2018)
9. A.N. Kiryukhina, E.A. Morozova, A.V. Mukhacheva, Manag Iss, **3(64)**, 125-137 (2020). https://doi.org/10.22394/2304-3369-2020-3-125-137
10. Kh.V. Arturovich, K.A. Hasanbievich, Gaps Rus Leg, **5**, 121–125 (2019)
11. D.V. Golovanova, O.M. Dudina, Creat Econ, **13(6)**, 1253-1268 (2019)
12. R. Urinboyev, *Uzbek Migrants' Everyday Encounters with Employers and Middlemen*, in Migration and Hybrid Political Regimes: Navigating the Legal Landscape in Russia, 62-80. (University of California Press, Oakland, 2021)
13. M.Yu. Markitanov, Devel Safety, **1**, 105-114 (2019)

14. T.N. Zhuravskaya, *"To the Far East": Internet, labor mobility and discrimination in the labor market*, in Fourth Industrial Revolution: Realities and Modern Challenges. The Tenth Anniversary Sociological Reading of Saint Petersburg. Proceedings of the International Scientific Conference, 38-40 (2018)
15. A.K. Kartseva, Econ Vector, **3(18)**, 61-64 (2019)

Pandemic inequality: new realities and prospects

Mayya Valerianovna Dubovik[1]

Plekhanov Russian University of Economics, Department of Economic Theory, Moscow, Russia

Abstract. This paper investigates the emergence of factors contributing to pandemic inequality. The pandemic itself is a completely novel factor, and it is associated with high uncertainty. Methodologically, this study builds upon sociological, economic, and psychological theories of inequality; these theories are elaborated by a combination of dialectical and causal analysis. The paper presents short overview of prior research into factors of pandemic and post-pandemic inequality in the context of the changing labor market; it further analyzes major trends and patterns of such inequality in order to identify the key causes and challenges of the situation today. It focuses on the need to help workers and businesses adapt to the irreversible processes of the post-pandemic world, to change the macroeconomic policy on many levels of the administrative hierarchy. The results of this study could lay the foundations for studying the behavior of workers in the labor market; they should also be accounted for by public agencies responsible for welfare programs intended to curb the pandemic inequality. The theoretical and practical contributions hereof lies in substantiation of the need to optimize the priorities of macroeconomic policies, to implement them competently in the fight against poverty, to balance the costs and benefits of labor market transformation, and to devise novel social policy mechanisms.

Keywords: socio-economic inequality, "pandemic" inequality, macroeconomic policy

1 Introduction

The shock that the pandemic brought to all aspects of life has polarized the society on all levels; the change that ensues can no longer be ignored by the macroeconomic pandemic policy makers. In Russia, 10% of the population has monthly income of or below 100 thousand rubles; the lower decile is ≤10 thousand rubles. The income gap is unacceptable. The pandemic made inequality and discrepancy even worse. The goal hereof was to identify new challenges and risks that far exceed all of the previous negative factors in gravity, contribute to polarization and to pandemic inequality. The paper analyzes global trends in the development of various forms of pandemic inequality, highlights the trajectories of their evolution, and proposes steps to mitigate their effects.

[1] Corresponding author: mydubovik@gmail.com

2 Materials and methods

At the heart of social inequality is economic inequality, although this is an egg-vs-chicken kind of relationship. These sets can be called fuzzy; however, their intersection creates novel forms of inequality that arise from today's macroeconomic shocks, whether endogenous or exogenous. Thus, today's reality is overshadowed by the so-called pandemic inequality, a new phenomenon that needs to be understood and calls for the creation of new tools to mitigate it.

Piketty claims that economic inequality is inherent to capitalism [1, 2]. Although he claims that this trend was suspended in the times of Word Wars, one can disagree. Apparently, today's conflicts and wars only exacerbate economic inequality, whether on the national, regional, or individual scale [3].

Economists and sociologists who pen papers on economics or work in business point to different causes of economic inequality, which might be due to the fact that these two groups pursue different interests. Thus, Ancient Greek philosophers would speak in defense of slaveowners and justify social and property-based differentiation [4, 5]. Nobel Prize winner and inequality researcher Joseph Stiglitz sees the primary cause of inequality in the pursuit of privileged rent ownership backed by market and political power. He highlights such political and institutional factors as staff remuneration systems, models of supply and demand in the labor market [6].

Authors [7, 8] believe inequality to be mainly attributable to the rapid change in various technologies, whether those involved in creating a product or used to promote it (logistic, marketing, and consumption applications). Change concerns the organization of production, business processes, and management.

Universal digitalization, one of the latest tech developments, exacerbates disparity in the labor market in the short and medium term, as demand for high-skill labor increases, whereas low-skill labor is falling out of grace, a situation that calls for novel, advanced programs for staff (re)training. Digitalization eliminates low-skill, mechanical labor; these workers lose their income and join the poor strata [9-11]. In [12] was reported that up to 86% of restaurant jobs, up to 75% of jobs in retail, and up to 59% of entertainment jobs can be automated as soon as by 2035.

Institutional factors of inequality include, according to [13-16], the weakening of trade unions. Trade unions have ever lesser influence on employee-employer relationships, which evolve; thus, trade unions can no longer mitigate economic inequality as extensively as they earlier could. Immigration rates and demand for workforce have declined amidst the pandemic; many jobs have been moved to home offices.

Some authors point to the mistakes that governments and public authorities made when applying a variety of macroeconomic policy instruments, which contributed to inequality [17].

Schwab and Malleret [12] note a new form of ownership as a factor of pandemic inequality: total control over the inner world of the people, pandemic slavery of a kind.

Economic inequality does not boil down to wage alone. Bureau of Economic Analysis reports that over the past decade, benefits and premiums payable by employers have increased more than twelvefold and come to account for most of employees' income. A broad range of allowances and reimbursements are payable as productivity incentives, overtime bonuses, work-on-vacation payments, insurance premiums, pension fund contributions, etc. This category of remuneration alone has grown by >10%.

Social inequality today manifests itself in health disparity and unequal access to medical care. There is no denying that at the core of any such disparity is unequal education and income. Some researchers believe that health quality and income / wealth correlate. Marmot [18] has shown every year of schooling to improve individual's health [19]. Similar but weaker correlation is found between income and health. Deaton A. describes inequality in the variety of its manifestations as affected by geographical, economic, and social status; the author presents a case study of two families to illustrate their position [20].

Kollier [21] describes inequality that has a geographic dimension to it, stems from the place of residence, correlates with education, views of morality, and cultural background. He describes the confrontation and opposition of people in the following dichotomies: residents of central locations vs residents on the periphery; poor, undereducated people with little to no opportunities vs educated, professional people who are tolerant to various social minorities, make up the ruling class, and trust in the government and in the state.

Inflation is an important driver of inequality and poverty. In Russia, inflation reached 6% in May 2021, whereas nominal average pension was indexed by 5.5% and totaled 15,806 rubles. Thus, real income, in particular pensions, dropped by 0.5%, which could be critical for retirees and other low-income people.

3 Results

Thus, whereas earlier one's origin and inherited wealth would make them part of the elite, now it is the professionalism and individual competitiveness of workers that enable them to join the ruling class.

Polarization of people working online and offline is another specific trait of pandemic inequality. Not everyone has the skills, equipment, mental resilience, and soft skills to work online. The stringent conditions of the pandemic contribute to income disparity and social differentiation. COVID-19 has brought significant gender changes to the labor market.

Any recession is associated with large-scale staff cuts, and unemployment usually hits women first. However, the pandemic-related crisis in the labor market does not discriminate by gender. Immigrant males have had it worst, as, e.g., in Russia, immigrants have less labor opportunities in lockdown and are more likely to lose their jobs. Similar downward trends in the employment of male immigrants are noted in [22]. Besides, low-skill immigrants are not needed online, and they are forced to go home. Men are known to be less tolerant to job loss and are more likely to show signs of deviant behavior as a result of being sacked [23].

4 Discussion

When exposed to technological change, companies take different spots in the global value chains (GVCs), which predetermines their domination or subordination in their respective industries, thus affecting their proportion of added value. Areas, regions, and countries where such companies are headquartered become prioritized. However, in light of the pandemic and associated lockdowns, GVCs have to be "undeployed", evacuated from their regions, and relocalized within a single country or locality, which does affect local development. One example is the United States, whose production chains were moved far to Asia and India only to be brought back home urgently. Such repositioning restructures the labor market and exacerbates inequality.

The availability of, and access to, digital tech in these difficult times depends on household income and varies by country or region, which only fuels this new digital divide. Discrepancy in education hinders any attempt to address the deformation of the labor market, as it has always been. Higher education does not necessarily make one a desirable employee. This mismatch of labor supply and productivity creates inequality pertaining to education.

The development of several vaccines against a variety of coronavirus strains is an important milestone of the 21st century's science and technology. However, access to vaccination varies from country to country, which is perhaps the most important highlight of what we call pandemic inequality. Interestingly, countries that offer multiple vaccines for free are also slow to progress in this respect. Russia is a vivid example of this trend. Duke University reports AstraZeneca to be the #1 vaccine manufacturer, and the EU and the US to be the major buyers that have purchased over 50% of all doses. China refers to this policy as "vaccine protectionism". Developed economies have allocated ~17% of GDP to fight the pandemic, middle-income countries are at 4%, whereas the least developed nations can only afford 1.6% [25].

For every three persons whose employer pays medical insurance in the upper quartile, there is only one recipient of such payments in the lower quartile, income-wise. That being said, in the lower income quartier, only 1 out of 4 persons eligible for such payments will actually receive any compared to 3 out of 4 in the upper quartile. Availability of bonuses and other supplementary remunerations varies from industry to industry. Bonuses are the way high-tech companies retain valuable employees. In reality, economic inequality is deeper, as it should be seen through the lens of total income, which includes not only salaries, but also income generated from assets, property, etc. FRS reports a 10:1 ratio of financial asset owners (upper vs lower decile) [24].

As for the instruments of fiscal and monetary policies, they have often been misused due to ignorance of national traits, the specifics of the phase of the cycle and time. On the one hand, global economic inequality was reinforced in the aftermath of WW2, as the world became split into two opposing parties once the winning nations were done splitting the spoils of war. On the other hand, further public regulations failed to create a fair state of welfare for all. For instance, a social state does not necessarily manage to address inequality. Middle class, deprived of welfare and benefits, becomes poor. Financial elites abuse the availability of the rent to accumulate wealth and assets; they only grow wealthier over time.

Research of the Institute of Health and Equity focuses on the root causes of inequality in health and how this gap could be bridged. However, there are a few contradictions here. Thus, long-lived residents of the Caucasus Mountains are not necessarily educated and/or rich. The rich can be just as powerless against diseases as the poor; Steve Jobs can attest to that. Apparently, people can succumb to COVID-19 regardless of their age and social status.

5 Conclusion

The pandemic left no country, no social stratum, no aspect of life unaffected. Richer nations have been funding the sectors that create and maintain human capital: health, education, and culture; they have thus been able to spend more to mitigate the effects of the pandemic. However, pandemic inequality has worsened the existing forms of inequality, exacerbated social tensions, and weakened the physical and mental capabilities of man.

The coronavirus shock has contributed to the gender-related and geographical manifestations of unemployment; immigration rates dropped, and those who have lost their jobs are facing ever greater psychological issues in light of such uncertainty.

The new experience of governments and health systems should help tackle the challenges of tomorrow and mitigate the risks of the upcoming waves of the pandemic. To that end, we need consolidated international effort and must focus on digitalization of public services, which in the long term will improve the well-being of a broader demographic while also enabling secure collection of people's data.

In pursuit of social security, governments must ensure access to medical insurance or universal healthcare in order to tackle pandemic inequality.

The reported study was funded by RFBR, project number 20-010-00578.

References

1. T. Piketty, Capital in Twenty-One Century. (Harvard University Press, Boston, 2014)
2. F. Novokmet, T. Piketty, G. Zucman, NBER Work Paper Ser, **23712** (2017). https://doi.org/10.3386/w23712
3. M.V. Dubovik, Int J Civil Eng Techn, **10(2)**, 1704–1711 (2019)
4. Plato, Gosudarstvo [The State]. (St. Petersburg, 2006)
5. Aristotle, Politika [The Policy]. (ACT, Moscow, 2010)
6. E. Seçilmiş, Ekonomik Yaklaşım, **27(100)**, 223-235 (2016). https://doi.org/10.5455/ey.35942
7. Y. Kalantzis, Y. Kambayashi, S. Lechevarier, Labour, **26(4)**, 514-541 (2012). https://doi.org/10.2139/ssrn.2153618
8. E. Brynjolfsson, A. McAfee, M. Spence, Foreign Aff, **93(4)**, 44-53 (2014)
9. E.L. Groshen, H.J. Holzer, The Russell Sage Found J Soc Sci, **5(5)**, 1-19 (2019). https://doi.org/10.7758/RSF.2019.5.5.01
10. C. Goldin, L.F. Katz, The Race Between Education and Technology. (Harvard University Press, Cambridge, 2008). https://doi.org/10.1086/ahr.114.5.1413
11. A.D. Skills, Science, **344(6186)**, 843-851 (2014)
12. K. Schwab, T. Malleret, *COVID-19: The Great Reset*, in World Economic Forum 91-93 route de la Capite CH-1223 Cologny/Geneva Switzerland (2020). https://doi.org/10.1007/s42413-021-00117-7
13. T.A. Kochan, W.T. Kimball, The Russell Sage Found J Soc Sci, **5(5)**, 88-108. (2019). https://doi.org/10.7758/RSF.2019.5.5.05
14. S. Machin, Res Labor Econ, **43**, 329-354 (2016). https://doi.org/10.1108/S0147-912120160000043017
15. H.S. Farber, D. Herbst, I. Kuziemko, S. Naidu, NBER Work Paper Ser, **24587** (2018). https://doi.org/10.3386/w24587
16. J. Sherk, Backgrounder, **2825**, 1-17 (2013)
17. L. Mishel, J. Bivens, E. Gould, H. Shierholz, The State of Working America. (Cornell University Press, New York, 2012). https://doi.org/10.7202/1024216ar
18. M. Marmot, et al., Health equity in England: the Marmot Review 10 years on the Health Foundation. (Institute of Health Equity, London, 2020)

19. D. Cutler, A. Leras-Muney, *Education and Health: Evaluating Theories and Evidence*, in R.F. Schoeni, J.S. House, G.A. Kaplan, H. Pollack (eds), Making Americans Healthier: Social and Economic Policy as Health Policy. (Russell Sage Foundation New York, 2008)
20. A. Deaton, Rev Austrian Econ **28**, 443-447 (2015). https://doi.org/10.1007/s11138-014-0269-5
21. P. Kollier, The Future of Capitalism. Facing the New Anxieties. (Harper Collins Publishers, 2020)
22. G.J. Borjas, H. Cassidy, NBER Work Paper Ser, **27243** (2020). https://doi.org/10.3386/w27243
23. V. Galasso, et al., NBER Work Paper Ser, **27359** (2020). https://doi.org/10.3386/w27359
24. Board of Governors of the Federal Reserve System. Accessed on: October 2, 2021. [Online]. Available: https://www.federalreserve.gov/econres/scf/dataviz/scf/chart/#series:Stock_Holdings;demographic:nwcat;population:1,2,3,4,5;units:median
25. J.L. Dieleman, et al., Lancet, **396**, 693-724 (2020). https://doi.org/https://doi.org/10.1016/S0140-6736(20)30608-5

Socio-economic and socio-legal aspects of labor market regulation in modern Russia

Oleg Muratovich Epkhiev[1] and *Anatolii Sergeevich* Kislyakov

Financial University under the Government of the Russian Federation, Department of Sociology, Moscow, Russia

Abstract. The object of this study is the problem of market regulation in modern Russia. The author has set a goal based on a comprehensive system analysis of the Russian Federation's scientific works and normative legal documents to uncover and characterize the socio-economic and socio-legal aspects of the problems of regulating the labor market at the federal and regional levels. In his work, the author relied not only on solidarizing but also on alternative theoretical approaches and conclusions. The scientific results obtained in the course of the study allowed us to understand and substantiate that at the federal and regional levels, the Russian Federation requires a very serious adjustment of institutional mechanisms for regulating the labor market, on which the development of its economy depends. The scientific novelty of the work lies in the conclusion that in current conditions, Russia should provide more in-depth fundamental research of the labor market, which would contribute to the formation of more effective public policy in the regulation of the labor market, employment of the population. This will entail the need to develop the implementation of social, economic, industrial policy, considering its impact on the labor market.
Keywords: labor market, Russia, state, employee

1 Introduction

Problems of functioning of the labor market, which can have a very negative impact on the situation in the social sphere of any state, naturally, at all times, have interested both science and the government, almost all states seeking to regulate it through certain economic, social, legal mechanisms. Suffice it to recall that as early as 1905, the British Parliament, in resolving the problems of employment, wage labor, including unemployment, passed the Unemployment Act, under which organizations were created in the industrialized cities of England, with the tasks of labor exchanges and statistical committees [1]. In 1911, the National Insurance Act was adopted, which provided compulsory unemployment insurance for workers over 18. The fund for insurance benefits was made up of contributions from the state, businessmen, and workers.

[1] Corresponding author: OMEpkhiev@fa.ru

In 1935, the United States passed the Labor Relations Act (Wagner Act), which enshrined workers' rights and labor unions. The Law on Social Insurance, adopted in the same year, provided social support measures for the unemployed and certain categories of employed citizens. Solving the employment problems and the labor market in 1946, the US Congress passed the Employment Act. As part of the anti-poverty program, laws were passed in the 1960s that were also designed to reduce social tensions in society and stimulate economic growth by providing high levels of employment and consumption, which in our view, laid the foundation for the mass consumption society [2].

It is believed that the labor market in Russia began to form much later than in Western Europe and the United States and refers to the period of the Great Reforms implemented by Alexander II in 1860-1870. In February 1861, Alexander II signed the Manifesto on the Abolition of Serfdom and the Statute on Peasants Departing from Serfdom, which consisted of 17 legislative acts that defined and regulated the issues of the liberation of peasants, the redemption of land by liberated peasants [3]. With the emergence of the working class and its growth, especially during the Peasant Reform or the abolition of serfdom, Russian cities became major industrial centers with a new labor market [4].

The history of the Russian labor market is very rich and of particular scientific interest since its development was influenced by different social systems. The interest in practical terms stems from the need for the state, as the main regulator of the labor market, effective measures of its regulation to achieve equilibrium.

At present, the state of the labor market as one of the most important conditions for the socio-economic stability of society is determined by many economic, demographic, political, legal, and other factors. Furthermore, if in economic science, economic theories, the labor market are considered only in the aspect of the relationship between the employer and the employee, selling their professional skills and competencies, sociology is more productive in its studies of the problem because of the multidimensional approach.

The study of the labor market in sociology has given rise to a special scientific tradition of considering it at both the organizational and social levels of society and, importantly, at the institutional level. The importance of the institutional approach in the study of labor market problems is obvious. It suggests a special focus of research attention on the social nature, content, factors, and in a sense, social process, considering the labor market as a system, primarily, social interrelations and relationships arising in this market between its main participants.

2 Methods

The base of sources was composed with a retrospective approach, which allowed to trace the evolution, since the mid-twentieth century, of labor market regulation in Russia and other foreign countries and to identify the main variables related to the subject of research. The methodological basis of the study was a system-complex approach of scientific papers [1-23], normative legal documents [24-27], which allowed to identify and substantiate the need to regulate the labor market in Russia. The epistemological potential of statistical and sociological methods of quantitative and qualitative research gave the opportunity to adequately interpret the results of scientific research on the subject of analysis. The novelty of the work and its results are determined by an original approach to identifying the characteristic features that reveal the socio-economic and socio-legal aspects of market regulation in Russia.

3 Results and discussion

The system-forming elements or subjects of the labor market are the state, the employee, and the employer. The labor market is a very important element of the state's economic system; if so, the state's primary task should undoubtedly be to regulate this market. The regulation of the labor market can and should be seen as a specific sector of the regulatory impact of the state on social and economic processes in society. State participation in the regulation of the labor market should be aimed at achieving a high level of employment; ensuring the most reasonable compliance of specialists' training with the structure of necessary jobs; leveling the effects of unemployment; creating sufficient and necessary social, economic, legal conditions for the functioning of workers. In essence, the achievement of an optimal social and economic state of the labor market, its stability and balance should be characterized by a combination of such factors as employment, decent wages, the implementation of measures stipulated by law for the social protection of workers in case of suspension of production or dismissal of workers, development, and implementation of measures to preserve and use the professional potential of workers, their social protection, improving working conditions and a number of other issues creating sufficient and necessary conditions for the balance of the labor market and the implementation of social stability policies in society. In modern Russia, the following regulatory components established by law can be distinguished:

1. The minimum wage, the size of which is established by the Federal Law "On the minimum wage". Since 2021, the ratio of the minimum wage and the median wage is set at 42%. The median wage is a value that divides all workers into two equal parts: half of the workers have wages higher than this amount, and the other half have lower wages. This indicator can be calculated on the scale of the entire state and separately in the region, locality, and even in individual enterprises or their structural divisions. This ratio is subject to revision at least once every 5 years based on the state's socio-economic development conditions. However, the minimum wage for the next year is not lower than the minimum subsistence level of the able-bodied population in the Russian Federation for the next year and not lower than the minimum wage set for the current year. Calculation of the median wage is based on the methodology determined by Rosstat.

Table 1. Federal minimum wage values in 2017-2021.

Date of the minimum wage introduction	1.07.2017	1.01.2017	1.01.2018	1.01.2019	1.01.2020	1.01.2021
Minimum wage amount, RUB	7,800	9,489	11,163	11,280	12,130	12,792 (new metrics)

2. The consumer price index for goods and services serves as the basis for a systematic increase in the monetary income of the population. This is a type of index created to measure the average level of change in prices of goods and services over a certain period. The consumer price index (CPI) is calculated following the official statistical methodology of the organization of statistical monitoring of consumer prices for goods and services and the calculation of consumer price indices, approved by order of Rosstat No. 734 dated December 30, 2014. It is an objective socio-economic indicator

calculated to measure changes over time in the general level of prices for consumer goods and services purchased for household consumption.

Table 2. Consumer price and tariff indices for goods and services.

	Total		**Including**					
			Food products		Non-food products		Services	
	To the previous period	By December of the previous year	To the previous period	By December of the previous year	To the previous period	By December of the previous year	To the previous period	By December of the previous year
2020								
Q1	101.3	101.3	102.2	102.2	100.8	100.8	100.7	100.7
Q2	101.3	102.6	102.1	104.3	101.0	101.8	100.7	101.4
Q3	100.2	102.9	98.8	103.1	101.3	103.1	100.9	102.4
Q4	102.0	104.9	103.5	106.7	101.7	104.8	100.3	102.7
2021								
Q1	**102.1**	**102.1**	**103.1**	**103.1**	**101.8**	**101.8**	**101.2**	**101.2**

Importantly, among other things, the CPI is also used to regulate wages and social security, and other benefits to partially or fully compensate for changes in the cost of living or consumer prices.

3. The Unified Rate Schedule for Employees was introduced by RF Government Decree No. 785 of October 14, 1992, "On differentiation of wages in the public sector based on the Unified Rate Schedule". It is a system of grades that determines the correlation between wages and qualification level of work of employees of the budgetary sphere. Consists of 18 grades dividing all types of work performed in organizations into 18 "types", depending on the level of qualification and degree of complexity. There are norms in current Russian labor law that provide for increased payment for difficult and hazardous working conditions. For example, extra pay for night work, overtime, work in the Far North and equivalent areas, other benefits and guarantees, and compensation for wages.

4. Norms of compensation for delayed wage payments. If the employer violates the deadlines for payment of wages, vacation pay, termination payments, and other payments to the employee, the employer must pay them with interest in an amount not less than one hundred and fifty percent of the key rate of the Central Bank of Russia in effect at that time. If wages and other payments to an employee are not paid on time, the interest on monetary compensation is calculated from the amounts not paid on time. The monetary compensation paid to an employee may be increased by a collective agreement, a local normative act, or an employment contract. The obligation to pay this monetary compensation arises regardless of the fault of the employer [25].

A very radical revision of institutional instruments of regulation of the modern Russian labor market is required, and the importance of this issue lies, above all, in the fact that the solution of this issue very seriously depends on the development of the socio-economic sphere of society.

The thing is that the improvement and development of socio-economic and socio-legal mechanisms of labor market regulation require the development and implementation of social, industrial, demographic, and other policies in terms of their impact on employment, with a reorientation from the solution of current problems to the development and implementation of important strategic directions of development and acceleration, while the process of forming a civilized competitive environment in the domestic market, given the real growing trends of digitalization of the economy, crisis trends both in the global economy and in the country, which in addition to other well-known factors are also caused by restrictive measures associated with pandemics, the varying level of socio-economic development of Russian regions, etc.

This gives us every reason to believe that today's Russian labor market is a systemic phenomenon. Therefore, the very concept, its essence, content, factors, and conditions of implementation of real mechanisms of regulation of modern Russian labor market refer to the key; we can say fundamental theoretical and practical research problems of such scientific directions as economic sociology, law, economics and of course, political science.

It is obvious that without strategic planning of the country's socio-economic development, a systematic approach to the regulation of the labor market, it is extremely difficult and impossible to talk about the development of economic, social, and political sectors of society and the state. Therefore, the formation and development of real labor market mechanisms, although not sufficient, is seen in modern Russia's state policy. For example, in 2008, the Russian government approved a Labor Market Action concept for 2008-2010 by its order in August 2008. Priority tasks of labor market development were the implementation of investment and innovation projects in high-tech areas of the economy; formation and development of production clusters, as well as the development of energy, industry, and transport infrastructure; development and modernization of education, health, housing, and agro-industrial complex; creation and improvement of market institutions, as well as the development of small entrepreneurship [26]. In November of the same year, 2008, the Government of the Russian Federation approved the Concept of long-term economic development of the Russian Federation for the period till 2020, where the development of the labor market section defines the goal of the state policy in the development of the labor market in the long-term perspective, which comes down to the creation of legal, economic and institutional conditions providing the development of flexible efficiently functioning labor market that allows to overcome the structural discrepancy between the labor supply and demand, to decrease the share of illegal employment, to increase the share of illegal employment.

In addition to the above factors of the labor market, it is obvious that in a market economy, the market mechanism of labor market regulation includes two important elements: the demand for labor and, of course, supply and the labor price on the labor market. The level of correlation between labor demand and supply on the market determines the real state of the labor market, its structure, and equilibrium.

If the essential understanding of the labor market in a broad sense is reduced and accepted as a system of relations between the employer, the employee, and the state, and related to these relations the whole system of socially important aspects, it is quite natural that the question of legal regulation of the labor market arises. No less important for deeper scientific understanding and practical implementation is the question of the functional structure of the modern Russian labor market, given such current global trends as globalization of the economy, the intensification of the process of convergence of economic relations and relations at the regional level, digitalization of the economy, sanctions wars, pandemics. Under these conditions, a well-considered state policy is required in crucial issues related to promoting and stimulating employment growth; the social partnership between employers, workers, and the state; ensuring competition in the labor market among both employers and workers; setting wage rates that would be equitable; and implementing effective social support measures for the unemployed.

Obviously, in regulating the labor market, the state must play and plays a special role, and the issues of legal regulation are laid down and permeate almost all the legal fields of modern Russian society and have their enforcement in the Constitution of the Russian Federation, international legal documents, the Labor Code, the Administrative Code of the Russian Federation and certainly in the Criminal Code, as well as in the decrees of the President of the Russian Federation, resolutions and orders of the Russian Government, industry and departmental regulations, laws and other regulatory legal acts of the Russian Federation. The basic principles of contemporary Russian state policy in employment promotion are set forth by law in Articles 7 and 37 of the Constitution of the Russian Federation, which states that "The Russian Federation is a social state, the policy of which is aimed at creating conditions that provide a decent life and free development of man... labor and health of people are protected, a guaranteed minimum wage is established...". [24].

The main national legal act in employment promotion and employment is the Law of the Russian Federation of April 19, 1991. No. 1032-1 "On employment of the population in the Russian Federation", which defines the economic and organizational foundations of the state policy in the field of population employment. The law on employment allows to establish economic and legal forms and methods of regulating employment relations (status of the unemployed, employment fund), to develop a system of social support: payment of temporary unemployment benefits and material assistance to disabled family members. Moreover, in March 2021, the Government adopted Decree No 362 of March 13, 2021, "On government support in 2021 to legal entities and individual entrepreneurs in job placement of unemployed citizens", which established rules for providing subsidies in 2021 to legal entities and individual entrepreneurs to stimulate job placements for unemployed citizens.

4 Conclusion

Thus, a rather brief description of the problem of socio-economic and socio-legal aspects of labor market regulation in modern Russia allows coming to understand that a very serious adjustment of institutional mechanisms of labor market regulation, on which economic development depends, is required at the federal and regional levels. In current conditions in Russia should provide more in-depth fundamental research of the labor market, which would contribute to the formation of more effective public policy in regulating the labor market and the employment of the population. This will entail the need to develop the implementation of social, economic, industrial policy, considering its impact on the labor market.

References

1. F.A. Rothstein, Ocherki po istorii rabochego dvizheniya v Anglii [Essays on the history of the labor movement in England] (State Publishing House, Moscow, Leningrad, 1925)
2. O.A. Zhidkov, N.A. Krasheninnikova, V.A. Saveliev, [Istoriya gosudarstva i prava zarubezhnykh stran, Chast 2] History of the state and law of foreign countries, Part 2 (Norma Publishing House, Moscow, 1996)

3. B.N. Mironov, Sotsialnaya istoriya Rossii v period Imperii (18 – nachalo 20 veka): Genezis lichnosti, Demokratii, Semi, Grazhdanskogo obshchestva i pravovogo gosudarstva [Social History of Russia during the Empire period (18th – early 20 century): Genesis of Personality, Democracy, Family, Civil Society and Legal State] (St. Petersburg, 2003)
4. V.S. Ilyin, *Promyshlennoe razvitie Rossii s kontsa 19 veka do nachala stalinskogo "Velikogo perelomnogo momenta"* [Industrial development of Russia from the end of the 19th century to the beginning of Stalin's "Great turning point"], in Collection of scientific works of the Peoples' Friendship University of Russia (Moscow, 2004)
5. Yu.P. Mador, Podem rabochego dvizheniya v Anglii v 1910-1913 godakh [The rise of the labor movement in England in 1910-1913] (Moscow, 1966)
6. P.A. Zayonchkovsky, Otmena krepostnogo prava v Rossii [The abolition of serfdom in Russia] (Moscow, 1954)
7. S. Baert, *Discrimination in the labor market: a register of (almost) all correspondence experiments since 2005*, in S.M. Gaddis (ed), Audit Studies: Behind the Scenes with Theory, Method, and Nuance, 63-77 (Springer, 2018)
8. S. Baert, B. Cox, N. Gheyle, C. Vandamme, Ind. Labor Relat Rev, **68**, 467-500 (2015)
9. H. Blumer, Pac Socio Rev, **1**, 3-7 (1958)
10. E. Bonacich, Am Socio Rev, **37(5)**, 547-559 (1972)
11. G.J. Borjas, Am Econ Rev, **77**, 531-553 (1987)
12. F. Dobbin, et al., Am J Sociol, **99**, 396-427 (1993)
13. D. Acemoglu, Rev Econ Stud, **64(3)**, 445-464 (1997)
14. W. Arulampalam, A.L. Booth, Brit J Indust Rel, **36(4)**, 521-536 (1998)
15. P. Cahuc, F.vPostel-Vinay, Labor Econ, **9(1)**, 63-91 (2002)
16. C. Dustmann, U. Schönberg, Rev Econ Stat, **91(2)**, 363-376 (2009)
17. H. Hopenhayn, R. Rogerson, J Polit Econ, **101(5)**, 915-938 (1993)
18. M. Picchio, J.C. van Ours, Labor Econ, **18(5)**, 712-722 (2011)
19. O. Åslund, D.O. Rooth, Econ J, **117(518)**, 422-448 (2007)
20. M. Barrett, *Gender and the division of labor Women's oppression today*, in The Marxist/feminist encounter, 152-186 (Verso Books, 2014)
21. D. Bhugra, M.A. Becker, World Psychiatry, **4(1)**, 18 (2005)
22. I. Kogan, M. Weißmann, Adv Life Course Res, **18(3)**, 185-198 (2013)
23. K.I. Paul, K. Moser, J Vocat Behav, **74(3)**, 264-282 (2009)
24. Konstitutsiya Rossiiskoi Federatsii (prinyata vsenarodnym golosovaniem 12.12.1993 s izmeneniyami, odobrennymi v khode obshcherossiiskogo golosovaniya 01.07.2020) [The Constitution of the Russian Federation (adopted by popular vote on 12.12.1993 with amendments approved during the all-Russian vote on 01.07.2020)]. Accessed on: October 14, 2021. [Online]. Available: http://www.consultant.ru/document/cons_doc_LAW_28399/
25. Trudovoi kodeks Rossiiskoi Federatsii ot 30.12.2001 No. 197-FZ (red. ot 28.06.2021, s izm. ot 06.10.2021) (s izm. i dop., vstup. v silu s 01.09.2021) [Labor Code of the Russian Federation dated 30.12.2001 No. 197-FZ (ed. dated 28.06.2021, with amendments. from 06.10.2021) (with amendments and additions, intro. effective from

01.09.2021)], 236. Accessed on: October 14, 2021. [Online]. Available: http://www.consultant.ru/document/cons_doc_LAW_34683/

26. Pravitelstvo Rossiiskoi Federatsii, Rasporyazhenie ot 15 avgusta 2008 g. No. 1193-r "Kontseptsiya deistvii na rynke truda na 2008-2010 gody" [The Government of the Russian Federation, Decree No. 1193-r of August 15, 2008 "The concept of actions in the labor market for 2008-2010"]. Accessed on: October 14, 2021. [Online]. Available: http://www.consultant.ru/document/cons_doc_LAW_79451/c28e8ae3291ba6db7098c2057be342fae6b5ca41/
27. Rasporyazhenie Pravitelstva RF ot 17.11.2008 No. 1662-r (red. ot 28.09.2018) "O Kontseptsii dolgosrochnogo sotsialno-ekonomicheskogo razvitiya Rossiiskoi Federatsii na period do 2020 goda" (vmeste s "Kontseptsiei dolgosrochnogo sotsialno-ekonomicheskogo razvitiya Rossiiskoi Federatsii na period do 2020 goda") [Decree of the Government of the Russian Federation of 17.11.2008 No. 1662-r (ed. of 28.09.2018) "On the Concept of long-term socio-economic development of the Russian Federation for the period up to 2020" (together with the "Concept of long-term socio-economic development of the Russian Federation for the period up to 2020")]. Accessed on: October 14, 2021. [Online]. Available: http://www.consultant.ru/document/cons_doc_LAW_82134/

Digital storytelling in the world economy classroom

Valerija Sergeevna Mincicova[1*], *Elizaveta Valentinovna* Ogloblina[1], and *Asem Hekimoglu* Naushabaeva[2]

[1]Financial University under the Government of the Russian Federation, Department of World Economy and International Business, Moscow, Russia
[2]Bitlis Eren University, Faculty of Economics and Administrative Sciences, Bitlis, Turkey

Abstract. This study aims to share the digital storytelling experience of higher education teachers including those using English as a Medium of Instruction. The methodology consists of giving a task to 130 students of bachelor course divided into 29 creative groups. Students created video stories on prearranged topics concerning the world economy agenda. We focused on analyzing the stages of project creation and the role of a teacher in the process, drivers of successful implementation of the digital storytelling, and compared the effectiveness of different types of classwork with digital storytelling. Thus, we conclude that such factors as the cohesion of the creative groups, ability to use video editing instruments and distribution of the tasks between members in the initial stages play the least important role in the successful implementation of the task, but at the same time can be most easily influenced and course-corrected by a teacher. The research also demonstrated that the potential to connect the topic with the own interests of the creative group, ability to present the result of work publicly and interest in the topic were the main drivers for success and involvement. Those are the characteristics of the generation Z, and we strongly believe that the teachers of higher education should consider the findings we present in the results section. It is also revealed that digital storytelling, despite being time-consuming, has a more emotional influence on students and gives them more satisfaction after a presentation.
Keywords: higher education, digital education, digital storytelling, world economy studies, teaching gen Z, student-centered pedagogy

1 Introduction

The modern teacher of higher education is faced with a difficult task. On the one hand, he or she has to develop in students the so-called classical skills that are reflected in the taxonomy of B. Bloom's pedagogical goals: remember, understand, apply, analyze, evaluate, and create [1]. On the other hand, a teacher has to lay the foundations and develop the competencies necessary for a fulfilling life in modern society [2, 3].

* Corresponding author: valerypavlova@rambler.ru

From the point of view of the evolution of oral communication, digital storytelling can be defined as a natural development of the oral tradition at a new stage [4].

From a technological point of view, digital storytelling is a new modern format for presenting information on the Internet using multimedia technologies [5].

From the sociological point of view, digital storytelling is a means of mediation, which has its social potential, and can even contribute to the strengthening of democracy [6].

Digital storytelling is the application of content methodology [7]. It helps to improve skills of navigating through the galore of information in today's world. It is the instrument of both absorbing and creating knowledge [8]. Formalized storytelling can encourage both reflection, a deeper understanding of a topic and stimulates critical thinking skills [9]. Holders and researchers experience transformative outcomes by crafting and reflecting on digital narratives [10]. It is also a proactive method of education, where one cannot stand aside from group work. And overcoming of isolation and of keeping quiet if you do not understand is one of the main objectives of a modern high school education [11]. Digital storytelling provides meaningful and permanent learning, is fun and motivational [12].

The hypothesis of the research: digital storytelling in education for generation Z promotes their interest, optimises the understanding of discipline. Its success mostly depends on psychological, social and subjective factors in the study group.

The research aimed to propose a successful mechanism of integration of digital storytelling into educational process to increase the intrinsic motivation of students and develop their higher-order thinking skills.

And thus, we found the objectives for the study:

- Provide proper task, forms of reports, criteria of evaluation;
- Distinguish the factors which make a digital storytelling task inspiring for the students;
- Evaluate the level of knowledge provided in stories, compare it with the knowledge and abilities gained from other forms of work in the classroom.

2 Methods

The study was undertaken at the Financial University under the Government of the Russian Federation, Moscow, Russia. The participants of the study are 130 students of the 2nd and 4th years in bachelor programme in World Economy and International Business studies. The storytelling was executed as a video, created by a group of students, explaining the set problematic situation. The age of the students varied from 19 to 21 years old in 2021, which makes them the paragons of the Generation Z. Students divided themselves into groups of 3-5 people on their own. Finally, 29 video stories were created, presented, analyzed, and discussed in the class.

The videos were created in both Russian (native) and English (foreign) languages. They were presented at the class or accumulated on the specially created website. All the videos were evaluated by the groupmates concerning the stipulated criteria. That evaluation became the basis for a teacher to put the final mark for the task.

3 Results

First of all, we define the place of digital storytelling in the educational process. It can be implemented within the framework of the study of disciplines according to the curriculum.

It can also be used for research activities, and the organization of extracurricular work of students.

The use of digital storytelling in the educational process can be implemented in various formats:

- a presentation prepared, for example, using the Microsoft PowerPoint application;
- a filmed video posted on a file hosting service;
- a digital publication with multimedia content.

The 5 stages of work on creating a digital story can be distinguished (see table 1). At each of the designated stages, the teacher organizing the students' work on creating a digital story will have different roles.

Table 1. Teacher's roles in different stages of digital story creating. Developed by the authors on the empirical basis.

Stage	Teacher's Role
Search for an idea and its development	Manager Observer Rapport builder
Working on the script	Facilitator Diagnostician
Storyboarding and editing	Observer
Publication of results	Manager
Presentation	Assessor

At the stage of searching for an idea, it is necessary to direct students as accurately as possible to formulate the topic of a digital story. At the initial stage, the abundance of information and topics can significantly complicate the work of students and lead them away from the goals and objectives set by the teacher. In this regard, it is recommended to provide extensive advisory assistance to students at the initial stage.

Further, students determine the purpose of the study and prepare a justification for the relevance of the research topic. Students collect information on the topic and carry out its primary analysis in the form of an annotated bibliographic list of reliable sources.

The work on the script also supposes the active involvement of a teacher in the process of its preparation, since students need help in finalizing it. The format for presenting the digital story are the following, each of them is accompanied with the list of programs and applications mostly used:

– Doodle video: VideoScribe, Powtoon, Sparkol;
– Animated storytelling: Powtoon, GoAnimate, Toon Boom;
– Comics & Storyboards: Pixton, StoryboardThat, Canva;
– Presentations: PowerPoint, Prezi, Sway;
– Video Editing: iMovie, YouTube Editor, Windows Live.

As for the actual programmes on PC students prefer Microsoft Windows Live Movie Maker, Microsoft PowerPoint, iMovie as has been noted for other European students [13]. Smart phone and tablet applications differ significantly.

The stage of storyboarding and video editing begins with a one-second storyboard and script adjustments. It can be presented in the form of a table, including the following mandatory columns:

– Sequential frame number;
– Frame time in seconds;
– Text on screen;
– Video;
– Sound.

Presenting digital stories involves evaluating them. The criteria for evaluating a digital story can be proposed as follows:

– Content of digital story (depth of topic disclosure and relevance of the content to the declared topic);
– Digital story structure (consistency of presentation);
– Use of technologies (video and sound).

As for the topics, the most interesting of them appeared to be those concerning the history of organizations, processes, stories of success (such as "BP ecology management" or "Flying geese paradigm in Asian Tigers rise"), and concerning the political cases influencing economics (such as new trends in pandemic-ruled world: "India's economy hit by COVID-19", "European tourist sector collapse" [14]). This trend is stipulated by the determinants of the Generation Z, highlighted as preferences for group-work, affirmation, clear rules, and for their opinions to be valued [15].

We evaluated how different characteristics of the study group drew interest to the task implementation (Table 2).

Table 2. Drivers of successful implementation of the digital storytelling task in a student's class. Developed by the authors on the empirical basis.

Which factor drew interest to the task?	Meaning of the factor for successful implementation of the task (1-10 points)	Can the factor be influenced by the tutor?
Cohesion of the creative group	3	Can be influenced while the teaching process
Strong leadership inside the group	5	Should not be influenced
Ability of video editing	4	Can be influenced while posing a task
Interest in the topic	8	Can be influenced while posing a task
Ability of public presentation of the video	7	Can be influenced while posing a task
Potential to connect the topic with the own interests of the creative group	10	Accidental, can be influenced if the topic is changed
Ability to distribute the tasks while preparation	4	Can be influenced prescriptively
The topic connected with the interests of Gen Z	10	Accidental, but can be corrected at initial stage

Student feedback confirms our conclusions on drivers of interest, the role of a student and a teacher, the mechanism of forming a group, etc. And it also confirms the actual hypothesis of the study.

Digital storytelling has its difficulties. It is quite labour intensive and time demanding (no less than 4 hours of work of the group per a video of 3 minutes) task type. But if we compare it with other types of work it wins in many criteria as shown in table 3.

Table 3. Comparison of effectiveness of different classical types of classwork with digital storytelling. Developed by the authors on the empirical basis.

	Digital storytellin g	**Q&A (answers prepared)**	**Discussion with scenario**	**Unprepared discussion**	**Text analysis with control questions**
Digestion of the discipline material	H (high)	H	H	M (medium)	M
Memorisation	M	H	M	L (low)	M
Interpretation	H	L	M	H	L
Application	M	L	H	H	M

Digital storytelling has a more emotional influence on students and gives them more satisfaction after presentation, although it lacks the effectiveness in memorization of all the material given for the rest of the group except the video authors.

Being a modern type of task, the digital storytelling has its certain demands for equipment. That establishes the demands for the university campus that should provide such equipment and software for everyone to be able to execute the task.

We also registered that the students experienced difficulties in scriptwriting, mostly while making short clear sentences, in finding the related images, videos and photographs. Despite the fact that they have a rich experience in such type of work, those difficulties are still present. But finally, the students concluded, that practically all topics are suitable for digital storytelling, even the ones demanding a sound scientific research.

4 Discussion

Our experience and study show that creating digital stories can be executed both with the strong participance of a teacher and without it. And if the group is motley enough it risks failing making a good product. So, we insist on the active position of a teacher at least at first stages. This viewpoint can be challenged, as generation Z, which we speak of in this study, appreciates their freedom and diversity. But modern higher school society is not ready for the education process without a supervisor, advisor, manager, facilitator, source of experience and knowledge, and mentor – a higher education teacher.

5 Conclusion

All new forms that provide effective communication and information presentation, which include digital storytelling, are in demand in various spheres of private, business and public life, including the academic environment. The digital storytelling format is a good example of how students rethink actual world economy topics competently and engagingly. This method of information presenting is a successful combination of traditional communication techniques and modern computer visualization capabilities. Modern programmes and services offer convenient development tools and resources that facilitate the creation of digital stories.

It should be noted that digital storytelling technology justifies the time spent, as it allows to consolidate and apply in practice a whole range of competencies related to the use of text, graphics, photos, audio, video, develops speech skills and stimulates internal motivation for learning.

References

1. Bloom's Taxonomy Is a Hierarchical Framework for Cognition and Learning Objectives. Accessed on: September 25, 2021. [Online]. Available: https://www.teachthought.com/learning/what-is-blooms-taxonomy-a-definition-for-teachers/
2. J. Soffel, Ten 21st-century skills every student needs (2016). Accessed on: September 25, 2021. [Online]. Available: https://www.weforum.org/agenda/2016/03/21st-century-skills-future-jobs-students/
3. Iniciativa "Navyki budushchego" [The "Skills of the Future" initiative] Accessed on: September 25, 2021. [Online]. Available: https://asi.ru/leaders/initiatives/education_leaders/future_skills/
4. Digital Storytelling for Communities. Accessed on: September 25, 2021. [Online]. Available: https://librarydigitalstorytelling.wordpress.com/why/
5. A.A. Kolosova, N.V. Poplavskaya, *Cifrovye tekhnologii v sozdanii novyh formatov m ediatekstov* [Digital technologies in the creation of new media text formats], in Medialingvistika. Vypusk 5: yazyk v koordinatah massmedia: sbornik materialov I Mezhdunarodnoj nauchno-prakticheskoj konferencii, 39, 6-9 September, 2016 (Varna, 2016)
6. N. Couldry, New Media Soc, **10(3)**, 373-391. (2008). https://doi.org/10.1177/1461444808089414
7. K. Daniels, J Sch Edu Tech, **9(1)**, 1–10 (2013)
8. L.M. Dreyer, South Afr J Edu, **37(4)**, 1-10 (2017) https://doi.org/10.15700/saje.v37n4a1475
9. M. Alterio, J. McDrury, Learning Through Storytelling in Higher Education: Using Reflection and Experience to Improve Learning (Routledge, 2003). https://doi.org/10.4324/9780203416655
10. C. Lenette, *Digital Storytelling*, in Arts-Based Methods in Refugee Research (Springer, Singapore, 2019). https://doi.org/10.1007/978-981-13-8008-2_5
11. H.L. Schnackenberg, C. Johnson, Preparing the Higher Education Space for Gen Z (IGI Global, Hershey, 2019). https://doi.org/10.4018/978-1-5225-7763-8
12. G. Gürsoy, Europ J Edu Res, **10(1)**, 97-113 (2021). https://doi.org/10.12973/eu-jer.10.1.97
13. G. Gürsoy, *Digital Storytelling*, in New Trends in Educational Sciences (SRA Academic Publishing, 2020)
14. K. Liuhto, The world economy in 2020 (2020). https://doi.org/:10.13140/RG.2.2.30689.94566
15. H. Raslie, Stud App Econ, **39(1)** (2021). https://doi.org/10.25115/eea.v39i1.4268

Empathy in teaching foreign languages in a distance format at universities

Svetlana Yurievna Dronova[1], and *Ainur Durbelenovna* Kurmangalieva[2]

[1]Financial University under the Government of the Russian Federation, Department of Foreign Languages and Intercultural Communication, Department of Political Science, Moscow, Russia
[2]Al-Farabi Kazakh National University, Department of Philosophy and Religious Studies, Alma-Ata, Kazakhstan

Abstract. Empathy is a key factor in teaching foreign languages, making it possible to establish an emotional connection both between the teacher and students and between students themselves, as well as between students and a target language, which has a positive effect on learning outcomes. The distance learning format is no exception, but it requires more attention to this factor, since there are objective obstacles to the emergence of empathy in a natural way, which often occurs in face-to-face training and requires certain awareness when it is introduced into the educational process by the teacher. The research purpose is to develop practical recommendations for language teachers at universities on the introduction of psychological-methodological elements that contribute to an increase in the empathy of all participants in distance and hybrid classes to improve the effectiveness of learning. The research methodology includes an experiment, observation, deduction, survey, post-analysis. As a result of the study, specific recommendations were formulated to build an emotional connection in distance classes in a foreign language. Thus, it is recommended to use methodological and psychological tools, gamification, and discussion format of classes, to control the dynamics of seminars. Establishing an empathic connection is facilitated by less formal and more benevolent communication at distance seminars, preferably a targeted appeal to students, to the whole group, keeping the camera on and enriching seminars with various emotionally charged and substantive elements, the teacher's active presence in social networks during off-hours, engaging all types of work in real time: interactive whiteboard, editing shared documents.
Keywords: teaching foreign languages, empathy, pedagogy, teaching methods

1 Introduction

The coronavirus pandemic has forced educational institutions and educators to take a fresh look at both full-time, distance and hybrid education. In addition to the learning format, the teacher's role has also changed in many aspects of the educational process [1]. In modern foreign language teaching, it is necessary to take into account several main aspects: active learning, motivation, feedback [2], the development of communicative competence [3, 4] and the interdisciplinary nature of modern education [5]. However, the most important challenge in these conditions, often due to the lack of a common physical educational space, is to build a stable teacher-student relationship [6]. The formation of this connection, as well as an increase in the effectiveness of classes, should be facilitated by the inclusion of an element of empathy in the educational material and teaching practice [7].

Empathy is considered one of the teacher's key characteristics, since it allows establishing an adequate dialogue and ensuring cooperation of all participants in the educational process [8]. The concept of empathy involves understanding the emotions of another person [9], the ability to put oneself in his/her shoes [10] and consists of two components: cognitive and affective ones.

Studies also show that the teacher's empathy affects the level of students' motivation to learn foreign languages in a distance learning format [11-13]. The level of students' empathy to the teacher can positively affect their progress in learning foreign languages [14]. The more empathy and collaboration arise between teacher and student, the higher the level of satisfaction, enthusiasm and interest in the subject [15]. Empathy can create a calm and friendly classroom atmosphere and students will be more willing to learn a new language [16].

However, despite a wide body of research on the influence of emotions and empathy on teaching foreign languages, there is a noticeable lack of practice-oriented research aimed at developing specific recommendations.

The research purpose is to develop practical recommendations for language teachers at universities on the introduction of psychological and methodological elements that contribute to an increase in the empathy of all participants in distance and hybrid classes to improve the effectiveness of learning. Research objectives are to identify the factors and tools affecting the empathy of participants in the educational process, determine the central and most effective tools, formulate recommendations for teachers on their basis on increasing empathy and introducing the emotional component into the process of teaching foreign languages at universities.

2 Methods

The study was based on an experiment, observation, deduction, and survey. During the experiment, for 14 months (2020-2021), elements that were supposed to increase empathy when learning foreign languages were introduced into seminars for 1st-3rd year undergraduates on the subject "Second Foreign Language (Spanish)" studying at the Financial University under the Government of the Russian Federation in distance and hybrid formats. The elements were selected based on the researcher's experience, observation and deduction. At the end of the seminars, students were asked to express their emotions in relation to the introduced elements or the material in which they were used orally or in the form of an anonymous snap poll. The seminars were followed by a detailed post-analysis with the identification of effective practices, which were subsequently refined and retested. As a result, a list of the 15 most effective psychological and methodological techniques was formed. This list formed the basis for a survey of 203 students of all courses at all levels of education from 17 universities, who studied fully or partially in a distance format, in order to identify the most effective ones in students' opinion. On the basis of the survey, a list of recommendations for language teachers' practical activities at universities was made.

3 Results

Table 1 shows the identified psychological and methodological tools in descending order according to the results of student surveys. The first column shows the percentage of students who voted for the effectiveness of this technique, while students could choose an unlimited number of points. It should be noted that all the practices identified during the first part of the study were assessed as effective.

Table 1. Recommendations for increasing empathy in distance learning of foreign languages at a university, based on survey materials.

Students who voted, %	Empathy tool in class	Recommendations for teachers
68	Humor regarding the distance learning attributes	You should not treat non-standard situations arising in distance learning as a violation of discipline, humor can help in establishing empathic connections and motivate students to study the subject.
63.1	Interactive real-time games (kahoot, quizlet, menti, etc.)	The use of gamification elements on various topics: both lexico-grammatical and country-specific ones.
56.2	Less formal communication between the teacher and students (compared to face-to-face communication in the classroom)	It is recommended to avoid harshness and take into account what words and actions in a distance format, somewhat distorting communication, may cause misunderstanding and, as a result, students' frustration [17]. It is necessary to more clearly formulate objectives and provide explanations, show emotional closeness with students and promote cooperation.
53.7	Frequent change of activities, dynamism of a seminar	It is recommended to enrich distance seminars with different forms of work within the framework of the educational goal and control the change of activities in time, so that students' attention is not scattered and the training retains its immersiveness, thereby maintaining an empathic connection.
52.2	Teacher's camera on	Before classes, you should check that the camera is working properly, lighting is sufficient (do not sit with your back against the window or other light source so that students can see your face), show your face in full, not part of it, and look directly into the camera as often as possible.
50.7	Discussions both in a general videoconference and in separate working conferences	Discussions on the outlined topics contribute to collaboration, which increases emotional connection.
47.3	Teacher's active presence in social networks and communication with students during off-hours	It is recommended to create common dialogues with students on social networks, promptly answer questions, send visual material not only to consolidate relevant educational material, but also to get

		acquainted with the culture, traditions and modernity of target-language countries.
43.8	Interactive whiteboard	Both the teacher and students can have access to the interactive whiteboard. Creative collaborative tasks are recommended.
40.9	Real-time surveys and voting (google forms, vk, ms teams, etc.)	Using voting tools to get feedback, select topics for discussion, task types, etc., as a tool to involve students in the organization of training.
39.4	The teacher should address the student, not the whole group; personal comments are recommended.	Due to the lack of a common physical space, the concept of a "study group" becomes blurred, and therefore it is recommended to address some of the students and then involve others in the discussion, give feedback as often as possible.
35	Students' cameras on	Asking students to keep their cameras on should not be imperative at training seminars, not control ones, however, you can introduce the "rule of five cameras", where students decide in advance who switches on the cameras, this will help avoid embarrassing situations and allow using all the psychological and emotional advantages of this tool.
31.5	Editing a shared document in real time	Organization of work on a shared document, which structure or specific template is emotionally charged (idioms, pictures, puzzles, etc.).
26.6	Home comfort in the video against the background: interior, paintings, books, things, pets (not a blurred background or the background replaced with a neutral one)	It is necessary to think ahead the position of the teacher's camera and the background so that it shows his/her interests or an interior element. However, in some situations, replacing the background with emotionally charged and meaningful ones can serve a pedagogical goal.
25.1	More time to discuss abstract topics at the beginning of the lesson.	In distance learning, it is recommended to pay special attention to "warming up" and "tuning up" before starting the lesson to establish an emotional connection and create the right atmosphere; do not start the lesson right away by checking homework or explaining new material.
24.1	The use of videos and audio recordings	Video and audio materials in learning foreign languages are an important source of knowledge; they also help to create an emotionally charged visual picture.

Thus, the presented tools are only an outer envelope, a carrier of the emotional part of information, while the recommendations indicate the content to be conveyed through these carriers, or a method of implementation.

4 Discussion

The results obtained show the possibility and effectiveness of introducing empathy in the distance learning of foreign languages at universities. Many of these recommendations have been mentioned in earlier studies. Thus, humor has a positive effect on the material assimilation, especially in foreign language classes [18, 19]. Researchers also come to conclusions about the important role of interactive games like Kahoot, Menti, Quizlet in bringing together the team of students based on the emotional factor, adding dynamics to studying a subject and elements of competition in distance and face-to-face training [20, 21]. According to a UNESCO study [22], students expect support from teachers in a difficult epidemiological situation that has forced the mass introduction of distance learning, which also affects the emergence of a desire for more informal and less strict communication in distance classes.

5 Conclusion

Empathy plays a key role in learning foreign languages in a distance format at the university. At first glance, this learning format suggests less involvement of emotions and empathy than face-to-face training, but when learning foreign languages, it is necessary to pay special attention to creating and maintaining empathy for effective material assimilation. Obviously, a teacher should use not only methodological, but also psychological tools in order to build and maintain an empathic connection in distance learning. Many of the recommendations considered are relevant both for distance learning and face-to-face training, such as gamification, discussion, dynamism. Building an empathic connection is facilitated by less formal communication in distance seminars, as well as keeping the camera on and enriching seminars with various emotionally charged and meaningful elements. A constant post-analysis of the conducted classes is required to identify the most effective practices for each group, maintain discipline and work on shortcomings.

The presented study can be continued by studying the possibilities of various computer programs and technologies, including immersive learning technologies, in terms of increasing students' empathy and, as a consequence, improving the efficiency of the educational process.

References

1. S. Bennett, L. Lockyer, Edu Media Int, **41(3)**, 231-248 (2004). https://doi.org/10.1080/0952398041000 1680842
2. I. Yengin, D. Karahoca, A. Karahoca, A. Yücel, Proc Soc Behav Sci, **2**, 5775–5787 (2010). https://doi.org/10.1016/j.sbspro.2010.03.942
3. I.I. Klimova, G.V. Klimova, S.A. Dubinka, XLinguae, **12(1)**, 207-218 (2019)
4. I.I. Klimova, N.E. Sharabarina, E.A. Tikhova, S.A. Dubinka, XLinguae, **11(1XL)**, 67-74 (2018)
5. T.A. Goroshnikova, E.S. Smakhtin *Interdisciplinary Curriculum Approach as a University Component for Large-scale Education Projects*, Proceedings of 2018 11th International Conference; Management of Large-Scale System Development, MLSD (2018) https://doi.org/10.1109/MLSD.2018.8551932
6. O. Khatsrinova, V. Bronskaya, S.V. Barabanova, et al., *Teacher Readiness for Distance Learning*, in M.E. Auer, T. Rüütmann (eds), Educating Engineers for Future Industrial

Revolutions. ICL 2020. Advances in Intelligent Systems and Computing, **1329**, 453-469 (Springer, Cham, 2021). https://doi.org/10.1007/978-3-030-68201-9_45

7. D.S. Melo-Solarte, P.A. Díaz, Inf Tecnol, **29(3)**, 237-248 (2018). http://dx.doi.org/10.4067/S0718-07642018000300237
8. S. Stojiljković, G. Djigić, B. Zlatković, Proc – Soc Behav Sci, **69**, 960-966 (2012). https://doi.org/10.1016/j.sbspro.2012.12.021
9. M. Davis, JSAS Catalog Sel Doc Psy, **10**, 85 (1980)
10. S. Baron-Cohen, S. Wheelwright, J Autism Devel Disord, **34**, 163-175 (2004). https://doi.org/10.1023/B:JADD.0000022607.19833.00
11. F.G. Escobar Fandiño, A.J. Silva Velandia, Heliyon, **6(8)**, E04630 (2020). https://doi.org/10.1016/j.heliyon.2020.e04630
12. S. Meyers, K. Rowell, M. Wells, B.C. Smith, College Teach, **67(3)**, 160-168 (2019). https://doi.org/10.1080/87567555.2019.1579699
13. T.M. Makoelle, *Teacher Empathy: A Prerequisite for an Inclusive Classroom*, in Peters M. (ed), Encyclopedia of Teacher Education. (Springer, Singapore, 2019). https://doi.org/10.1007/978-981-13-1179-6_43-1
14. Y. Hao, Comp Human Behav, **59**, 295-303 (2016). https://doi.org/10.1016/j.chb.2016.01.031
15. D. Goleman, Inteligencia Emocional. (Editorial Kairós, Barcelona, 1997)
16. H. Gardner, Estructuras de la mente. Inteligencias multiples. (FCE, Santafé de Bogotá, 2001)
17. F. Borges, Digithum, **7(7)**, (2005). http://dx.doi.org/10.7238/issn.1575-2275
18. S. Bolkan, D. J. Griffin, A. K. Goodboy, Comm Edu, **67(2)**, 144-164 (2018). https://doi.org/10.1080/03634523.2017.1413199
19. M. Azizinezhad, M. Hashemi, Proc – Soc Behav Sci, **30**, 2093-2098 (2011). https://doi.org/10.1016/j.sbspro.2011.10.407
20. K.E. Cameron, L.A. Bizo, Resn Learn Techn, **27**, (2019). https://doi.org/10.25304/rlt.v27.2225
21. M. Curto Prieto, L. Orcos Palma, P.J. Blázquez Tobías, F.J.M. León, Edu Sci, **9(1)**, 51-56 (2019). https://doi.org/10.3390/educsci9010055
22. S. Duggan, AI in Education: Change at the Speed of Learning. (UNESCO IITE Policy Brief, 2020)

Gadgets and new educational technologies: real assistance or an obstacle?

Svetlana Alekseevna Molochkova, *Marina Valentinovna* Gromova, *Valery Alekseevich* Bartenev[1], *Aleksander Viktorovich* Borodkin, *and Aleksander Mikhailovich* Salnikov

Financial University under the Government of the Russian Federation, Yaroslavl Branch, Department of Management and General Humanitarian Sciences, Yaroslavl, Russia

Abstract. The rapid spread of modern computer and communication technologies resulted in some global problems in society, including the adoption and rejection of such technologies in education. The COVID pandemic forced a widespread use of modern-style communications between tutors and students but unsolved problems still exist and will exist in the post-COVID era. The main goal of the paper is to find the potential borders of using modern technologies for students learning economics and business management in Russia. The authors conducted the survey in March of 2021 in the Yaroslavl Region, Russia. The main finding of the paper is the necessity of limiting the use of computer technology. The main reason for such limits is a possible decrease in students' health and the quality of education. All the results gained by the authors are applicable in the practice of organization of educational processes for students involved in economics and business management. Probably, the results may be inapplicable for students of other fields such as medicine, art, engineers, etc.

Keywords: distance learning, survey, higher education, technology implementation

1 Introduction

It is generally accepted that the Internet, electronic mail, and digital television took a significant part in people's lives. "The modern development of the society was marked by the rapid development of modern technologies", Yurov wrote in 2018 [1].

According to a survey kept by analytics of the online school "Skysmart" in 2020, traditional mass media move to the back stage of contemporary life, and teenagers are leading in the processes of refusing classic TV, radio, and printed media. According to the research results, all respondents spent at least 4 hours a day using their gadgets, primarily smartphones and tablets. Twenty-nine percent of the interviewed teenagers declared that they used gadgets for up to 6 hours a day. "Considering that the future of the Internet looks very promising and cloudless, it is obvious that the number of students who have modern communication devices will rapidly increase", the authors conclude [2, 3].

[1] Corresponding author: VABartenev@fa.ru

Antony M. Townsend notices that the appearance of the World Wide Web in 1993 resulted in the cardinal change in human behavior. Another leading communication technology of the early 1990s was mobile phones. Both of them caused a transformation of reality perception and gave humankind brand new abilities to communicate worldwide regardless of distance [4]. Bringing the Internet to the top of communication technologies allowed people to find any information easily. Researchers denote the following advantages of using the Internet in education:

- obtaining the required information in an easy and fast way;
- comfortable visualization of the information;
- pre-programmed solving of problems and cases;
- interaction with the tutor any time the student needs;
- simplifying the control over educational results;
- instant access to libraries and storages of educational materials;
- ability to store huge volumes of textual and graphical information for educational purposes [5, 6].

On the other hand, theorists of education suspect that the widespread use of modern computer technologies in education may cause some global problems that cannot be solved right now. First of all, it is needed to develop a brand new paradigm of learning. Since gadgets have changed the lifestyle, all approaches to education should be changed too. Probably new ways of learning should use both synchronous and asynchronous communication between tutors and students, as well as increase the volume of work in virtual teams without direct tutor's supervision [7].

Another problem of implementing new technologies into educational processes concerns the readiness of tutors to use gadgets and communicate with students using the Internet. In fact, there are too many older tutors who resist any new things due to their age. It was too difficult for them to start using electronic mail two decades ago, and it is bigger trouble to start lecturing through video communication systems and review students' papers existing in electronic form only. Universities tried to re-learn their tutors for so-called "electronic education" but these attempts mostly have failed. On the other hand, new generations of tutors are ready for using modern educational technologies. According to some studies, one-third of interviewed tutors are ready to use any new way to learn their students [8].

The third problem found by researchers is the high differentiation in the availability of the high-speed Internet between urban and rural territories. This problem is not so relevant in European countries but it really becomes an unbreakable wall on the way to full electronic learning in ex-USSR countries. Most villages in Russia (especially in Siberia) are deprived of the ability to use an optical or wired connection to the Internet; mobile networks are unable to transmit data at the high speed needed for real-time face-to-face communication. This causes rural students to use asynchronous communication technologies only and probably decreases the quality of educational results [9].

The fourth major problem of using gadgets and new technologies in education concerns legal aspects. Some countries (e.g., Russia) have an education branch deeply regulated by public administration. Due to the full absence of regulation legislature, any modern electronic technologies are in the "grey area": they are not prohibited to be used and not allowed at the same time. Universities use them at their own risk. The "grey" status of new technologies brings up troubles related to academic fraud. Since testing is kept asynchronously without the direct supervision of a tutor, some students can use books, other materials during answering test questions or simply can find proper answers on the Internet. There are still no solutions to such problems [10, 11].

So theorists of education and higher-school tutors came to the following conclusion. Electronic educational technologies appeared too recently to be studied properly. The COVID pandemic in 2020 forced the implementation of these technologies and it is not clear if it was implemented correctly or it was a life ring dropped to educational organizations to survive. Obviously, it is needed to deeply study all new technologies separated by their technological nature, as well as students' specializations [12]. It is more than likely that the results of such studies will reveal a large amount of ambiguities due to the high differentiation in the nature of learning for different professions. "Any innovation in education is a very controversial issue", concluded [13].

To answer the questions mentioned above for a specific university, the authors conducted a study. The main goal of the study was to decide are new technologies and gadgets suitable at the Financial University or not. The main goal determined the two particular tasks:

1. To determine the positive and negative influence of using modern computer technologies and gadgets on the educational processes in the fields of economics and business administration.
2. To determine students' attitudes toward using gadgets in learning.

2 Methods

The authors have created an electronic questionnaire to anonymously interview full-time students of the Yaroslavl branch of the Financial University. The questionnaire had six questions:

1. How frequently do you use gadgets (smartphones and/or tablets)? (The specific number of hours a day that the respondent spends using gadgets).
2. Did you increase or decrease the time spent using gadgets during the COVID pandemic in 2020? (The specific number of hours a day that respondents spent on gadgets during the pandemic).
3. Do you use gadgets for educational purposes after the COVID pandemic? (Answers are supposed to compare the intensity of using gadgets for educational purposes and for other ones).
4. What software/websites do (did) you use for educational purposes? (It is a semi-closed question; every answer is educational software/a website widely known in Russia).
5. Do you want to continue distance learning even if the COVID pandemic stops or do you prefer to learn in university classrooms? (Two available answers: "distance learning" or "classroom learning").
6. Why do you prefer distance learning or learning in classrooms? (It is semi-closed question that describes the most probable reasons to continue distant learning or return to the classroom; the respondent could type in his/her own reason).

To get results for future students that could enter the Financial University in 2021 and 2022, the authors expanded the survey and asked to answer the questionnaire under-graduate schoolchildren of a municipal school in the Myshkin town. The survey included 35 persons. According to the last census results, the population of the Yaroslavl Region aged 15–24 is 10,1509 persons. Assuming them as the universe, one can conclude that the survey has ±12.17% error with 85% trust probability. Thus, the sample has a required and sufficient size and the survey is statistically justified.

3 Results

The following results were obtained after processing the survey data.

As for the first question, the majority of respondents (more than half) answered that they used gadgets more than six hours per day (see Fig. 1). There were no schoolchildren and students who did not use gadgets. Only one respondent answered that he (or she) still used gadgets with the same intensity; others answered that the usage of gadgets increased (see Fig. 2). At the same time, the majority of interviewed persons declared that they began to spend less time on gadgets after the COVID pandemic and returning to classroom learning, and only one person answered that he (or she) increased the time spent on gadgets (see Fig. 3).

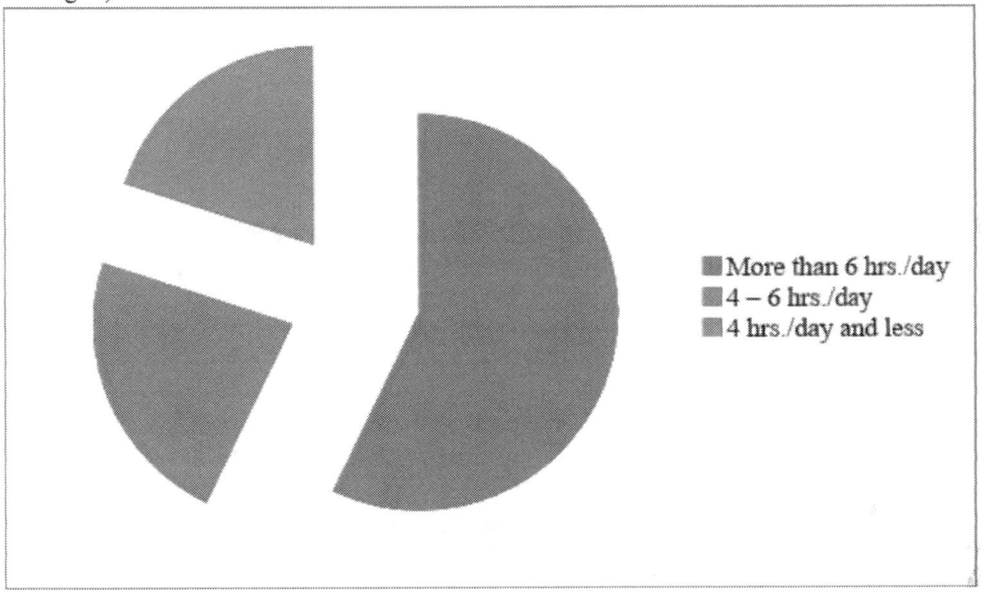

Fig. 1. Distribution of answers to the first question of the survey questionnaire: how frequently do you use gadgets?

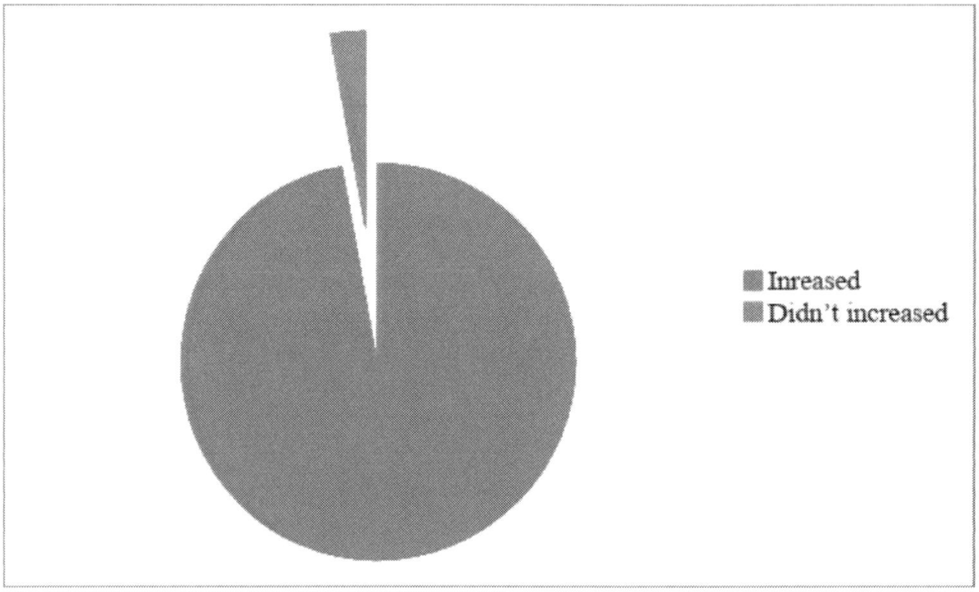

Fig. 2. Distribution of answers to the second question of the survey questionnaire: Did you increase or decrease the time spent using gadgets during the COVID pandemic in 2020?

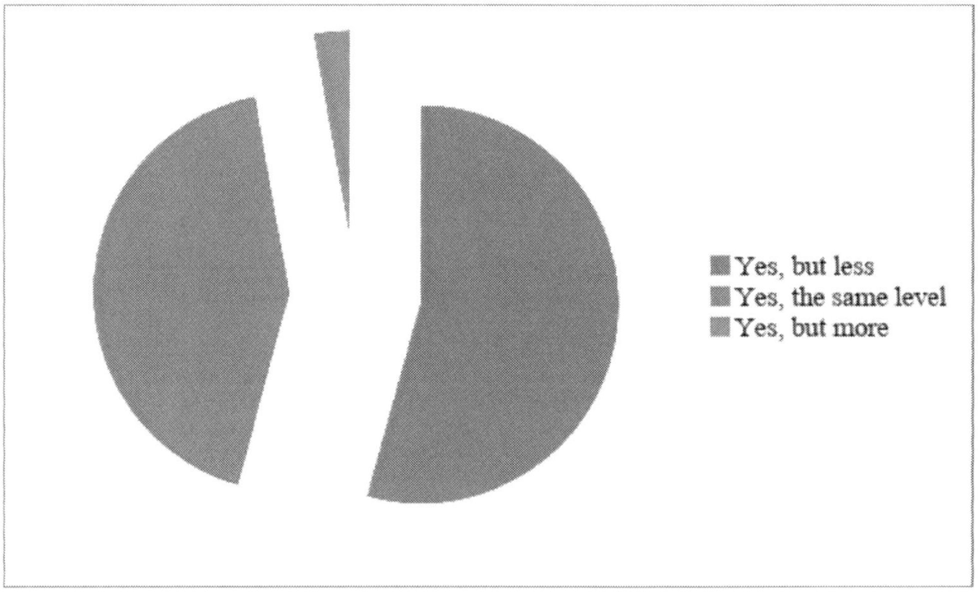

Fig. 3. Distribution of answers to the third question of the survey questionnaire: Do you use gadgets for educational purposes after the COVID pandemic?

The fourth question gave the rating of software and web services that were used during the COVID pandemic (see Fig. 4). Electronic mail, the oldest communication technology, took the first place. Discord and the FU portal – the technologies that have been chosen at the Yaroslavl branch of the Financial University took the second place. The Regional Diary and Yaclass, services intended for use at schools, not at universities, took the third place. Besides, 11.43% of respondents answered that they used something else, e.g. VKontakte, the most popular social network in Russia and some ex-USSR countries.

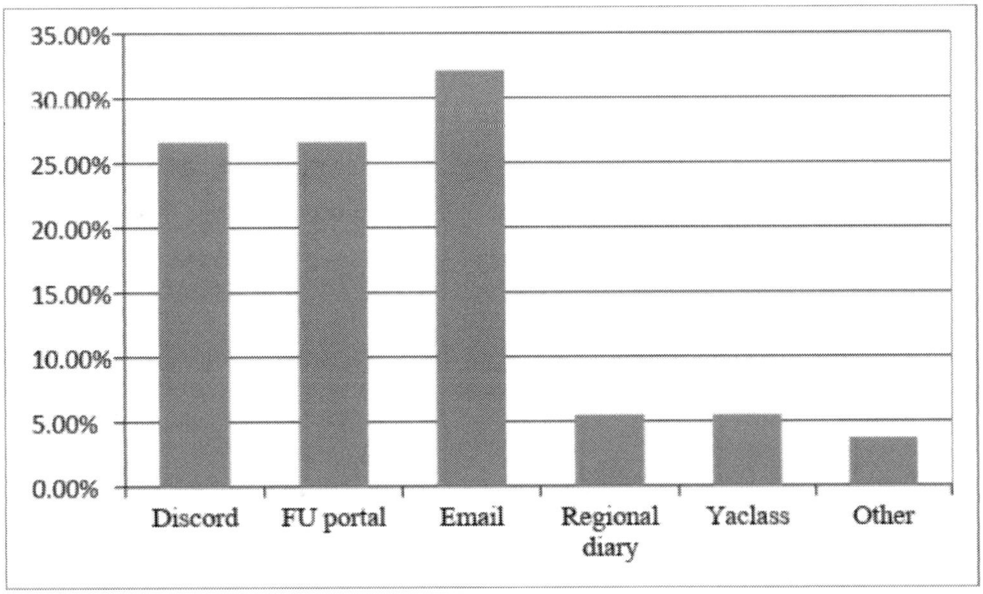

Fig. 4. Distribution of answers to the fourth question of the survey questionnaire: What software/websites do (did) you use for educational purposes?

Answers to the fifth and sixth questions were surprising (see Fig. 5). More than 71% of students wanted to return to classroom learning instead of continuing with distant classes. Everyone had at least two reasons why he/she wanted it.

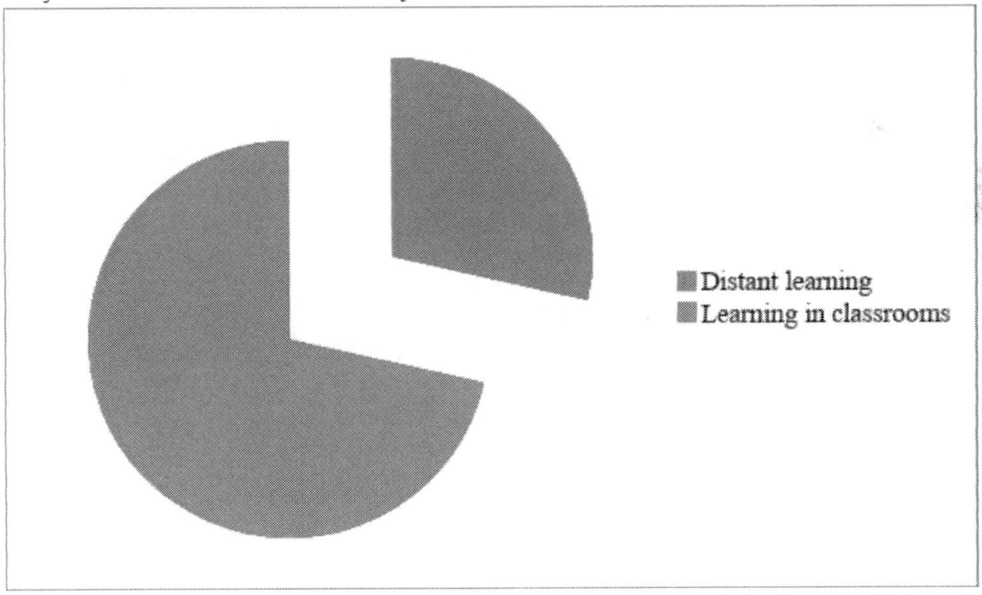

Fig. 5. Distribution of answers to the fifth question of the survey questionnaire: Do you want to continue distance learning even if the COVID pandemic stops or do you prefer to learn in university classrooms?

In general, the survey was kept perfectly. There were no compulsions to take part in the survey or any compulsion to choose specific answers, so one can assume that the respondents really expressed their will and opinion.

4 Discussion

The survey showed that most teenagers and students in the Yaroslavl Region spent more than six hours a day with gadgets. It is more than the mean value in Russia. This result could be explained by the proximity of the Yaroslavl Region to the Moscow city and the high availability of mobile networks even in rural lands. The same explanation could be applied to answers to the second question. On the one hand, one can conclude that the Yaroslavl Region is a leading territory and local universities and schools probably will not face technical troubles on students' side during the total implementation of distant electronic learning. On the other hand, spending more than six hours a day with a computer or gadget is a very bad result from a medical point of view. One can assume that this enormous use of gadgets will lead to a massive decrease in the health of the generation born about 2000.

The same conclusion can be made after analyzing answers to the third question. Traditional schoolchildren's or students' schedule assumes classes from 8:30 a.m. to 3:00 p.m., amateur sport or art activities after classes and execution of home assignments in the evening without electronic learning technologies. During the COVID pandemic, schoolchildren and students were deprived of physical activities; all day between bed-times was dedicated to electronic learning. It was one of the negative consequences of the COVID pandemic, and it is obvious that implementing new computer technologies into education during the post-COVID era must be limited by the time spent with computers or gadgets [14].

Answers to the fourth question did not bring any unexpected results. Schoolchildren and students were forced to use software and websites chosen by educational organizations. Moving to distance learning in March of 2020 was so fast, and there was no time to discuss with students the software they prefer. System administrators were forced to recommend tutors the software and services that could be used on the majority of devices. In the future, the question of software and web service potentially could become one of the key questions. It is obvious that software must be comfortable to use and able to be run on a variety of devices to ensure high-quality results of education; students and tutors should not solve technical difficulties instead of the learning process [15, 16].

The fifth and sixth questions gave unexpected survey results. Based on the high level of gadget use, it could be assumed that students prefer to study electronically without a presence in classrooms. However, the reality is quite another. Students want to use gadgets for common purposes (communicate between themselves, obtain news, get fun, and so on), but they do not want to use computers and gadgets for education. They prefer to learn in classrooms with limited use of electronic learning technologies. The main benefits of learning in classrooms noted by students are:

1. Face-to-face communication between students and tutors.
2. A higher level of the skills obtained in the classroom with the tutor's assistance rather than those obtained during distance learning.
3. Stable conditions of classroom learning (including schedule, deadlines, etc.).
4. Internet outages during online classes that may result in a low level of understanding some topics or home assignments.
5. Most of older tutors are unable to communicate with students through video communication services, to process and rate home and class assignments written electronically.

The minority of students who preferred distance learning could explain their choice by comfortable conditions at home and saving time to get to the campus (due to high traffic jams in Yaroslavl, students and tutors spend up to 3 hours a day to get to campus and back by the municipal transportation system). The same results have been obtained by other studies in different countries [17-22].

Based on the result mentioned above, the Yaroslavl branch of the Financial University decided to recover classroom learning with the restricted use of distance learning. The problem of implementing the new educational technologies is too versatile to be solved in a year. Each of the participants in the learning process has reasons to use new technologies or decline them; there is and will be no universal solution that satisfies everyone. Also, it should be kept in mind that educational processes in other fields rather than economics and business administration (e.g., medicine, engineering, art) could have peculiarities not mentioned in this study; the results of a similar survey of the students could be completely different [23, 24].

5 Conclusion

Summarizing the study results, one can conclude that total moving to distant electronic-based technologies is inappropriate in current economic, social, and technological conditions. Such transformation of educational processes can result in a significant decrease in students' health, as well as a decrease in the quality of education. Another negative consequence that could be met after electronization of the educational process is an increase in the number of unsatisfied students.

Nowadays, a lot of problems of electronic education are still unsolved. These problems are related to insufficient learning materials, the low-level skills of tutors and the probable impossibility to teach students in some specific fields. Gradual solving of such problems will allow spreading new educational technologies more and more but it is suspected that the process will be stretched for years and years.

References

1. I.A. Yurov, Young Sci, **1(15)**, 101-106 (2018)
2. A. Balandina, "Sidyat ponikshie": pedagogi o problemakh obucheniya onlain ["Sitting drooping": teachers about learning problems online]. Accessed on: October 1, 2021. [Online]. Available: https://www.gazeta.ru/social/2020/08/24/13208521.shtml
3. Ya.S. Medvedeva, Young Sci, **15(119)**, 473-476 (2016)
4. A.M. Townsend, J Urban Techn, **7(2)**, 85-104 (2000). https://doi.org/10.1080/713684114
5. A.V. Tatarintsev, Young Sci, **25(211)**, 317-319 (2018)
6. X. Chen, D. Zou, H.Xie, G. Cheng, Edu Tech Soc, **24(1)**, 205-222 (2021)
7. R.M. Mukhamadeeva, Euroasian Sci Union, **2-1(48)**, 57-59 (2019)
8. E.A. Shefer, Young Sci, **16(358)**, 22-25 (2021)
9. I.S. Kargopolov, Young Sci, **12(250)**, 268-269 (2019)
10. T.F. Kryaklina, Econ. Prof. Bus, 2(2), 77-82 (2016)
11. Z.Z. Kodirov, Young Sci, **15(149)**, 585-586 (2017)
12. N.P. Kiselev, Young Sci, **4(346)**, 340-344 (2021)

13. M.Sh. Bozorova, Int J Human Nat Sci, **2**, 39-41 (2018)
14. O.N. Ragozin, et al., Bul Nizhnevartovsk State Univ, **2**, 130-135 (2020). https://doi.org/10.36906/2311-4444/20-2/17
15. A. Ellahi, Edu Techn Soc, **21(4)**, 64-75 (2018)
16. Ch.K. Siu, Edu Techn Soc, **22(1)**, 28-43 (2019)
17. G.G. Karhanyan, Cross-Cult Stud: Edu Sci, **5(2)**, 92-96 (2020). https://doi.org/10.24411/2470-1262-2020-10080
18. N. Guzachchova, Bul Sci Prac, **6(5)**, 457-460 (2020)
19. F. Manganello, C. Falsetti, T. Leo, Edu Techn Soc, **22(1)**, 44-58 (2019)
20. N.L. Schroeder, Edu Techn Soc, **20(4)**, 138-147 (2017)
21. J. Zilvinskis, A.A. Masseria, G.R. Pike, Res Higher Edu, **58(8)**, 880-903 (2017). https://doi.org/10.1007/s11162-017-9450-6
22. M. El-Masri, A. Tarhini, Edu Techn Res Devel, **65**, 743-763 (2017). https://doi.org/10.1007/s11423-016-9508-8
23. I.S. Grigoriev, S.A. Belyaev, Software J: Theory App, **1**, 2 (2020)
24. A.M. Tokesheva, et al., Sci Healthcare, **22(6)**, 80-85 (2020)

Construction of the Ishikawa diagram for the incorporation into employment of graduates of the Ministry of emergency situations of Russia

Vladimir Olegovich Mikryukov[1*], *Sergey Anatolievich* Melkov[2], *Andrey Yurievich* Lyabakh[2], *Sergey Nikolaevich* Ilyushin[3], and *Aleksandra Olegovna* Boeva[2]

[1]Financial University under the Government of the Russian Federation, Department of Sociology, Moscow, Russia
[2]Civil Defense Academy of the Ministry of Emergency Situations of Russia, Department of State and Municipal Administration, Khimki, Russia
[3]State Budgetary Educational Institution of the City of Moscow "School No. 1576", Moscow, Russia

Abstract. The paper presents the results of the author's study of the problem of employment of universities graduates of the Ministry of Emergency Situations (EMERCOM) of Russia for vacant posts in the EMERCOM of Russia. The purpose of the study is to develop a draft "road map" aimed at solving the management problem of insufficient active participation of EMERCOM universities in the employment of graduates for civil service in the EMERCOM of Russia. The authors have used the following methods: traditional document analysis, expert sociological survey in the form of a questionnaire; method of studying statistical data; method for constructing the Kaworu Ishikawa chart. As a result of processing statistical data and experts' answers, the paper presents the author's view of the draft "road map" the implementation of which will solve the problem of effective employment of graduates of EMERCOM universities. The paper presents for the first time the results of the causal relationship between the need for employment of graduates of EMERCOM universities and the low percentage of this employment presented in the form of the Ishikawa chart.
Keywords: Ishikawa chart, employment, EMERCOM of Russia, expert survey, road map

1 Introduction

In our opinion, there are at least three aspects that determine the relevance of the topic of this study.

Firstly, in modern Russian society, the employment of young experts is the main way not only to organize their own life and self-expression but also to apply the professional knowledge and skills acquired during the educational process.

* Corresponding author: mikryukov.v.o@gmail.com

Secondly, the research carried out by the authors is aimed at identifying a management problem that affects the efficiency of the functioning of the student employment system existing in the universities of the EMERCOM of Russia and, subsequently, the effectiveness of the activities of the EMERCOM structural units.

Thirdly, an in-depth study of the chosen topic, as well as the formation of a "road map" with proposals for their implementation at the universities of the EMERCOM of Russia are designed to improve the existing processes of employment for vacant posts in the ministry.

Thus, the management problem that is solved in this study can be formulated as follows: low efficiency of assistance to graduates of universities of the EMERCOM of Russia in the process of employment of students in the EMERCOM structural units.

A consistent solution to this problem can be achieved by posing and solving the following research questions:

1. Is it possible, with the help of the traditional analysis of documents on the employment of graduates of universities of the EMERCOM of Russia, as well as using the internal organizational documents of the Civil Defence Academy of the EMERCOM of Russia, to find out the specificity of employment of graduates and the existing problems?

2. Will it be effective to use methods such as the construction of the Ishikawa chart to develop proposals for solving the problem of low efficiency of assistance to graduates of the Civil Defence Academy of the EMERCOM of Russia in the process of employment students in the structural units of the EMERCOM of Russia?

3. What blocks should the "road map" consist of to solve the problem of employment of graduates of the EMERCOM of Russia?

Thus, the hypothesis of the study sounds like this: assistance of the EMERCOM universities of Russia to students in employment in the structural divisions of this ministry in the early stages will allow the selection of applicants who are most interested in employment for vacant positions in structural bodies among prospective candidates.

Let us turn to the analysis of the research problem. The authors Mikryukov, Melkov and Lyabakh have already analyzed the skills that a university graduate should have for employment in the article [1]. With regard to the application of the Ishikawa chart for various areas of activity, it should be noted that there is a conspicuous lack of articles on its application in Russian. It is possible to name some authors who use the Ishikawa chart: Logunova [2], Gavrilova, Novikova, Petelyak, Nazarova and Agdavletova [3], as well as Voronin, Goryacheva and Kokhova [4], Sheromova and Yanchenko [5]. For this paper, the approach used by Aronov is especially important. He uses it in his scientific article "Modifications of the Ishikawa chart for managing the process of developing regulatory legal acts" [6].

In detail, but in plain language (which is important for young researchers), Ilenkova [7] has conveyed the essence of the Ishikawa chart. We would also note the merits of the Ukrainian researcher Ivanov in popularizing the SWOT analysis and the Ishikawa chart [8, 9]. We also relied on the experience of foreign researchers who use the Ishikawa chart in their articles [10-14].

2 Materials and methods

The purpose of the author's research is to develop a draft "road map" aimed at solving the management problem of insufficiently active participation of EMERCOM universities in the employment of graduates for civil service in the EMERCOM of Russia. Research objectives directly intersect with research questions, which are indicated in the first section of the paper.

The authors used the following methods during the study: traditional document analysis, expert sociological survey in the form of a questionnaire; method of studying statistical data; method for constructing the Kaworu Ishikawa chart.

Methodological limitations: the authors have conducted a survey of experts from among the employees of the EMERCOM of Russia. A total of 6 people were interviewed. This is enough to solve the problems of this study. However, in order to scale up and develop recommendations at the level of the EMERCOM of Russia, it is advisable to interview other experts (for example, among the employees of the central office of the EMERCOM of Russia).

3 Results

The most important current global issue and in the Russian Federation as well is the problem of employment of working-age population. Its importance is determined by the possible positive prospects for the stability of the State. However, the pace and trends of development of modern society require constant monitoring and analysis of the current status. The EMERCOM of Russia has a regulatory framework that regulates the procedure for entering the state civil service in the structural bodies of the department:

1. Federal Law No. 79-FZ of 27.07.2004 "About the State Civil Service of the Russian Federation" [15];
2. Federal Law No. 58-FZ of 27.05.2003 "About the civil service system of the Russian Federation" [16];
3. Decree of the President of the Russian Federation No. 112 of 01.02.2005 "About the competition for the vacant position of the state civil service of the Russian Federation" [17];
4. Order of EMERCOM of the Russian Federation No. 174 dated 05.04.2012 "About qualification requirements to the professional knowledge and skills necessary for the performance of official duties by federal state civil servants of the central office and territorial bodies of the Ministry of the Russian Federation for Civil Defense, Emergency Situations and Elimination of the consequences of Natural Disasters" (registered in the Ministry of Justice of Russia on May 14, 2012 No. 24140) [18].

Let us analyze the content of each of the above-mentioned documents using the method of traditional document analysis. Federal Law No. 79-FZ of July 27, 2004, fully explains the system of state and municipal service in the Russian Federation highlighting the most important issue for this study on the admission of a candidate to the federal state civil service in the Russian Federation.

Decree of the President of the Russian Federation of 01.02.2005 No. 112 regulates the procedure and competition for filling vacant posts in the state civil service. Considering this document, the authors have revealed that an adult applicant who meets the requirements of an employer can apply for the position. The competition itself is held in two stages.

At the first stage of the competition, information is posted about the search for applicants to fill vacant posts in a state body and basic information about the competition (position, qualification requirements, conditions for civil service, etc.); then an applicant collects some relevant documents At the second stage, a meeting of the recruitment panel takes place, at which applicants are assessed using the documents they have provided; their professional and personal qualities are also estimated.

When considering this decree, the authors revealed an interesting fact regarding the conditions of the second stage of the competition. The meeting can be held only if there are at least two candidates for the vacancy. Although clause 3 of this regulatory legal act does not specify the conditions for canceling the competition. If one applicant applies for the post, apparently, the competition should still be held.

The Order of the EMERCOM of Russia No. 174 dated 05.04.2012 gives the right to include the listed requirements in the official regulations. The system of admission to the civil service in the Russian Federation is conditioned by certain professional skills and knowledge that are special for each of the groups of relevant posts.

Thus, having analyzed the most important regulatory legal acts for the research using the traditional document analysis method designed to regulate the process of employment of applicants for vacancies of the federal state civil service in the structural bodies of the EMERCOM of Russia, the authors have drawn the following conclusions:

- Normative legal acts have the main provisions governing the process of employment of applicants in the structural bodies of the EMERCOM of Russia are presented;
- The analyzed regulatory legal acts prescribe the basic professional knowledge and skills that are necessary for passing the civil service for different categories of job groups;
- For the purposes of this study, the procedure presented in the regulatory legal acts data for holding a competition to fill a vacancy is interesting.

From the data available about the labor market's need for experts for the EMERCOM of Russia departments, the authors conclude that the structural divisions have a significant staffing reserve, which can be used by any graduate of EMERCOM universities. Having considered some vacancies for the employment of students, which are offered by the personnel structures of the EMERCOM of Russia, the authors have identified some features of the labor market in the conditions of seeking posts in the structural divisions of the EMERCOM of Russia. They are presented in table 1.

Table 1. Features of the labor market for full-time students of the Civil Defence Academy in the context of seeking vacancies in the EMERCOM of Russia. Source: compiled by the authors.

The identified feature of the labor market for job seekers	**Substantiation of the identified features of the labor market for applicants to the structure of the EMERCOM of Russia**
Geography	There is no association with a specific location. Vacancies in the structural bodies of the EMERCOM of Russia are open in every region of the Russian Federation. Most vacancies for civilian experts in the humanitarian and engineering fields can be found outside the Central Federal District.
Group of posts	This feature is due to the fact that for applicants for filling vacancies in the structural divisions of the EMERCOM of Russia, a group of positions not higher than the main group of positions are offered.
Technical experts	Most of the posts (discarding territorial linkages) have a technical nature in the conditions of training "State and Municipal Administration" (hereinafter – SMA), which makes graduates of the State Medical University less employable (meaning the EMERCOM of Russia) than graduates of engineering training areas.
Conditions for filling vacant posts	The procedure for employment in the structural divisions of the EMERCOM of Russia is a long and rather complicated procedure for holding a competition for a vacant post.
Labor potential	In the context of the dynamic socio-economic conditions in the country, the low percentage of graduates employed in the structural units of the EMERCOM of Russia is caused by a lack of their work experience or by their unwillingness to work in that ministry at all.

Therefore, we can conclude from the analysis of Table 1 that employment in the specialized structures of the EMERCOM of Russia occurs with significant difficulties. In addition to these factors, job market in specialized organizations lays down strict disciplinary requirements, which also affect the motivation of an applicant. Accordingly, quite often, after graduating from the EMERCOM universities prospective applicants choose employment in commercial organizations where an employer offers flexible working arrangements and the employment process itself does not take as much time as in the structural units of the EMERCOM of Russia.

Having considered the system of employment in the Department for the Eastern Administrative District of the Directorate General of the EMERCOM of Russia in Moscow, the authors consider it appropriate to turn to the analysis of statistical data taken from official sources of one of the leading universities – the Civil Defence Academy of the EMERCOM of Russia. If we consider the overall picture of the engineering and humanities faculties of the EMERCOM of Russia (civilian students study at these faculties), then the total number of graduates from 2017 to 2019 is about 254 people [19]. Of all the graduates (254 people), only 49 people are employed in the EMERCOM of Russia, while there are a total of about 64 unemployed or employed in non-governmental organizations (36 are employed in non-governmental organizations; 28 are not employed at all) [19]. General information on the employment of full-time students at the expense of the federal budget in 2017, 2018, and 2019 from the Civil Defence Academy is shown in Figure 1.

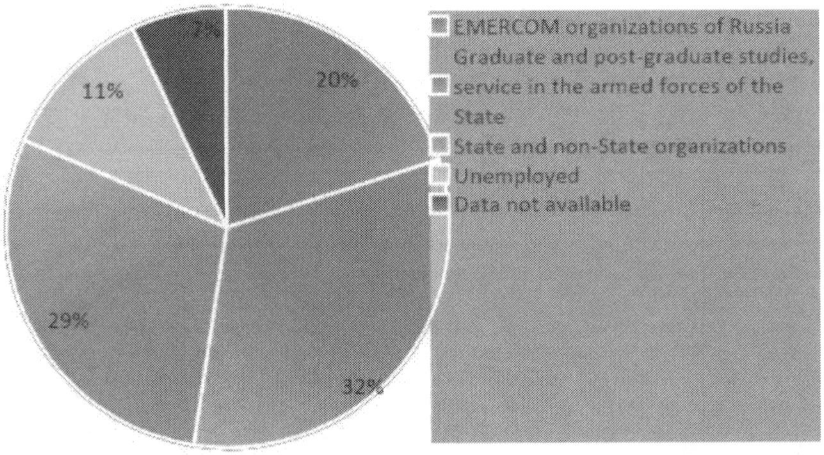

Fig. 1. General information on the employment of students enrolled in full-time education at the expense of the federal budget in 2017, 2018, 2019, graduated from the Civil Defence Academy of the EMERCOM of Russia. *Source:* compiled by the authors.

Figure 1 shows that the share of employed graduates, taking into account employment in governmental and non-governmental organizations, is 32% (71 people) of the total number of all experts graduated from the Academy (254 people). At the same time, the share of graduates employed by the EMERCOM of Russia is 20% of the total number of graduates (49 people), which is even less in comparison with the percentage of those applicants who got government and non-government jobs. We can state the unstable dynamics of the employment rates of graduates of the Civil Defence Academy of the EMERCOM of Russia and a small percentage of employment of graduates of this university. It indicates the ineffectiveness of meeting the goals and objectives of the

Development Strategy of the Civil Defence Academy of the EMERCOM of Russia until 2030, which means the management problem of low efficiency of assistance to graduates of the Civil Defence Academy of the EMERCOM of Russia in the process of employment students in the structural divisions of the EMERCOM of Russia.

According to the author's intention, in order to solve this problem, it is necessary to construct a cause-and-effect diagram or an Ishikawa chart [20]. To do this, the authors have conducted an additional expert survey in the form of a questionnaire. The experts were 6 managers of the Department for the Eastern Administrative District of the Directorate General of the EMERCOM of Russia in Moscow. The introduction of the cause-and-effect diagram is undertaken with the aim of graphic display the relationship between the problem being solved and the reasons that influence its occurrence. The Ishikawa chart itself is constructed in four stages. As a result, it looks very impressive (see Figure 2).

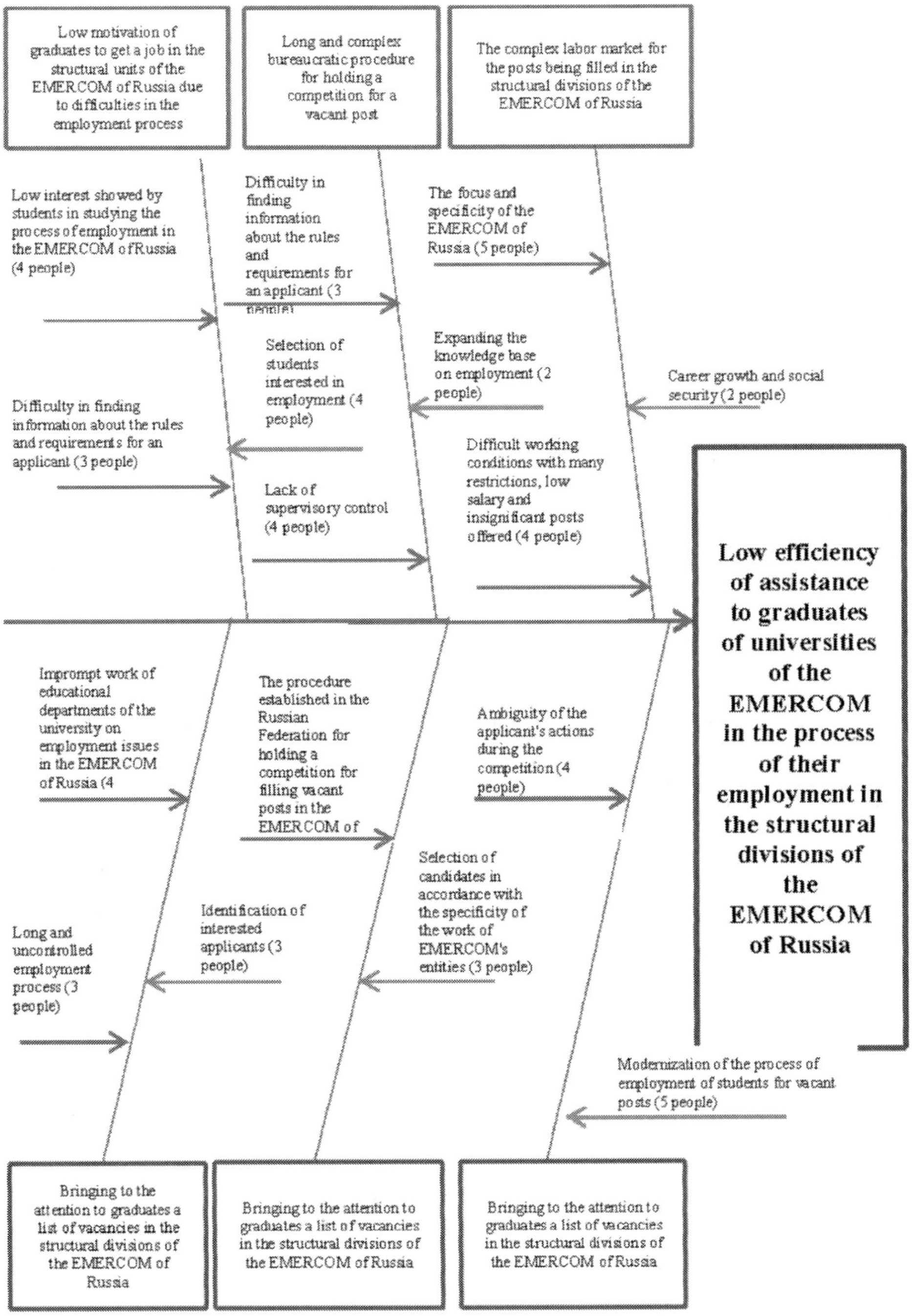

Fig. 2. The result of the graphical analysis on the research problem is the Ishikawa chart. *Source:* compiled by the authors.

The Expert Group has also identified 3 reasons that, in their opinion, most affect the problem, namely:

- bringing to the attention to graduates a list of vacancies in the structural divisions of the EMERCOM of Russia;

- independent solution of problematic issues by graduates when applying for employment in the structural bodies of the EMERCOM of Russia;
- visible activity of universities of the EMERCOM of Russia in the process of employment of graduates.

Next, the experts were asked to determine the significance of factors affecting the employment of graduates of the EMERCOM of Russia in this Ministry. As a result of the calculation, it turned out that the most important factors are a long and complex bureaucratic procedure for holding a competition for a vacant post, low interest of students themselves in studying the employment process, selection of candidates in accordance with the specificity of work in the structural bodies of the EMERCOM of Russia, and modernization of the process of student employment for vacant post.

4 Discussion

Having constructed the Ishikawa chart, based on the data of an expert survey, the authors have justified the key issues that affect the solution of the problem of employment of graduates of the EMERCOM of Russia in this ministry.

We believe that the measures proposed in the chart (see Figure 2) should be taken to fruition, that is, to propose specific management actions that will be able to optimize employment processes making them more effective. We accordingly get a "road map" to solve the problem of employment of graduates of the EMERCOM of Russia for vacant posts in this ministry. The proposals set out below were also verified during an expert survey (the survey of the expert group was conducted in mid-February 2021, 5 people were also interviewed – experts of the personnel bodies of the EMERCOM of Russia). Therefore, the authors consider it necessary to propose the following activities:

1. Creation of a structural unit in each university of the EMERCOM of Russia, where students are enrolled on a paid basis of the organizational unit responsible for coordinating the employment of graduates in the structural bodies of the EMERCOM of Russia;
2. Reforming of the chapter on employment on the official website of the universities of the EMERCOM of Russia, so that graduates always have access to an updated database of vacant posts and can easily get this information;
3. Organization of mentorship activities in the universities of the EMERCOM of Russia (in the structural divisions of the EMERCOM of Russia (Directorate Generals among the constituent entities of the Russian Federation, rescue centers of the EMERCOM of Russia, etc.). You should have a person who would be in permanent contact with the student council of the University of the EMERCOM of Russia and would explain the specificity of employment and the algorithm of rapid selection for vacant posts;
4. Add a new introductory discipline, where students will be able to study in detail the entire process of employment in the EMERCOM of Russia;
5. Use of means and methods of informing students about the prospects of work/service in the EMERCOM of Russia (social videos about the professions of the EMERCOM of Russia, posters, trainings, master classes of the best professionals of the EMERCOM of Russia, etc.).

5 Conclusion

In conclusion, we note that based on the conducted research, the Ishikawa chart was constructed to identify problematic issues affecting the employment of graduates of the

EMERCOM of Russia in this ministry. With the help of an expert survey, specific measures were identified to solve this problem.

Because of the conducted research, the stated research goal was fulfilled – a draft "road map" was developed. This project consists of five specific activities with those responsible for their implementation, persons and the expected usefulness for both parties: for students of universities of the EMERCOM of Russia and for the leadership of the ministry. In fact, the developed "road map" can form the basis of a management decision. The practical implementation of this project can bring the solution of the problem of employment of graduates of the EMERCOM of Russia as close as possible.

The research tasks set within the framework of the ongoing work have been completed in full.

References

1. S.A. Melkov, V.O. Mikryukov, A.Yu. Lyabah, *Dolzhen li vypusknik rossiiskogo vuza vladet analiticheskimi navykami? (na primere napravleniya "gosudarstvennoe i munitsipalnoe upravlenie")* [Should a graduate of a Russian university have analytical skills? (on the example of the direction "state and municipal administration")], in Applied scientific research and experimental developments based on the results of fundamental and exploratory research, Conf. Proc. Moscow, 06–07 December. 2016, 145-150. (Publishing house: Plekhanov Russian University of Economics, Moscow, 2017)
2. O.E. Logunova, *Primenenie prichinno-sledstvennoj diagrammy Ishikawy v reputacionnom menedzhmente* [Application of the Ishikawa causal diagram in reputation management], in Science problems, **1**. Proc. I International Scientific and Practical Conference "Scientific research: key problems of the III millennium" Moscow, 30 December, 2015, 54-56. (2015).
3. I.V. Gavrilova, et al., Fund Res, **8**, 247-251 (2015)
4. M. Voronin, N. Goryacheva, L. Kokhova, Mod High Tech. Reg App, **3(43)**, 48-56 (2015)
5. I.A. Sheromova, A.V. Yanchenko, Fund Res, **8(2)**, 464-469 (2017)
6. I.Z. Aronov, Meth Qual Manag, **8**, 18-20 (2012)
7. V. Ilenkova, Diagramma Isikavy: zachem primenyat i kak postroit [Ishikawa Diagram: Why to use and how to build]. Accessed on: October 4, 2021. [Online]. Available: https://pmclub.pro/articles/diagramma-isikavy-zachem-primenyat-i-kak-postroit
8. S. Ivanov, *Modeling company sales based on the use of SWOT analysis and Ishikawa charts*, in Selected papers of the 8th International conference on monitoring, modeling & management of emergent economy (M3E2-EEMLPEED 2019), 385-394 (2019)
9. S. Ivanov, SHS Web Conf, **65**, 04018 (2019). https://doi.org/10.1051/shsconf/20196504018
10. R. Cahyana, IOP Conf. Series: Mater Sci Eng, **434**, 012050 (2018). https://doi.org/10.1088/1757-899X/434/1/012050
11. N.I. Idris, *A Case Study of Coffee Sachets Production Defect Analysis Using Pareto Analysis, P-Control Chart and Ishikawa Diagram*, in M.S. Bahari, A. Harun, Z. Zainal Abidin, R. Hamidon, S. Zakaria (eds) Intelligent Manufacturing and Mechatronics.

Lecture Notes in Mechanical Engineering, 1295-1305 (2021). https://doi.org/10.1007/978-981-16-0866-7_115

12. F. Ikwan, et al. *Intelligent Risk Prediction of Storage Tank Leakage Using an Ishikawa Diagram with Probability and Impact Analysis*, in K. Arai, S. Kapoor, R. Bhatia (eds) Intelligent Systems and Applications. IntelliSys 2020. Advances in Intelligent Systems and Computing, **1252**, 604-616 (2021). https://doi.org/10.1007/978-3-030-55190-2_45
13. G. Ilie, N.C. Carmen, Manag Res Pract, **2(1)**, 1-20 (2010)
14. M. Suárez-Barraza, F. Rodríguez-González, Int J Qual Serv Sci, **11(2)**, 302-316 (2017). https://doi.org/10.1108/IJQSS-12-2017-0113
15. Federalnyi zakon "O gosudarstvennoi grazhdanskoi sluzhbe Rossiiskoi Federatsii" ot 27.07.2004 N 79-FZ [Federal Law of the Russian Federation "About the State Civil Service of the Russian Federation" 27.07.2004 No. 79-FZ]. Accessed on: October 4, 2021. [Online]. Available: http://www.consultant.ru/document/cons_doc_LAW_48601/
16. Federalnyi zakon "O sisteme gosudarstvennoi sluzhby Rossiiskoi Federatsii" ot 27.05.2003 N 58-FZ [Federal Law of the Russian Federation "About the civil service system of the Russian Federation" 27.05.2003 No. 58-FZ]. Accessed on: October 4, 2021. [Online]. Available: http://www.consultant.ru/document/cons_doc_LAW_42413/
17. Ukaz Prezidenta RF ot 01.02.2005 N 112 (red. ot 31.12.2020) "O konkurse na zameshchenie vakantnoi dolzhnosti gosudarstvennoi grazhdanskoi sluzhby Rossiiskoi Federatsii" [Decree of the President of the Russian Federation 01.02.2005 No. 112 (ed. 31.12.2020) "About the competition for the vacant position of the state civil service of the Russian Federation"]. Accessed on: October 4, 2021. [Online]. Available: http://www.consultant.ru/document/cons_doc_LAW_51515/
18. Prikaz MChS Rossii ot 05.04.2012 № 174 "O kvalifikatsionnykh trebovaniyakh k professionalnym znaniyam i navykam, neobkhodimym dlya ispolneniya dolzhnostnykh obyazannostei federalnymi gosudarstvennymi grazhdanskimi sluzhashchimi tsentralnogo apparata i territorialnykh organov Ministerstva Rossiiskoi Federatsii po delam grazhdanskoi oborony, chrezvychainym situatsiyam i likvidatsii posledstvii stikhiinykh bedstvii" (zaregistrirovan v Minyuste Rossii 14.05.2012 № 24140) [Order of the Ministry of Emergency Situations of Russia dated 05.04.2012 No. 174 "On qualification requirements for professional knowledge and skills Necessary for the Performance of Official Duties by Federal State Civil Servants of the Central Office and Territorial Bodies of the Ministry of Civil Defense, Emergencies and Elimination of Consequences of Natural Disasters" (registered with the Ministry of Justice of Russia on 14.05.2012 No. 24140)]. Accessed on: October 4, 2021. [Online]. Available: https://www.mchs.gov.ru/dokumenty/normativnye-pravovye-akty-mchs-rossii/608
19. Report on the employment of graduate students for 2017, 2018, 2019. Internal document of the Center (training and methodological) [text] // Civil Defence Academy of the EMERCOM of Russia, 2021
20. K. Ishikawa, What is Total Quality Control? The Japanese Way. (Prentice Hall, London, 1988)

Reserves of teaching staff's productivity growth based on lean technologies

Alena Alexandrovna Borisova[1], *Victoria Alexandrovna* Vinichenko[1,2*], *Ekaterina Vladimirovna* Spiridonova[1], and *Elena Nikolaevna* Kirichenko[3]

[1]Novosibirsk State Technical University, Management Department, Novosibirsk, Russia
[2]Siberian State University of Water Transport, Department of Transport Economics and Finance, Novosibirsk, Russia
[3]Novosibirsk State Technical University, Department of Foreign Languages, Novosibirsk, Russia

Abstract. The purpose of this study is to find and justify reserves for increasing the productivity of teachers on the example of the Novosibirsk State Technical University. Some of the methods presented in the paper are well-known and are of interest from a practical point of view. These include a survey, observation, experiment, and in-depth interview. Some methods are specific and are more often used by experts in the field of lean manufacturing. The key set of lean tools was the kaizen umbrella, which includes the philosophy of continuous improvement, the method of mapping the value stream, the "five why" method. Diagrams of various purposes are widely used in the article: Ishikawa, Pareto, spaghetti. The main problem solved in this paper is a large number of errors in syllabuses, a long time of their creation and excessive control. The hypothesis about the simultaneous elimination of these shortcomings by automating the process itself was tested. The experiment proved the validity of the hypothesis. The duration of the value creation process during the development of syllabuses was reduced by 15%, and the overall satisfaction of the participants in the process increased by 27.5%. The total savings of the working time fund amounted to 0.5 million man-hours across the entire university, accounting for almost 100 thousand syllabuses.
Keywords: lean manufacturing, labor productivity, syllabuses, automation of the process

1 Introduction

The steady interest of business in the search and improvement of management tools is explained by the need to ensure the competitive advantages of the company. The ways of increasing competitive advantages are differentiated and will be actively expanded: diversification [1], speed, focus orientation, technologies, innovations [2], as well as rational use of limited resources [3]. Methods of ensuring competitive advantages are implemented in two directions: external, aimed at finding the best business practices in

* Corresponding author: vika_06.07@mail.ru

comparison with competitors, and internal, involving the search for reserves within the company itself.

The search and updating of the company's internal reserves is implemented in the concept of rational business management and the introduction of lean technologies. The concept of lean technologies appeared due to the founders of practice-oriented regular management (A. Gasteva [4], P. Kerzhentseva [5], T. Ono [6] and was tested on empirical objects – complex production systems. Until now, manufacturing companies prevail in terms of the frequency of implementing lean technologies. Eliminating irrational losses and setting up a balanced system of regular improvements provide a significant increase in labor productivity. After manufacturing enterprises, trade organizations and institutions of the service sector (healthcare [7], transportation carriers [8] and education [9]) joined. Federal programs are being created to support companies that implement measures to update internal reserves, as well as associations that accumulate and promote the practice of regular process improvements.

In general, we can note an increase in the number of followers of the concept of lean thinking and the search for internal reserves of development. But the involvement of a larger number of Russian participants is slower compared to non-CIS countries [10]. Therefore, it is relevant to request proof of the effectiveness of the system of lean and rational organization of activities, as well as an assessment of the effectiveness of the transfer of lean tools from the production sector to others [11]. The availability of such evidence can contribute to the spread of the practice.

The purpose of this study is to test a set of lean production tools for a specific field-education – and to substantiate the possibility of improving (efficiency) the implementation of the current process based on the application of the concept of "lean thinking".

The choice of the object of observation – the organization of the educational process is due to several arguments:

firstly, the education sector is at the stage of active testing of changing formats for organizing activities and the educational process. This stage is resource-intensive and requires the search for internal resources to eliminate inefficient activities and operations;

secondly, regular changes in the field of education have led to an increase in the measures of resistance and protection of personnel, as well as to the underutilization of professional and qualification potential. It is possible to increase the index of trust and loyalty of personnel to the management system and the decisions initiated by it on the basis of evidence-based experiments of participants' benefits. Therefore, the implementation of a real project with the justification of the increase in improvements and leveling the problems of the labor subject contributes to reducing resistance to changes and increasing confidence in the management system;

thirdly, positive results can be replicated to a sufficiently large number of institutions, since the regulatory system is set up.

Therefore, it is necessary to test the technology and its instrumental support for educational institutions.

The achievement of the research goal indicated the fulfillment of a number of tasks:

- substantiation of the prerequisites for the need to regulate the implementation of the educational process on the example of a single element – development and updating of curriculums;
- identification and analysis of sources of process losses;
- selection and justification of a set of lean production tools necessary for the implementation of the improvement project;

- evaluation of the effectiveness of the implementation of the improvement project and justification of the impact of the used tools.

Such an organization of the research will allow: first to justify the need to regulate the current organization of the educational process and to digitalize the problems factually (to fix the state as it is); then to develop and implement an improvement project that includes a set of lean production tools and determine the necessary resource support for its implementation (experiment) and finally, to assess the effectiveness of the changes introduced (to characterize the state as it has become).

2 Materials and methods

The logic of achieving the set research goal indicated the choice of research methods. The experiment was chosen as the main method of proving and evaluating the effectiveness of the use of lean production technologies.

For the experiment, one element of the educational process was selected, that is the preparation and updating of curriculums by teachers. Criteria for selecting a process element are as follows: high labor intensity of work on the process and the interest of participants in reducing labor costs, as well as the possibility of introducing instrumental solutions into the process of regulating the process.

The subjects of observation were the direct participants of the process. Initially, the analysis of the studied process was carried out by the method of observation, questionnaires and in-depth interviews. The assessment of the types of losses identified during the experiment was carried out according to the traditional classification proposed by Tahiti Ono [12]. To process the received information material, both traditional methods of analysis and specific methods typical for the implementation of projects for the implementation of improvements using lean technologies were used. The selection criteria are related to the frequency and prevalence of tools in the business environment. The arsenal of tools includes mapping the flow of value creation, the method of highlighting problems and their analysis. The root causes were determined by constructing an Ishikawa diagram [13]. The Pareto diagram was used to select the root causes that affect the occurrence of the identified problems. Confirmation of the objectivity of the analysis was carried out by the "five why?" method, which was used in relation to the heads of departments and directly participants in the process.

After selecting the key problems, a pool of recommendations was formed to eliminate them. The expert assessment made it possible to exclude difficult to perform or difficult to achieve tools, focusing on those that can form a mechanism for increasing the productivity of the teaching staff in the foreseeable future. The chosen solution did not require significant investments and at the same time, it allowed to eliminate several types of losses at the same time.

Then a project group was formed and a roadmap for the implementation of the experiment was drawn up. Finally, after a while, the research team characterized and evaluated the results of the experiment.

3 Results

The key reasons that significantly determine the need to initiate this experiment were the low level of satisfaction with the speed and quality of the process of creating work programs of the discipline. The assessment of the level was carried out by interviewing participants. 175 people were interviewed, including heads of structural divisions performing supervisory functions in the process. The survey results are shown in Figure 1.

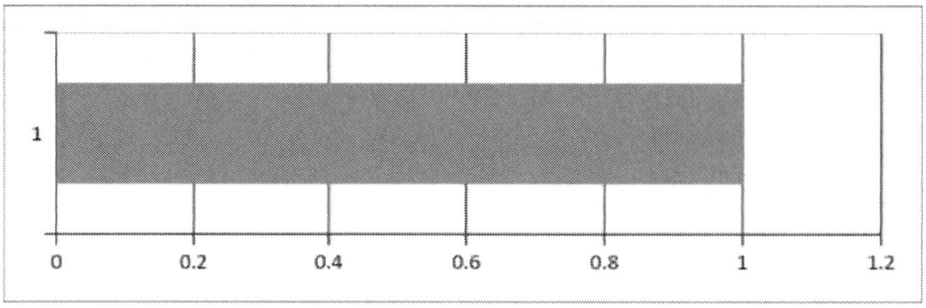

Fig. 1. The level of satisfaction of the participants in the process. *Source:* compiled by the authors.

The lowest level of satisfaction was revealed by the parameters "duration" and "organization of the process". The same conclusion was confirmed by the results obtained when constructing the Pareto diagram (Fig. 2)

Fig. 2. The Pareto diagram to identify the root causes of the poor quality of the process of creating work programs. *Source:* compiled by the authors.

Based on the analysis of the Pareto diagram, the most significant losses were a high level of expectation and frequent defects in the work. Focusing on the received losses allowed us to determine which of them are of the greatest interest from the point of view of optimizing the analyzed process (Table 1).

Table 1. Analysis and diagnosis of problems. *Source:* compiled by the authors.

Type of loss	The key reason	Direction of measures to eliminate
Overproduction	The difficulty in signing the title pages, as there are often no managers on the ground	Unification of the electronic template with the inability to correct and include information from the General description of the educational program
Stocks	A large number of errors and a frequent need for reprinting the Syllabus	
Unnecessary processing	The template can be changed by the performer, different versions automatically transform the document	
Expectation	There is no automatic notification of receiving a document on the table. Often you have to call, write and visit several times	Automatic pull-up of information from databases

Unnecessary movements	Duration of the process of physical movement of information	
Unnecessary transportation		
Some reject	The human factor	Creating internal control at all stages of syllabus development
Unrealized human potential	Excessive control at several stages leads to an irrational distribution of potential	

In most cases, there was a human factor, since the program template was easily changed the teacher working in it can happen to make spelling and punctuation errors. The hours periodically did not coincide with those specified in the curricula. The duration of the process was increased by additional control (often unnecessary). However, the further down the value stream the error was discovered, the more the process was delayed. Sometimes the error occurred already in the process of placing the agreed program on the site. Thus, it was necessary to completely go through the entire procedure again.

If we structure the key reasons that cause losses, we can talk about the existence of three key areas of countering these causes – automation of the process, unification of the template with the impossibility of changing it, and quality built into the process.

As a result of the experiment, it was decided to create a software product that allows you to automatically create syllabuses and build the necessary connections with the requirements of educational and professional standards.

The created system allowed to build the process in the personal account, access to which was provided by login and password in the electronic information educational environment of the teacher. The duration of the process of creating a syllabus was reduced by 227 minutes (from 1558 minutes), and the number of errors per syllabus was 1 (instead of 9 previously noted).

The process has been significantly simplified by embedding quality in it. The system did not allow the teacher to specify irrelevant data, since it synchronized several databases simultaneously. Thus, databases were created with up-to-date sources of educational, methodological and scientific literature (electronic library system), which compiled a list of 8279 titles. An additional database was created using software, which the teachers indicated in the work programs when describing the material and technical equipment of the educational process. The database on the teaching staff included about 1.5 thousand teachers. In total, the system made it possible to integrate syllabuses in more than 180 specialties (areas of training). Currently, the software contains about 100 thousand syllabuses only for full-time training.

A repeated survey of the participants in the process showed that the overall increase in satisfaction was almost 30% (Figure 3).

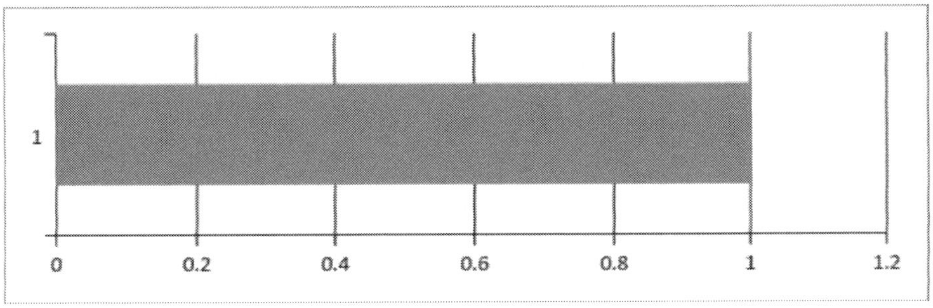

Fig. 3. The level of satisfaction of the participants in the process after the transformations. *Source:* compiled by the authors.

The total savings of the working time fund amounted to almost 0.5 million man-hours during the preparation of syllabuses for one period. The map of the value creation flow is not presented in this paper due to the limited amount of information available for placement. However, it should be noted that during the mapping process, the number of participants decreased from 7 to 3, and the total number of touch points of the process decreased by 37%. Control functions were integrated into the system, thus removing the burden of verification from the heads of departments, deans and educational management.

4 Discussion

The inclusion of software in the electronic educational environment allows you to strengthen several bottlenecks of the process at the same time. Facilitating the process of creating syllabuses in general increases the motivation of the teacher. The creation of a syllabus is considered routine work and teachers often boycott this process. The head of the structural unit has to shift this responsibility to a less qualified employee who does not deliver the discipline (for example, an assistant of the department), which leads to a formal approach to performing the task of methodological support of the discipline. Automation allows the teacher not to be distracted by technical issues and fill the program cells exclusively with meaningful content that is directly related to the discipline being taught. The embedded templates (for example, the type and form of classes) set the teacher a vector of the educational process that does not contradict the requirements of educational standards, which takes into account the interests of the modern labor market (for example, the content of competencies).

5 Conclusion

According to the lean production methodology, the key losses were identified within the framework of the project under consideration – some reject, expectation, unnecessary movements. The developed software product was focused on their minimization. The results obtained made it possible to significantly reduce the cycle time for preparing syllabuses and increase the satisfaction of the participants in the process. At the same time, lean production involves building a system of continuous improvements.

The first direction of improving the project is the context menu (pop-up windows with hints). The survey of the participants of the process both before and after the introduction of the software product revealed that the quality of support and service is one of the bottlenecks in the process of preparing syllabuses.

The second area of improvement is the spell checking module. One of the main losses in the process was some reject. The implemented program allowed to reduce the average number of errors in the program from 9 to 1. The main attention was paid to the moments that violate the restrictions of external (federal) and intra-university standards. Along with these fundamental deviations, spelling and punctuation errors were also detected, which are not yet controlled in the author's texts.

The third direction is timely interaction with customers. When building a value stream map, external users were identified as the main customers of syllabuses: students, applicants, employers. Employers are interested in ensuring that graduates master all the necessary competencies, receive the required knowledge and skills. In this regard, when preparing syllabus of specialized disciplines, a dialogue with the employer is extremely important. The remote format of interaction will reduce the losses when organizing this dialogue. The prepared (corrected) syllabus should be automatically placed on the site so that users can see its current version. Thus, the modules "examination by the employer" and "automatic upload to the site" are the third additional step.

The planned directions for improving the software product require additional resources, but it seems that they will be able to significantly increase the value of the product for the consumer.

The research was carried out with financial support within the framework of the NSTU development program, scientific project No. C21-20.

References

1. L.L. Klein, M.S. Tonetto, L.V. Avila, R. Moreira, J Cleaner Prod, **286**, 125386 (2021). https://doi.org/https://doi.org/10.1016/j.jclepro.2020.125386
2. C. Tasdemir, K. Gazo, J Cleaner Prod, **265**, 121759 (2020). https://doi.org/10.1016/j.jclepro.2020.1217590959-6526
3. K. Marx, Kapital. K kritike politicheskoi ekonomii. Pervyi tom. Protsess proizvodstva kapitala. Pervyi razdel. Tovary i dengi. Pervaya glava. Tovary [Capital. To the criticism of political economy. First volume. The process of production of capital. First section. Goods and money. First chapter. Goods] (Book laboratory, Lenizdat, 2018)
4. A.K. Gastev, Model Philosophical Encyclopedia. 3. V (Moscow, 1964)
5. A.J.H. Enthoven, J.V. Sokolov *Accounting Practices and Education in Russia and Other Republics of the C.I.S.*, in J.O. Burns, B.E. Needles (eds), Accounting Education for the 21st Century, Pergamon, 35-41 (1994). https://doi.org/10.1016/B978-0-08-042405-7.50010-1
6. R. Shah, P.T. Ward, J Oper Manag, **25(4)**, 785-805 (2007). https://doi.org/10.1016/j.jom.2007.01.019
7. D. Paul-Eric, P. Rafael, S. Cristiane., Ch.Jr Joao, Procedia Manuf, **51**, 1657-1664 (2020). https://doi.org/10.1016/j.promfg.2020.10.231
8. B. Villarreal, J.A. Garza-Reyes, V. Kumar, M.K. Lim, Int J Logist, **20(2)**, 163-180 (2017). https://doi.org/10.1080/13675567.2016.1170773
9. N. Sremcev, et al., Procedia Manuf, **17**, 595-602 (2018). https://doi.org/10.1016/j.promfg.2018.10.101
10. J.K. Liker, J.M. Morgan, Acad Manag Perspect, **20(2)**, 5-20 (2006)
11. S. Zargun, A. Al-Ashaab, Adv Mater Res, **845**, 668-681 (2013). https://doi.org/10.4028/www.scientific.net/AMR.845.668
12. M. Holweg, J Oper Manag, **25(2)**, 420-437 (2007). https://doi.org/10.1016/j.jom.2006.04.001
13. T. Varzakas, *HACCP and ISO22000: Risk Assessment in Conjunction with Other Food Safety Tools Such as FMEA, Ishikawa Diagrams and Pareto*, in B. Caballero, P.M. Finglas, F. Toldrá (eds), Encyclopedia of Food and Health, 295-302 (Academic Press, 2016). https://doi.org/10.1016/B978-0-12-384947-2.00320-2

Risk factors of the Volgograd labor market of university graduates

Inna Viktorovna Vasilenko[1], *Ekaterina Olegovna* Belikova[1*], and *Erkaiym Kozuyeva* Sharipova[2]

[1]Volgograd State University, Institute of History, International Relations and Social Technologies, Department of Sociology and Social Technologies, Volgograd, Russia
[2]Osh State University, Department of Public and Municipal Administration, Osh, Kyrgyzstan

Abstract. This article deals with the problem of risk accumulation in the regional labor market of university graduates as a specific socio-demographic group. Risks are investigated on the basis of a synthetic concept that takes into account the mutual influence of subjective (personal) and objective (contextual) factors. The article aimed to study the risk factors of subjective and objective nature in the regional labor market of university graduates. The analysis of the available scientific experience has shown that there is no holistic concept of risks in the labor market of graduates, although there are works that study professional risks in the context of employment, analyzing the situation of young people in the labor market. Risk factors were investigated based on four characteristics describing the risk society: the uncertainty of the situation, uncertainty of the environment, unpredictability of the development trend, and inadequacy of personal resources. The inadequacy of resources is the level of theoretical and practical preparedness, the correspondence of the performed functions to the received specialty, and the desire to change jobs. The information was collected using an online questionnaire. The analysis of each characteristic of the regional labor market showed the presence of enlarged risk factors influencing the decision to choose a job: lack of information and knowledge about the situation in the labor market and insufficient professional potential of graduates, who are motivated only by a prestigious, well-paid job, rather than by creativity and self-realization in their profession.
Keywords: risky nature, synthetic concept of risk, risky situation, labor market, difficulty of orientation, unpredictability of resources

1 Introduction

1.1 A Subsection Sample

The famous Russian sociologist Yanitsky claimed that Russia was a risk society [1]. Risks are created, spread, overlap, permeate the whole social system, are included in social life,

* Corresponding author: eo_belikova@volsu.ru

creating conditions for new risks. Risks interfere with the normal functioning of all subsystems and institutions; the control over them is gradually being lost. The labor market is no exception: risks emerge and constantly multiply here as well. The market system is the basis of economic and social development, so the risky nature of its functioning negatively affects all social actors, from the micro level – individuals, their families, organizations, ending with the macro level – institutions, including the state. This situation does not contribute to the development of the state and society but inhibits their positive dynamics. This article will focus not only on the risks in the labor market but also on the dangers in the labor market for graduates of higher education institutions, which are created both by objective conditions and by the young subjects of the market themselves. It should be noted that the nature of risks in this labor market is conditioned by the peculiarities of these subjects. For example, the risks in the labor market for graduates of the humanities will be different from those for graduates of technical specialties. Risks for young subjects who have not graduated from a higher education institution will be different from those for former students.

1.2 Analysis of the theoretical groundwork for the study of risks in the labor market of graduates and young people

To begin the analysis of the scientific literature on this or similar topics, one can consider the keywords present in this topic. Talking about the risks in the labor market of university graduates, it should be noted that this combination of key concepts is rarely found in publication titles. The research by Yekaterinburg sociologists is devoted to identifying the risks and threats that arise for young people in the modern regional labor market [2], such as unemployment [3].

The greatest threat in the regional labor market is the mismatch between places of employment and the qualifications and skills obtained. This not only leads to the loss of professional skills and competencies but also increases the probability of other risks in the labor market. There is scientific literature devoted to occupational risks, which in a certain way relates to this study as well [4-8]. Three main approaches to the study of risks were considered above: objective, subjective, and synthetic. The authors listed above are supporters of either the first or the second approach. Some authors consider the contradiction between the desire of young people for self-development and the dysfunctions of contemporary Russian society [5, 8]. At the same time, other authors pay more attention to the role of young people's professional behavior, professional choices that lead to risks [4, 6, 7].

The main part of scientific research is devoted directly to the issue of employment, without conceptualizing risks and their universality in Russian society [9, 10]. A number of works do not consider risks, but investigate the issues of the socially safe youth labor market and the impact of casual employment on it [9]. Some scientists consider the situation in the labor market as a crisis, due to the persistence of social stereotypes and non-compliance of the profession with the requirements of the labor market [10]. In the authors' opinion, these views refer to the production of risks in the labor market, but the labor market is not considered as part of the risk society, there is no analysis of the content side of the risks.

1.3 Research methodology

It is known that there are two consolidated approaches to the study of risks: objective and subjective. According to the first one, risks are presented as objectively existing dangers, threatening the loss of something valuable for a person and society. "Hazards are treated as independent variables, and people's reactions to them are treated as a dependent one" [11]. The second approach emphasizes the subjective side of risks, that is, their perception. Therefore, on the basis of this approach, the risk does not exist if one is not aware of it. Risks are constantly produced by society because they are consequences of social actions such as decision-making [12, 13]. In the authors' opinion, it is not possible to limit the study of risks to one side of reality, but it is necessary to consider both of them, especially since in everyday life, subjects' perception of the situation depends directly on the context, that is, the established everyday reality, which is perceived by subjects in different ways. It is not by chance that Giddens pointed out the existence of an ever-increasing dependence between everyday decisions of individuals and the transformation of structural elements of society, the consequence of which is the risks [14].

There is a very good classification of characteristics of this context of risk: unrecognizability, uncertainty, unpredictability, and inadequacy of personal resources [1]. These features allow describing the basic conditions of risk society functioning, in this case, it is the graduate labor market.

2 Materials and methods

The labor market includes not only those who look for work and provide vacancies but also those who work, but in the current conditions of instability and uncertainty of the labor market situation can lose their jobs. The article aims to describe the risk characteristics of the regional labor market of Russian regional university graduates and to fill them with the real empirical content of social facts.

Unrecognizability occurs when the usual course of life is disrupted, when stereotypes and social dispositions break down, and when individuals do not have the necessary information about the situation. In relation to work, this characteristic can be given meaning through graduates' attitudes toward their permanent (formal) employment, the identification of the ways and agents, by which job search was carried out, and the presence of professional continuity in the family [15].

Uncertainty is the most important property of the modern era. The risk in this situation is the result, including in the labor market [16]. Uncertainty as a basic feature of the risk society, as a context surrounding a graduate in the labor market, will be defined by the indicator – the main difficulties in finding a job.

Unpredictability occurs because most people may not be aware of most of the risks in most cases [17]. The unpredictability of the situation in the labor market for graduates in general and specifically for this individual was clarified by such indicators as the content of future plans and the characteristics of future work.

The inadequacy of resources makes it possible to define the presence of risks through the potentials of protection of the subject [18]. The potential is identified through the level of theoretical and practical competence, compliance of the functions performed with the received specialty, and the desire to change/not to change the specialty.

Based on the above, below is a methodological scheme of the situation of graduates of higher education institutions in the labor market and the risks they face in one of the regions of modern Russia (Volgograd and the Volgograd Region) (see Fig. 1):

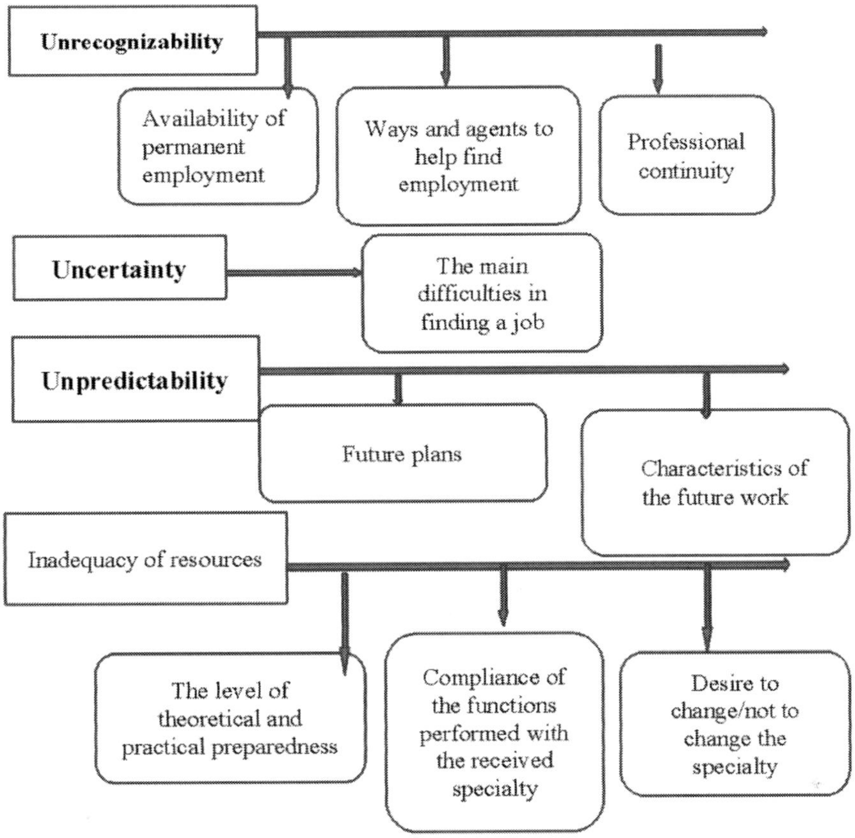

Fig. 1. Methodological model of the study of risk factors in the labor market for graduates of regional universities.

Based on this methodological model, a questionnaire was developed that included three blocks of questions. The purpose of the first block was to get information about the contradictions between the existing stereotypes and the real situation. The questions of the second block were aimed at finding out the uncertainty in the context of graduates' labor market and their awareness of this uncertainty. The third block included questions to investigate the unpredictability of the situation in the labor market and the respondents' knowledge about it. The fourth block of questions made it possible to determine the respondents' real and potential resources for protection in the labor market.

This questionnaire was used to conduct an online survey (the Google platform was used). The sample size was 376 people, and the survey was conducted in September-October 2020. The data obtained were processed using the SPSS Statistics 21 application package.

The objects of the study were residents of the city of Volgograd (n=200) and the Volgograd Region, in particular residents of the Kamyshinsky, Frolovsky, and

Mikhailovsky Districts (n=176). The number of male respondents was 43% and the number of female respondents was 57%. The age distribution of the respondents was as follows:

respondents aged 21–22 were 30.0%;

respondents aged 23–24 were 42%;

respondents aged 25 years and older were 28.0%.

The study involved graduates who had obtained social professions: economists (21.0%), lawyers (16.0%), managers (34.0%), sociologists and social workers (9.0%), and psychologists (20.0%), which allowed obtaining the opinion of representatives of different categories of former graduates of regional universities about the labor market and risks there.

3 Results

The first characteristic is unrecognizability. According to the results of the survey, 45.0% of graduates have permanent (formal) employment, and 55.0% do not have it. In this case, permanent employment denotes not those characteristics of work that existed before: stability, sustainability, and security, but simply a variant of formal, rather than informal, work. The data indicate that risks exist for both groups of graduates, as some do not have any protection from the actions of employers at all, and others, although they do, perhaps not for long. At this point, no protection from the situation and no change for the better is in sight for this group of graduates.

Unawareness of the situation in the labor market makes former graduates ask their acquaintances (41.0%) and relatives (9.0%) for help in finding a job. The other half got a job by sending out CVs – 26.0% or by choosing a job from a list of vacancies (21.0%). Three percent stayed on after the training practice. In other words, respondents reduce the level of unrecognizability of the situation at the expense of those who know it better, at least in some part. These are acquaintances, friends, and relatives, these are vacancies provided by employers in mass media and searching for workers with their help. At the same time, the first examined indicator shows that even after employment, the risk of the environment does not decrease.

The next indicator concerns the continuity of the profession received. It is no secret that a significant part of parents try to get their children into the sphere of activity in which they themselves work. The survey showed that the complete continuity of the profession was observed in 12.0% of the respondents, partial continuity, defined as a specialty close to the one their parents had – 26.0%. The rest had no such connection of professions. In other words, more than 60.0% did not find out the continuity of the profession. This means that for the majority of the respondents, the unrecognizability of the situation in the labor market is very high, and hence the riskiness as well.

The uncertainty of the situation in the labor market was determined by analyzing the main difficulties in finding a job. The study showed that the main difficulties in finding a job are: the lack of decent places in the labor market (high wages, interesting work, prospects of promotion) – 50.0% of respondents; lack of assistance of universities in employment – 18.0%, the lack of vacancies in the specialty – 18.0%, and the requirement of practical experience by employers – 16.0%. The uncertainty of the situation in the labor market often leads to inadequate requirements for the professional potential or makes graduates look for agents who can help them become oriented in the labor market – these are universities. The university has a big role to play as an organization that can help graduates enter the labor market according to their specialty and on good terms. Forty-four percent of the graduates counted on universities and considered such assistance necessary,

40.0% were less demanding, but assessed such assistance as rather necessary. Only 8.0% of respondents said they did not need it, and 8.0% found it difficult to answer. Thus, the majority of former graduates considered the assistance of the university in employment necessary.

The next feature characterizing the riskiness of the situation in the labor market is unpredictability. It was revealed with the help of the information about who from the nearest environment most influenced the formation of future plans, and about the criteria for evaluating the future place of work. The first aspect of unpredictability arises from the fact that it is impossible to foresee how the situation in the labor market will develop, although the image of a promising job exists in the minds of respondents. The second aspect concerns the personal potential of graduates and its adequate assessment. So, the main features of perspective work, which were highlighted by respondents, are: the reputation and fame of the organization – 64.0%, the stable position of the organization – 45.0%, the level of salaries – 37.0%, a favorable atmosphere in the team – 27.0%, the prospective development of the organization – 23.0%, decent social package – 22.0%, the availability of career prospects – 15.0%. Respondents are much less interested in: the duties performed – 12.0%, the possibility of professional development – 12.0%, the convenient location of the office – 8.0%.

Naturally, the respondents named more than one trait at a time, so it is very difficult to predict the situation in which there will be vacancies in organizations of this kind. However, graduates do not research labor markets and do not have accurate information about the development of the situation in it. At the same time, information is drawn from relatives and acquaintances, whose knowledge is also minimal or even non-existent. At this, it is known that the risks are reduced by the available information, but it should be small in volume and rich in content. When asked who influenced the formation of plans for the future, the respondents provided the following answers: 48.0% made plans for the future on their own, 30.0% were guided by the opinion of parents and other relatives, 8.0% took into account the opinion of friends and acquaintances, 8.0% listened to teachers, and 6.0% trusted the media. The data obtained indicate that the majority of respondents either make independent decisions about their future or are influenced by their parents, but nowhere in the answers does the employment service or data about the situation in the labor market in the region appear. The lack of accurate information undoubtedly increases the riskiness of the graduate labor market.

The last characteristic of the labor market riskiness is the inadequacy of the respondents' resources. The assessment of the situation in the labor market according to this characteristic was determined by indicators concerning the level of theoretical and practical preparedness (self-assessment), correspondence of the functions performed to the received qualification, and the desire to change the place of work.

The survey showed that 65.3% of the respondents evaluated their level of theoretical knowledge as average, 23.1% evaluated it as high, 9.6% evaluated it as low, and 2.0% found it difficult to answer. Besides, 46.2% of the respondents assessed their level of practical training as average, 30.7% assessed it as low, 19.2% assessed it as high, and 3.9% had difficulty answering. The findings suggest that graduates rate their level of professional training as average. In terms of risk, this means that, with average potential, graduates are also moderately protected against the risk of losing their jobs or the risk of performing uninteresting duties.

The next indicator of the inadequacy of resources for the modern labor market is aimed at identifying the correspondence of the functions performed to the received specialty. Only 9.1% of the graduates believe that their work fully corresponds to the received qualification. For 59.1% of the respondents, the job corresponds partially to the specialty. For more than a third of the respondents, it does not correspond to the specialty at all. Thus, the possibility to use professional potential at the average level of training leaves much to be desired. It means that the potential of protection of graduates of universities from a change of situation, from risks of loss of a job or positive results of its search, is utterly low.

Finally, the last indicator is a subjective assessment of one's resources related to the acquired specialty. This indicator shows that the situation is assessed and the respondent hopes/does not hope that he/she will be able to use his/her resources. The survey shows that 31.8% of respondents are quite satisfied with their specialty and would not want to change it, 13.6% of the respondents are less sure about it, but still would not want to change their specialty either. This group makes up 45.4%. Besides, 13.6% of respondents are sure they would like to change their specialty. Another 13.6% of respondents, though not so sure, but still would like to change their specialty. Together this group makes up 27.2%, i.e. about a third. Also, 9.1% found it difficult to answer the question (see Fig. 2). Thus, about one-third of respondents-graduates are aware of the risk of insufficiency of their professional potential in the modern labor market, related to the obtained specialty. However, there is an unconscious risk of inadequate assessment of their resources and additional expenses for getting another specialty.

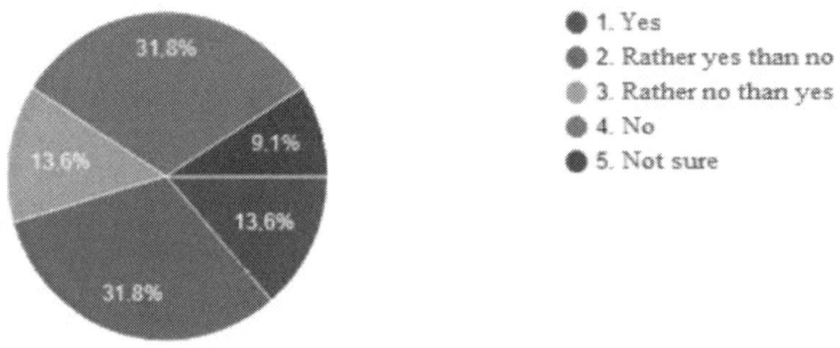

Fig. 2. Respondents' decision to change/not to change their specialty, %.

An important addition to the subjective assessment of the potential of one's specialty in the labor market is the respondents' disposition with respect to the specialty as the main professional path. There were no respondents with a positive, stable disposition. A positive, but not stable disposition, expressing that working in a specialty should rather be a professional path than not, was characteristic of 58.0% of the respondents. Six percent of respondents had a negative stable disposition, and 26.0% of them had negative, but not stable disposition (the specialty should rather not be the main professional path). In the dispersion of dispositions, the problem of resources and adequate estimation of them is traced, which is formed initially, still at the solution of a problem: what university to enter and what specialty it is better to receive and with what purpose.

4 Discussion

The most important problem in the context of the universality of risks is the riskiness of the youth labor market in general, and the labor market of graduates of higher education institutions, in particular. This problem is most acute at the regional level. Certain knowledge in this field has been formed in the scientific literature. Scientists studying regional labor markets identify, first of all, the threat of unemployment; non-compliance with labor legislation; work not in the specialty and "leaving" the profession, as well as salary risks; involvement in informal employment [9]. A number of authors consider the core of all risks to be professional risk, as a result of the non-compliance of the profession with the requirements of the modern labor market and the lack of necessary jobs [2, 7, 10]. These social facts are a consequence of the spread of risk factors in the labor market: the lack of information among young people about its condition, the low level of theoretical and practical competence of graduates, and the lack of clearly expressed motivation to obtain a specific specialty.

In this study, the riskiness of factors affecting the regional labor market was considered differently.

First, it was done based on the concept of the universality of risk in Russia.

Second, it was done with the help of the following characteristics: unrecognizability, uncertainty, unpredictability, the inadequacy of resources describing a risky situation.

Third, the research was guided by a system of indicators that describe each characteristic of the risk situation in the labor market.

Fourth, the fullness of these characteristics with real social facts of subjective and objective reality was analyzed.

Fifth, it was done identifying on this basis the range of risk factors of the regional labor market.

Of course, some results of this article echo the conclusions of the scientists listed above. For example, the one-sided motivation of graduates only for a prestigious job, without taking into account their own potential, which they themselves estimate as average.

It is necessary to continue further work on the study of risk factors in the regional labor market. It is planned to get information about the market situation from employers, parents, as well as graduates themselves, but also with the help of qualitative methods, if the coronavirus pandemic does not interfere with the collection of information.

5 Conclusion

Currently, in the regional market of graduates of higher education institutions who have received quite popular professions, risk factors of a subjective and objective nature are being formed and continue to be produced. Subjective risk factors are created by the results of graduates' decision-making in the conditions of limited information, poor knowledge of the market situation, and low professional potential. Objective factors are manifested in the risky social context surrounding the graduates and embodied in the current conjuncture of the labor market, parents' advice, friends' ideas, information and images transmitted by the media.

The situation in the regional labor market was described based on four characteristics of the risk society: unrecognizability, uncertainty, unpredictability, and inadequate resources.

The unrecognizability of the situation is formed due to the existence of formal and informal employment. According to this indicator, more than half of the respondents do not have formal employment. The risks within this characteristic increase due to the fact that some graduates cannot use acquaintances and personal relations during employment, which means they do not have information about their future work. Half of the respondents are such graduates. The lack of occupational and parental continuity for 60.0% of graduates also increases the risk of a lack of knowledge of the regional labor market situation for graduates.

The uncertainty of the situation was clarified with the help of an indicator of the presence of difficulties in finding employment. Risks are created in the labor market by the overestimated motivation of graduates for a prestigious job, the imbalance between the received specialty and vacancies in the labor market, and the desire to solve their employment problems at the expense of the university.

The unpredictability of the situation is manifested in the idealized image of the future place of work and for half of the respondents in the absence of "guides" in the labor market who can not only clarify the situation but also help with employment.

Finally, the inadequacy of resources is manifested in the average level of theoretical and practical readiness of graduates, indifference to the choice of specialty, employment not in the specialty, when the duties performed do not correspond to the profession received, and as a result, the desire to urgently change one specialty to another without a guarantee of a change in the situation.

Reference

1. O.N. Yanitsky, Sotsiologiya riska [Sociology of risk]. (LVS Publ, Moscow, 2003)
2. A.N. Tarasova, A.N. Chernyh, Bul Tyumen State Univ. Soc-Econ Leg Stud, **5(4(20))**, 52-75 (2019)
3. M. Choudhry, E. Marelli, M. Signorelli, Int J Manpower, **33(1)**, 76-95 (2016)
4. E.V. Kombarova, Central Rus Bul Soc Sci, **1**, 44-49 (2010)
5. E.V. Zinchenko, *Professionalnaya karera talantlivoi molodezhi: perspektivy i vozmozhnye riski* [Professional career of talented youth: prospects and possible risks], in A.V. Chernaya (ed), Materials of the All-Russian scientific and practical conference "Personality in culture and education: psychological support, development, socialization", Southern Federal University, 11-13 December, 2018, **6**, 51-56. (Foundation, Rostov-on-Don, 2018)
6. A.B. Fokina, News Higher Edu Inst. Soc. Econ. Pol, **4**, 88-90 (2014)
7. Yu.M. Pasovets, Innov Expert: Sci Works, **1(26)**, 166-176 (2019). https://doi.org/10.35264/1996-2274-2019-1-166-176
8. N.Yu. Priyatkina, Mod Prob Sci Edu, **4**, 118 (2019)
9. L.F. Gurbaeva, Forum. Ser: Human Econ Sci, **2(17)**, 192-196 (2019)
10. M.A. Gnatiuk, A.A. Horovinnikov, S.I. Samygin, Sci Notes, **2**, 157-163 (2018). https://doi.org/10.22394/2079-1690-2018-1-2-157-163
11. M. Douglas, Risk Acceptability According to the Social Sciences. (Russel Sage Foundation, New York, 1985). https://doi.org/10.4324/9780203708781
12. U. Beck, Risk Society. Toward a New Modernity. (SAGE, London, 1992)
13. U. Beck, Ecological Enlightenment. Essays on the Politics of the Risk Society. (Humanities Press, New Jersey, 1995)
14. A. Giddens, Living in a Post-Traditional Society, in Beck U., Giddens A., Lash S. Reflexive Modernization. Politics, Tradition and Aesthetics in Modern Social Order. (Stanford University Press, Stanford, 1994)
15. R. Sennet, The Corrosion of Character. (Norton, New York, 1998). https://doi.org/https://doi.org/10.5172/jmo.1999.5.2.53
16. Z. Bauman, Liquid Modernity. (Polity Press, 2000)

17. M. Douglas, A. Wildavsky, Rick and Culture: An Essay on the Selection of Technological and Environmental Dangers. (Univ. of California Press, Berkley and Los Angeles, 1982)
18. R. Reich, The Work of Nations. (Vintage Books, New York, 1991)

Professional objectives of higher education: imbalances and adaptation

Elena Konstantinovna Voronkova[1] and *Mariya Vladimirovna* Dolgova

Plekhanov Russian University of Economics, Finance and Prices Department, Moscow, Russia

Abstract. This article analyzes the systemic issues of personnel education in higher education, reveals their causes and consequences. Qualitative and quantitative characteristics of university graduates are one of the conditions for the state economic development. The goals and objectives of higher education should meet the needs of the state, industries, regions, and business. The lack of strong signals from consumers of educational services creates an imbalance between the higher education system and the market requirements and results in the conservatism of higher education. In this context, the purpose of this article is to develop a conceptual model for assessing the contribution of higher education to the national economy as one of the drivers of its transformation. The need to optimize the professional structure of education, with reference to the potential and prospects of the national economy, is emphasized. Research methods involve analysis and synthesis, induction and deduction, system and logical methods. The study result is a model that takes into account a number of indicators of employer satisfaction in university graduates, as well as indicators of regional and industrial security of specific professionals with higher education. An effective motivation for employers to take a more active part in the organization of higher education is the system of educational loans. The state's management of the credit sphere creates opportunities for consumers of educational services to direct financial flows primarily to education in priority specialties. The effectiveness of this tool requires ensuring its accessibility to employers and differentiating interest rates.
Keywords: higher education, professional education, financing

1 Introduction

Higher education (HE) is the climax of professional education of specialists, and it should correspond as closely as possible to national strategies. This is possible if the goal of managing the higher education system is to form appropriate schemes for the interaction of the provision and the consumption levels of educational services [1, 2].

Educational processes in society have a fundamental impact on the national economy [3]. There is a direct correlation between the qualitative and quantitative education characteristics and the indicators of the state economic development. The number of

[1] Corresponding author: Voronkova.EK@rea.ru

students and graduates, the structure of graduates by specialties and territories are among the indicators of such influence, etc. In turn, the effectiveness of higher education is strongly determined by state targets [4].

The number of students in the Russian higher education institutions is gradually decreasing. The number of students studying under contracts for the provision of paid educational services is increasing, mainly at the expense of state and municipal institutions of higher education, despite a significant increase in the cost of education in this group of institutions. State and municipal higher education institutions steadily remain on the top in terms of the number of students, but the number of students receiving education at the expense of budget allocations is decreasing. These trends were especially significant after 2010 (Table 1).

Table 1. Dynamics of indicators of the higher education development in the Russian Federation. *Source:* compiled by the authors based on the materials of the statistical digest "Education in numbers" [https://www.hse.ru/primarydata/oc/].

Indicator	Years					
	2000	2005	2010	2017	2018	2019
1. The rate of change in average consumer prices for education per term in higher education institutions:						
private	1.00	1.82	2.77	5.69	6.18	6.96
in state and municipal	1.00	2.28	3.63	8.20	9.71	10.62
2. The rate of change in the number of students (at the beginning of the academic year)	1.00	1.49	1.49	0.90	0.88	0.86
including at the expense of:						
budget allocations	1.00	1.07	0.94	0.68	0.68	0.68
under contracts for the provision of paid educational services in institutions, including	1.00	2.09	2.28	1.21	1.16	1.12
state and municipal	1.00	2.03	2.20	1.3	1.28	1.26
private	1.00	2.29	2.55	0.89	0.79	0.69
3. The share of students studying in state and municipal educational institutions, % including	90.29	84.72	82.96	90.04	90.88	91.84
under contracts for the provision of paid educational services	73.43	72.69	72.90	82.19	83.45	85.07
4. The share of students in state and municipal educational institutions studying under contracts for the provision of paid educational services, %	34.39	49.83	55.22	50.34	49.64	49.60

In recent years, the professional demand has gradually changed in favor of such technical specialties as engineering, technology, and technical sciences. At the same time, the number of graduates in economics and management is decreasing, although this area of education remains among the leading. The structure of professional education in higher education is shown in Fig. 1.

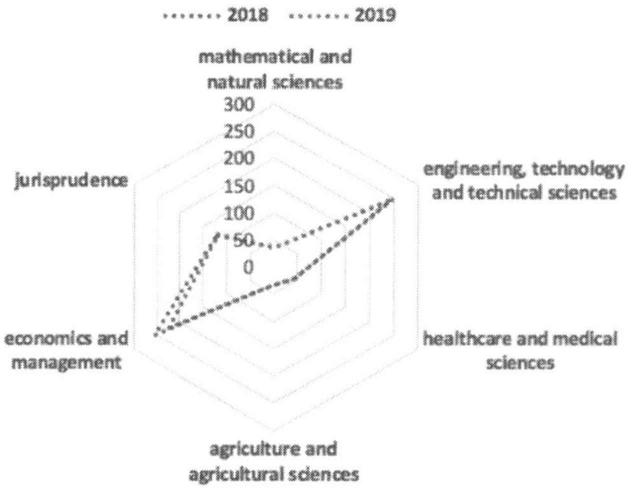

Fig. 1. Structure of professional education in higher education, the number of graduates, thousand people. *Source:* compiled by the authors based on the materials of the statistical digest "Education in numbers" [5].

2 Materials and method

The purpose of this work is the presentation of the author's view of promising models for assessing the contribution of the higher education to the national economy, as one of the incentives for its transformation. The research methodology is based on the principles of scientific objectivity and a systematic approach. It is based on general scientific methods, such as analysis and synthesis, induction and deduction, system and logical methods. This study uses official regulatory documents, statistical and analytical reviews to identify current trends in the higher education development, the features of its financing. Extensive factual database of state and international institutional structures has facilitated the reliability of the scientific provisions of this study.

3 Results

The current state of higher education in the Russian Federation represents some connection with the market requirements [6]. Thus, due to the development of the banking sector, trade, foreign trade, active integration processes, professions related to finance, commerce, and law became particularly popular at a certain historical stage. Today, there is an increased demand for technical professions. As was previously noted, the higher education institutions enhance the education of specialists in this area. At the same time, there is obvious conservatism and inertia in the organization of the higher education system according to the structure of professional personnel education.

Higher education should have such features as flexibility, the ability to quickly and adequately adapt to modern realities. The severe conditions of the Russian modern development require the creation and implementation of a higher education model, which would make it possible to "pinpoint" the imbalances in the interaction of higher education and the market [3]. At the same time, the higher education specialties should, first of all, attract by the demand for the market rather than by the unusual "names". This requires increasing the role of employers representing both the state and corporate levels. Their task is to form an understanding among consumers and sellers of educational services in the society interest in specific professions and specialties. Otherwise, the market of educational services will increasingly break away from the economy realities. It is fair to say, that having obvious difficulties with finding professionals in some areas, employers weakly give a warning of such problems to higher education.

Insufficient involvement of employers in the forming a professional structure in higher education appropriate for the market results in failures in education in certain professional areas. Already today, there is a shortage of some specialties due to the fact that universities have stopped such education due to a lack of applicants' interest. Thus, now we do not educate specialists in pricing, but they were greatly educated in the number of leading universities until the mid-90s. During the period of economic reforms, during the transition from a planned to a market economy, the point of view prevailed that only the market will determine prices and, therefore, education of these specialists is not relevant. Pricing began to be considered, first of all, as one of the marketing directions and the corresponding disciplines were excluded from the curricula of the vast majority of economic specialties, and education of specialists in pricing was terminated.

The consequences of this decision are currently a severe shortage of highly qualified specialists in state structures and companies who are able to comprehensively analyze the relationships in the economy at the macro, meso and micro levels, to identify the influence of various (at first glance, unrelated) factors on specific prices and pricing processes. Belated and ambiguous decisions related to keeping the prices for socially important products under control, the processes of forming and determining price dynamics – this result was particularly pronounced this year. Business needs pricing specialists who possess a wide range of competencies and skills, as evidenced by the data from the websites of recruiting companies. At the same time, the difference between the minimum and maximum salaries offered to "price makers" in the country is tenfold, in Moscow – fourfold, which indicates, among other things, a very different level of professionality of specialists employed in this field.

In recent years, a number of professional standards for pricing specialists have been developed and approved, as well as professional standards that include specific labor functions for pricing. Their introduction has resulted in an explosive increase in demand for seminars and advanced training courses on pricing, especially in construction and procurement. A similar pattern was after the entry into force in 2012 of changes in the Tax Code of the Russian Federation concerning transfer pricing.

A number of top universities have scientific schools and experienced personnel to educate specialists in pricing. But potential employers do not set specific tasks for higher education about the education of bachelors and masters of the appropriate profile, are not ready to accept students for industrial and pre-graduate internships, as well as conclude agreements on targeted financing or preferential lending to students. Thus, employers, experiencing the need for personnel in certain specialties, do not send clear requests to the higher education system for their education.

It is possible to eliminate the described imbalance and promote the activities of employers in the HE system through the financing system. The development of conditions and financing schemes for Russian higher education is constant [7]. However, the most important moment problem of the state is to optimize the processes of allocating financial resources in accordance with consumer demand, which is supported both by applicants and employers [8]. It is necessary to strengthen professional specifics more strongly when developing state programs of financial support for higher education. For example, a group of higher education institutions is financed by a "blend pool", not identifying critical points of expenditure of funds allocated by the state, even within the framework of the strategic Project 5-100, aimed at maximizing the competitive position of top Russian universities in the global market of educational services and research programs [9].

In our opinion, as priorities when making managerial decisions concerning the financing of higher education, it is necessary to take into account the long-term consequences of the higher education quality in the national economy. Its key guideline should be the degree of satisfaction of employers' interests detailed by profession: more than just economists, but specialists in prices or corporate finance, etc. [10]. At the same time, regional and sectoral monitoring of business satisfaction is particularly important for our country.

So here is a task to develop a model for assessing the contribution of higher education to the national economy, turning it into a key reference point when choosing a professional orientation at the higher education level [11, 12]. Traditional approaches to this assessment are based mainly on GDP growth rates and the level of wages. The latter indicator, as for the authors, quite logically reflects the degree of the economy interest in workers with higher education. But it is only if the ratio of wages of qualified and unskilled labor is appropriate [13, 14], taking into account the importance of a particular specialty for the national market. It is necessary to exclude such situations typical for modern times when the average salary of a driver in Saint Petersburg is a quarter higher than that of an engineer, which the domestic economy lacks, or 10% higher than that of a doctor [15].

In the integral assessment of the contribution of higher education to the economy, there should be indicators of the detailed level of employers' satisfaction in the needed professions in general, as well as indicators of regional and industrial security in higher education graduates in the professional context.

$$K_{HEC} = \{B1, B2, B3\}, \tag{1}$$

where K_{HEC} – integral indicator of the contribution of higher education to the national economy;
$B1,2,\dots\dots n$ – relevant indicators that characterize the degree of satisfaction of employers, regions and industries in specialists with higher education by profession. If necessary, each element of the formula can be argued depending on the significance of the indicator, which may change with changes in the state of the economy.

4 Discussion

Finally, the system of educational credit should become a real motivation for employers to pay more attention to the organization of educational processes in higher education. The state manages this sphere, allowing consumers of HE services to realize their desires, thus directing financial flows to universities. An educational loan should be considered as a fine-tuning tool that adapts to the market situation, primarily the domestic market. There is a need for this in the Russian Federation. For example, the step to reduce interest rates on educational loans in 2020 was very timely. However, it is necessary to further promote the dynamics of educational credit, including in the context of the problem of attracting employers to renovate the scheme of higher professional education.

Educational loans make it possible to find the right way to structure the system of higher professional education. For this purpose, firstly, it is possible to differentiate the interest rates on educational loans depending on the specific specialization of higher education, depending on the requests of employers, which will increase the motivation of the entire "employer-university-applicant" bundle. Secondly, it is necessary to manage the availability of employers to an educational loan in order to provide financial support for the future specialists of the company, now students, paying for their education and providing regular living conditions. Then universities, receiving additional funding from employers for education in certain specialties, will be more willing to switch to educating students "on request". The differentiation principle of interest rates on an educational loan can also work in this case, for example, depending on the complexity of the educational process or the level of the profession scarcity.

It should be noted that the practice of financing education, guaranteeing a workplace and a certain level of income after graduation by the future employer of a student in the higher education system should become wider. This is extremely important, especially for emerging markets. The analysis of the relevant indicators on the EAEU market shows the direct dependence of personal expenses for higher education on the real disposable income of the population in a vast majority of the EAEU member states [16]. Otherwise, if the disposable real income of the population falls, the plans for educating specialists that employers rely on may be disrupted [17, 18].

5 Conclusion

Thus, the study shows that the balance of interests of all participants in the process of educating specialists in the Russian higher education system can be ensured only with an increase in the role of employers, professional adaptation of educating to territorial and industrial peculiarities, and management decisions on the scale and cost of paid education should become more flexible from the position of social orientation.

Renovation of approaches to the formation of the professional structure of personnel education in the higher education system, involving the strengthening of the employer role, will increase both the competitiveness of the higher education system itself and the national economy as a whole, the efficiency of public and private spending on higher education, the degree of satisfaction of needs in the education market.

References

1. F.J. Altbach, Global perspectives of higher education, A. Ryabova (ed) (Ed. house of the Higher school. Economics, Moscow, 2018). https://doi.org/10.17323/978-5-7598-1712-3
2. H. Hall, Econ Law, **17(1)**, 33–42 (2018). https://doi.org/10.12775/EiP.2018.003
3. A. Hamdan, A. Sarea, R. Khamis, M. Anasweh, Heliyon, **6**, e04046 (2020). https://doi.org/10.1016/j.heliyon.2020.e04046
4. E.E. Lehman, M. Meoli, S. Paleari et al., Eurasian Bus Rev, **8(1)**, 37–54 (2017). https://doi.org/10.1007/s40821-017-0091-7
5. Education in Figures. Accessed on: September 25, 2021. [Online]. Available: https://www.hse.ru/en/primarydata/oc/
6. L.V. Ávila, et al., J Cleaner Prod, **164**, 1268-1278 (2017). https://doi.org/10.1016/j.jclepro.2017.07.025
7. T.N. Sedash, E.A. Kameneva, Rus Foreign Econ J, **5**, 50-58 (2021). https://doi.org/10.24411/2072-8042-2021-5-50-58

8. O.P. Ovchinnikova, N.E. Ovchinnikova, Fin Credit, **23(38)**, 2305–2316 (2017). https://doi.org/10.24891/fc.23.38.2305
9. Project 5-100. Accessed on: September 25, 2021. [Online]. Available: https://5top100.ru/documents/regulations/97332/
10. E.K. Voronkova, M.V. Dolgova, Edu. Sci. Sci Pers, **4**, 197-200 (2020). https://doi.org/10.24411/2073-3305-2020-10239
11. E.Ya. Kogan, N.Yu. Postalyuk, T.G. Kuteinitsyna, Higher Edu Rus, **28(7)**, 9–18 (2019). https://doi.org/10.31992/0869-3617-2019-28-7-9-18
12. B. Stensaker, J.J. Lee, G. Rhoades, S. Ghosh, et al., J Higher Edu, **90(4)**, 539–562 (2019). https://doi.org/10.1080/002215 46.2018.1513306
13. B.S. Potanin, Stud Rus Econ Devel, **30(3)**, 319-326 (2019). https://doi.org/10.1134/S1075700719030146
14. E.V. Romanov, The Edu Sci J, **21(9)**, 9-48 (2019). https://doi.org/10.17853/1994-5639-2019-9-9-48
15. Statistika zarplat v Rossii [Salary statistics in Russia]. https://gorodrabot.ru/salary?l=%D1%80%D0%BE%D1%81%D1%81%D0%B8%D1%8F
16. Obrazovanie v tsifrakh [Education in Figures]. http://www.eurasiancommission.org/ru/act/integr_i_makroec/dep_stat/econstat/Documents/Education%20Yearbook/EducationEAEU2020.pdf
17. X. Sheng, British J Soc Edu, **38(5)**, 721-737 (2017). https://doi.org/10.1080/01425692.2016.1158638
18. M. Shnarbekova, Rus Edu Soc, **60(4)**, 70-380 (2018). https://doi.org/10.1080/10609393.2018.1473700

Transformation of legal education in modern conditions

Natalia Nikolaevna Bocharova[1*], *Nikolai Viktorovich* Veretennikov[2]

[1]Southern Federal University, Department of Financial Law, Rostov-on-Don, Russia
[2]Southern Federal University, Department of Criminal Procedure and Criminalistics, Rostov-on-Don, Russia

Abstract. The article examines the main trends in education, including legal education. The authors studied the Federal Law "On Education", educational standards, and a number of industry laws; identified the positive and negative sides of digitalization of education, and also separately studied the impact of distance learning in the conditions of coronavirus restrictions as a side of digitalization. It is concluded that the transition to a new paradigm of education corresponds to the constitutional values of Russia. The competency-based approach creates prerequisites for the creative development of a person based on his/her activity and involvement in the learning process. Digitalization in education makes it possible to optimize and improve a number of educational procedures; however, as the authors showed in the study, it also has negative sides in legal education. The distance learning mode as part of digitalization also received a mixed assessment. The authors concluded that universities should carefully comprehend the experience of digitalization and distance learning in the conditions of coronavirus restrictions in order to eliminate problems and improve the educational process.
Keywords: education, competencies, educational standards, digitalization, distance learning

1 Introduction

The processes of transformation today have affected many spheres of the life of the state and society, including education. It was impossible to build a new statehood in the post-Soviet space without changing the basic paradigm of education, which was developed through the Federal Law "On Education" and educational standards [1]. The implementation of the educational concept is also influenced by digitalization, which was most clearly manifested in the conditions of coronavirus restrictions in 2020–2021. Studies conducted on this topic do not give an unambiguous assessment of these processes: Babinova notes the prerequisites for activating the creative development of the individual; experts of the Higher School of Economics talk about innovative tasks of education; Klochkova introduces the concept of "digital competencies" that should be formed in the

* Corresponding author: nbocharova@sfedu.ru

modern educational environment. In the presented article, the authors tried to critically comprehend the impact of the main trends on legal education, highlighting the positive and negative aspects.

2 Materials and methods

In the presented article, the hypothesis is formulated that modern education is influenced by several factors. Based on this, the goal was set to study the transformation of education in the current conditions through the example of legal education. To achieve this goal, it is necessary to solve a number of tasks: to study the prerequisites and implementation of the competence approach in legal education; to analyze the main trends of digitalization in education and the impact of the period of "coronavirus" restrictions on this process. The main method of research was the formal-logical one – the study of the Federal Law "On Education", as well as federal state educational standards. The authors also relied on the experience of teaching.

3 Results

Today, higher education is influenced by three main trends: the competence approach, the introduction of information technologies (digitalization), and coronavirus restrictions. Let us take a closer look at each of these areas.

In accordance with the Federal Law "On Education", participants in educational relations are students, parents (legal representatives) of underage students, teaching staff and their representatives, and organizations engaged in educational activities. The interaction between these participants is based on the principle of subject-subject relations (S → S).

Subject-subject relations (S → S) are a type of relations that develops in the educational process, consisting in creating parity participation of students and teachers in the organization and implementation of joint activities [2]. Subject-subject relations imply the identification of the student as a subject, the recognition of the key value of the entire educational process, the development of a student's abilities based on individual capabilities as the main goal of education [3], in contrast to the Soviet education system built on the subject-object system of relations (S → O), where the teacher is the subject, that is, an active person, and the student is the object of the teacher's influence. In fact, learning in subject-subject relations in education is a purposeful two-way process of transferring and assimilating knowledge, skills, and skills carried out by both the teacher and the student.

In this context, the experience of Finland is interesting, where education is focused on the student. According to the research conducted within the framework of the International Program for the Assessment of Educational Achievements of Students (PISA), Finland has been showing very good results over the past ten years, and Finnish schoolchildren are at the top of international rankings. Finland is one of the rare countries where a high level of student achievement is combined with life satisfaction (PISA 2018). Finnish schoolchildren feel more comfortable and study more effectively compared to other European children [4].

This educational paradigm is primarily focused on the development of the individual, increasing his/her activity and creative abilities. It also correlates with one of the main constitutional values enshrined in the Constitution of the Russian Federation [5] – civil society. In the works of modern researchers, civil society appears as a system of private and interpersonal relations of a social, political, ideological, cultural, religious, family, or other nature, directing people's activities to ensure conditions for the comprehensive development of the individual [6]. For this purpose – to activate the creative and psychological potential of a student – the use of methods of students' independent work, self-control, active and interactive teaching methods is increasing.

This trend has been consolidated and developed in educational standards (federal and specific universities). For example, one of the universal competencies of a bachelor in the field of training 40.03.01 Jurisprudence is the ability to manage one's time, build and implement a trajectory of self-development based on the principles of lifelong education [7]. The following universal competency is fixed in the Federal State Educational Institution in the direction of training 40.04.01 Jurisprudence: the ability to determine and implement the priorities of one's own activity and ways to improve it on the basis of self-assessment [8]. At both levels of training (bachelor's and master's degree), these competencies are located in the section "Self-organization and self-development (including health care)", which confirms the conclusion about the orientation of the modern education system, including legal education, to train specialists with a high level of creative potential development, who are able to systematically solve various tasks, and shows the continuity of tasks between levels of training during their development.

Other universal competencies are also aimed at solving the same problem at the Federal State Educational Institution: bachelor's degree – the ability to carry out social interaction and realize one's role in a team (teamwork and leadership); the ability to determine the range of tasks within the set goal and choose the best ways to solve them, based on existing legal norms, available resources and restrictions (project development and implementation) [7]; master's degree – the ability to organize and manage the work of the team, developing a team strategy to achieve the goal (teamwork and leadership); the ability to manage the project at all stages of its life cycle (development and implementation of projects) [8].

Here it is impossible not to mention another interesting trend in education, which is that in accordance with the Federal Law "On Education", as well as within the framework of the concept of the "Bologna" education system, there is no requirement for the coincidence of bachelor's and master's degree courses. On the one hand, there is a plus in this – the student can independently form his/her own trajectory of professional development. On the other hand, there are a number of professions that involve the assimilation of basic knowledge in the bachelor's degree and their consolidation and development at the master's level in one specialty. For example, prosecutors may be citizens of the Russian Federation who have received a higher legal education in the specialty "Jurisprudence", or a higher education in the field of training "Jurisprudence" with the qualification "master" in the presence of a bachelor's degree in the field of training "Jurisprudence", or higher education in the specialties included in the enlarged group of specialties "Jurisprudence", with the qualification "lawyer", having the necessary professional and moral qualities, capable of performing the official duties assigned to them for health reasons (Article 40.1) [9]. The same thing is seen in the Law of the Russian Federation "On the status of judges in the Russian Federation": a judge can be a citizen of the Russian Federation who has a higher legal education in the specialty "Jurisprudence" or a higher education in the field of training "Jurisprudence" of the qualification (degree) "master" if he/she has a bachelor's degree in the field of training "Jurisprudence" (Article 4) [10]. These legislative specifications can be considered fair and justified.

It is also impossible not to say about the impact on legal education of one of the main trends of the present time – digitalization. Digitalization in education is the introduction of information technologies into the educational process for the transition to a personalized and result-oriented educational process [11]. Not everyone positively assesses this process, believing that this is nothing more than another "fashionable topic" that will pass, while the "eternal values" of education as the most stable public institution will remain the same. However, the authors believe that it is useless to deny the obvious things – digitalization has already become part of the education system, which is also confirmed by the state: the National Project "Education" notes that "accessibility of all types of education is impossible without creating a modern digital environment" [12]. Accordingly, the introduction of this process into education poses a number of tasks for it – not only material and technical support but also the development of new educational technologies, as well as a system for monitoring training and improving the digital literacy of all parties to this process. However, it is worth noting a trend among teachers – not everyone is ready to actively participate in this process, and in a number of areas, there is an outflow of teachers from educational institutions [13].

The driver of digitalization in education was 2020, when, under the conditions of coronavirus restrictions, education switched to a distance learning mode. However, it is important to understand here that the digitalization of the educational environment and the distance learning mode are not the same things: digitalization is a broader concept that includes some kind of automation of educational processes; the distance mode has become more of a part of digitalization and at many universities has been reduced to the possibility of conducting classes in video conferencing mode.

Some authors believe that the introduction of artificial intelligence as the main trend of digitalization will deprive lawyers of work [14]. Indeed, in a number of countries, one can observe such an experience: in China, artificial intelligence makes decisions when considering simple cases. However, the authors do not share this point of view and believe that the legal profession is very individualized and requires a thoughtful and professional approach.

4 Discussion

Without detracting from the positive features of both digitalization and the distance mode for legal education, one nevertheless should note a number of aspects of legal education, where it is important to maintain the mode of personal interaction between the teacher and law students:

a) an important direction of training a high-level specialist at the Faculty of Law is the expansion of cultural horizons, the formation of a personal worldview position, the development of students' legal awareness, a value-normative system with a predominance of socially significant values and priorities. In this regard, one can mention an interesting study that was conducted in 2013 by the Laboratory of Socio-Legal Research and Comparative Law "New educational standards in the assessments of students of the Faculty of Law of the Lomonosov Moscow State University". These studies have shown that for students of the Faculty of Law of Moscow State University, it is necessary to introduce such a lecture course as psychology into the curriculum [15]. A lawyer should be able to communicate with people, establish psychological contact, analyze motivation, etc. Mastering the skills of communicative competence, public speaking, self-presentation, knowledge and understanding of people cannot be taught online. For example, at Harvard Law School, there are courses and training based on experience, involving personal intensive participation [16].

b) the same can be said about teaching one of the basic disciplines – criminology: in accordance with the Federal State Educational Standard, the minimum list of material and technical support necessary for the implementation of the bachelor's program includes a specialized classroom equipped for conducting classes in criminology. The most obvious set of tools in this regard is the so-called "investigative briefcase": for future lawyers, classes on the use of fingerprinting techniques are of the greatest interest, which allows students to come into contact with practical activities, feel like specialists, experts.

Another serious problem of the distance learning mode has been the decline in the quality of education: not all teachers are ready to switch to a new learning mode that does not involve feedback from the student; students, for their part, are not always ready to take greater responsibility for learning outcomes, quick access to information generates a false sense of competence. In continuation of this thought, it is worth evaluating the value of the system of point-rating evaluation of students' work: on the one hand, this system has become a powerful stimulating factor – the student can see the results of his/her work in real time; on the other hand, students are guided not by obtaining new knowledge, but by obtaining an appropriate number of points, the achievement of the established limit of which contributes to the loss of interest in learning.

Thus, it can be stated that the scientific and technological progress that has taken place, as everywhere else – in all spheres of activity, makes it easier to work, leads to the need to rethink the ongoing processes and make the necessary decisions.

5 Conclusion

As a result of the conducted research, the authors conclude that legal education in modern conditions is changing, responding to the realities and challenges of the time: the training of high-quality lawyers should be based on a deep understanding of the specifics of this profession, which is reflected in the relevant educational standards and federal legislation. At the same time, it is impossible to deny the impact of digitalization on the educational process and the legal profession, which should be taken into account by universities and implemented in the process of preparing for the legal profession.

References

1. Federalnyi Zakon Ob obrazovanii v Rossiiskoi Federatsii [Federal Law" On Education in the Russian Federation"] Accessed on: October 4, 2021. [Online]. Available: http://pravo.gov.ru/proxy/ips/?docbody=&nd=102162745
2. N.V. Babinova, Pedag Edu Rus, **4**, 45-50 (2015)
3. V.S. Bezrukova, Osnovy dukhovnoi kultury (entsiklopedicheskii slovar pedagoga) [Fundamentals of spiritual culture (an encyclopedic dictionary of a teacher)]. (Ekaterinburg, 2000)
4. S. Eriksson, Obuchenie, orientirovannoe na uchenika. Na primere finskoi sistemy obrazovaniya [Student-centered learning. On the example of the Finnish education system]. Accessed on: October 4, 2021. [Online]. Available: https://education.forbes.ru/authors/obuchenie-orientirovannoe-na-uchenika
5. Konstitutsiya Rossiiskoi Federatsii [The Constitution of the Russian Federation]. Accessed on: October 4, 2021. [Online]. Available: http://www.kremlin.ru/acts/constitution
6. K.A. Strus, E.R. Arakelyan, Quest Mod Jurisp, **27**, 136-139 (2013)

7. Federalnyi gosudarstvennyi obrazovatelnyi standart vysshego obrazovaniya – stepen bakalavra po napravleniyu podgotovki 40.03.01 Yurisprudentsiya [Federal State educational standard of Higher Education-Bachelor's degree in the field of training 40.03.01 Jurisprudence"]. Accessed on: October 4, 2021. [Online]. Available: http://fgosvo.ru/uploadfiles/FGOS%20VO%203++/Bak/400301_B_3_07092020.pdf
8. Federalnyi gosudarstvennyi obrazovatelnyi standart vysshego obrazovaniya – magistratura po napravleniyu podgotovki 40.04.01 Yurisprudentsiya [Federal State educational standard of higher education-Master's degree in the direction of training 40.04.01 Jurisprudence"]. Accessed on: October 4, 2021. [Online]. Available: http://fgosvo.ru/uploadfiles/FGOS%20VO%203++/Mag/400401_M_3_11032021.pdf
9. Federalnyi Zakon O prokurature Rossiiskoi Federatsii [Federal Law "On the Prosecutor's Office of the Russian Federation"]. Accessed on: October 4, 2021. [Online]. Available: http://pravo.gov.ru/proxy/ips/?docbody=&nd=102014157
10. Zakon Rossiiskoi Federatsii O statuse sudei v Rossiiskoi Federatsii [The Law of the Russian Federation "On the status of judges in the Russian Federation"]. Accessed on: October 4, 2021. [Online]. Available: http://pravo.gov.ru/proxy/ips/?docbody=&nd=102017065
11. Sem zadach tsifrovizatsii rossiiskogo obrazovaniya [Seven tasks of digitalization of Russian education]. Accessed on: October 4, 2021. [Online]. Available: https://trends.rbc.ru/trends/education/5d9ccba49a7947d5591e93ee
12. Opublikovan pasport natsionalnogo proekta "Obrazovanie" [The passport of the national project "Education" has been published]. Accessed on: October 4, 2021. [Online]. Available: http://government.ru/info/35566/
13. M. Shirrell, Amer J Edu, **127(3)** (2021). https://doi.org/10.1086/713828
14. S. Egorkin, Ischeznovenie yuridicheskoi professii. Mysli posle prokhozhdeniya garvardskogo kursa po iskusstvennomu intellektu [The disappearance of the legal profession. Thoughts after taking a Harvard course in artificial intelligence]. Accessed on: October 4, 2021. [Online]. Available: https://zakon.ru/blog/2020/06/09/ischeznovenie_yuridicheskoj_professii_mysli_posle_prohozhdeniya_garvardskogo_kursa_po_iskusstvennomu
15. A.V. Klochkova, A.A. Sharonova, Bul Moscow Univ. Ser. 11. Law, **1**, 43-52 (2014)
16. D. Levina, Yuridicheskoe obrazovanie v SShA. Lichnyi opyt studentki iz Rossii [Legal education in the USA]. Accessed on: October 4, 2021. [Online]. Available: https://education.forbes.ru/authors/yuridicheskoe-obrazovanie-usa

Digitalization in the system of methods of scientific knowledge

Anatoly Vasilyevich Denikin[1,*], *Aleksandr Valentinovich* Sablukov[2], *Viacheslav Leonidovich* Primakov[2], *Valery Alexandrovich* Lapshov[2] and *Pirmagomed Shikhmagomedovich* Shikhgafizov[3]

[1]Financial University under the Government of the Russian Federation, Department of Humanities, Moscow, Russia
[2]Moscow State Linguistic University, The Department of Social Sciences at the Institute of International Relations and Social and Political Sciences, Moscow, Russia
[3]Financial University under the Government of the Russian Federation, Department of Sociology, Moscow, Russia

Abstract. The study aims to clarify the methodological status of digitalization in the framework of the philosophical theory of knowledge. The subject of the study: explication of the instrumental and methodological nature of digitalization. The problem is solved by employing the categorical and conceptual apparatus of philosophy of science, particularly in paradigm analysis. A comparative analysis of the concepts of "information" and "digitalization" is carried out. Information within the framework of classical scientific rationality coincides with the phenomenon of absolute-relative knowledge. In non-classical rationality, it represents a movement of meaning and expresses the value-valued nature of information technology. In postnonclassical science, the phenomenon of information is associated with intrasystemic communications. Digitalization as a mode of data transmission is incorporated into the deductive and inductive logic of classical science and leads to an essential limitation of the source base. In modern science, data selection entails the creation of systematic factual meaning, and digitalization coincides with analytical activity. In addition to general logical functions, digitalization points to new perspectives on such types of systems methodology as modeling and design. In relation to the information as a methodology, digitalization acts as a meta-methodology. That is, the digital world is defined as an observing reality. In the framework of digitalization, information is identical to knowledge in its postnonclassical meaning. In digitalization, the process of cognition is incorporated into reality itself, forming an intersubjective field of meaning. In this context, digitalization serves as an inter-paradigmatic method of scientific knowledge. Digitalization refers to the rational side of cognition and is a type of systematic methodology and general logical methods.

* Corresponding author: andenikin@yandex.ru

Keywords: information, digitalization, information data, data selection, information technology, digital world

1 Introduction

The topic of the methodological status of digitalization is very relevant for philosophical, sociological, and interdisciplinary analysis [1, 2] and has significant epistemological and practical significance. It is difficult to deny that modern technical and technological innovations need to be understood from the perspective of social knowledge (discussions about the information society, the knowledge society) [3-11] and from the perspective of the philosophical theory of knowledge [12-15]. It is known that scientific knowledge is a developing system of knowledge, including empirical, theoretical, and metatheoretical levels. Collection of facts, their primary generalization, description of observed and experimental data, systematization, and classification are essential features of empirical knowledge. The theoretical level of scientific cognition is characterized by the predominance of the rational moment – concepts, theories, laws, and other forms of thought; the metatheoretical level serves to explain the lower levels. There are "eternal" methodological constants that form the substratum of empirical and theoretical cognition, but in the course of the evolution of modern science, the forms of their expression are transformed.

The initial hypothesis is that digitalization refers to the rational side of cognition and is a type of theoretical and metatheoretical methodology. To understand the logic of this transformation, let us compare the concepts of "information" and "digitalization". In our view, within the framework of classical scientific rationality, information is understood as the process of producing, storing, and transmitting knowledge in order to approach the absolute-relative truth.

Another concept of information is associated with a non-classical type of rationality, the desire for a consensus of heterogeneous discourses. Information is the process of producing, storing, and transmitting meanings. Unlike the phenomenon of knowledge, the meaning has a subjective content and expresses the value-evaluative nature of information technologies. Information becomes a cognitive and practical resource of the subject if communication closes between sender and receiver and meaning is formed. In other words, information is symbolic communication.

In postnonclassical science, the interpretation of information as an intra-system communication is actualized, affecting the degree of system self-organization with decreasing or increasing chaos. It seems that this understanding expresses the need to engage a paradigmatic approach to the methodological tools of epistemology. Broadly speaking, information technology contributes to an effective functional state of social being. At the same time, the scientific literature is increasingly asserting the idea that social life is already saturated with information tools and information products, that the informatization of social space has already taken place. Based on these perceptions, we can talk about the completion of the informatization phase [16].

Today's world is transitioning to a new technological state: digitalization. In a substrate sense, the basic technologies, the carriers of digitalization, are the Internet and mobile communications. Currently, the sociological definition of digitalization prevails in Russian science. Wiktionary defines "digitalization" as "a digital way of communicating, recording, transmitting data using digital devices" [17]. Marey sees digitalization as a paradigm shift in communication and interaction with each other and society [18]. Vartanova, Makseenko,

Smirnov expand the content of this concept, referring to digitalization as a complex solution of infrastructural, managerial, behavioral, cultural nature [19]. The prerequisites of digitalization include globalization, the development of Internet technologies, the growth of computing power of processors, the ubiquity of mobile devices, and the deep integration of international information networks into the cognitive process [20].

2 Materials and methods

Let us adopt the concept of digitalization to the conceptual and categorical apparatus of social philosophy and philosophy of science. The fact that digitalization is a way of communicating data indicates the practice-oriented nature of the technology, and the fact that digitalization serves as a way of communicating that data signals the intellectual nature of the technology. The term "data", which dominates definitions of digitalization, is interesting in terms of a family of ontological and epistemological concepts.

First, the data are neutral for the characteristics of empirical and theoretical knowledge. These can be empirical facts, theoretically loaded facts, formalized objects, abstract objects, theoretical constructs, etc. Second, the data are neutral for the subject-object nature of the cognitive process; they equally reflect both subject and object reality. Third, data is ontologically indifferent to objective and virtual reality, both of which act as a source of cognition (online games, social networks, etc.).

Digitalization removes the ontological opposition of finite and infinite reality. Now digitalization includes information resources, information data arrays, "limiting" information infinity (photos, audio, etc.). The deductive and inductive logic of classical science presupposes an essential limitation of the source base. The order of ideas follows the order of things, that is, the essentially finite reality.

3 Results

In modern science, the evidence base of digitalization is not limited, but data not selected, not digitized, do not make systemic sense, and do not yet become facts. Choice data processing is a methodological technique, a way of limiting an infinite procedural reality. In terms of postnonclassical science, selecting data from the flow of information and searching for relevant content is embedded in the evidence base.

Digitalization coincides with the technical support of a specific analysis procedure; it acts as a type of analytical activity. Digitalization is a kind of mechanism for entering information about the phenomena of reality into the subject's consciousness, providing its processing and certain processing. As an analysis, digitization performs the function of "sorting" the information received by the subject. In this sense, digitalization is a type of general logical method of analysis and synthesis, induction and deduction, because the search and selection of information are carried out under conditions of certainty and uncertainty.

In our view, digitalization should also be classified as a type of systematic methodology. It seems that digitalization has provided new epistemological opportunities for such types of systems methodology as modeling and design [21]. Digitalization acts as a meta-methodology since the information world is an observable reality for digital technology with regard to informatization. The digital world is an observing reality that models the objective world and the world of information data.

4 Discussion

Undoubtedly, digitalization has a targeted practical orientation [5, 22-26]. The effectiveness of this procedure depends largely on the result of the selection operation, the comprehension of the data stream. In this respect, the concepts of "technology" and "method" are related as method and technique. It seems that in the case of digitalization, there is a postnonclassical epistemological effect of "recycling" of theory and method and ontologization of methodology. How the data is connected is the project activity and the reality with which the communication subjects subsequently work. Digitalization is often defined as an instrumental method [26] and as an environment of existence [27].

The epistemological function of digitalization is the discovering of postnonclassical knowledge (not in Popper's sense). In science, Popper adheres to the separation of discovery and the situation of justification [28, 29]. Discovery means an intuitive, logically not reasoned proposal of a new theory. If a theory is not falsifiable, it tells us nothing about the world, the entire process of scientific cognition is the selection and interpretation of facts.

On the other hand, digitalization implies a logically reasoned offer of information since digitalization constitutes the object of knowledge, arranges knowledge about that object, and determines its practical value. In our view, within the framework of digitalization, information is identical to knowledge in its postnonclassical meaning. Knowledge in any form of intrasystem reflection contributes to the self-organization of the system. In digitalization, traditional methods of scientific knowledge are modified, the system methodology acquires a new sounding.

5 Conclusion

Let us summarize. Explicating the phenomenon of digitalization as a method of scientific knowledge expands the categorical apparatus of the description of the object and means of knowledge. An analysis of digitalization as a cognitive parameter reveals the watershed between sociological and philosophical views of this process and, at the same time, indicates points of convergence between them.

Thanks to digitalization, the process of cognition is incorporated into reality. The uniformity achieved in digitalization in the receipt, processing, storage, and use of information is enshrined meaningfully and functionally, forming an intersubjective field of meaning. Digitalization does not reflect a reality independent of the subject; rather, digitalization contributes to a holistic picture of the world, serving a metatheoretical function. The subject of cognition is within reality, which is by no means ephemeral; the selection and processing of information data point to the rational nature of the subject's constructive actions [30].

Digitalization belongs to the rational cognition, theoretical and metatheoretical levels of knowledge, the class of general logical methods, and is a single-order method along with modeling and construction. Digitalization can be effectively applied in classical, non-classical, and postnonclassical types of scientific rationality; that is, digitalization is an inter-paradigm method of scientific cognition. Like any other method, it has undoubted advantages and disadvantages. The task of the subject of cognition, the modern researcher, is to realize the benefits and minimize the negative consequences of the use of digitalization in scientific cognition.

Acknowledgments

The article is based on the results of research carried out at the expense of the Scientific Fund of the Financial University under the Government of the Russian Federation. Research work "Strategy of humanitarian education".

References

1. D.I. Dubrovsky, V.A. Lectorsky, *Ot redaktorov* [From the editors], in D.I. Dubrovsky and V.A. Lectorsky (eds), Artificial intelligence: an interdisciplinary approach (Moscow, 2006)
2. M. Castells, The End of the Millennium, The Information Age: Economy, Society, and Culture, Vol. III (Cambridge, Oxford, Blackwell, 1998)
3. N. Behmann, Quest Philos, **2**, 113-126 (2010)
4. A.A. Chernov, Formirovanie globalnogo informatsionnogo obshchestva: problemy i perspektivy [Formation of the global information society: problems and prospects] (Publishing and Trading Corporation "dAshkov and K.", Moscow, 2003)
5. *Osnovnye kontury kontseptsii obshchestva znanii* [The main outlines of the concept of the knowledge society], in V.V. Vasilkova, J.A. Verbitskaya (eds), The Society of Knowledge: from idea to practice, in 3 parts. Part 1 (Scythia-print, St. Petersburg, 2008)
6. A. Bardi, S.H. Schwartz, Pers Soc Psychol Bul, **29(10)**, 1207-1220 (2003)
7. W. Bilsky, M. Janik, S.H. Schwartz, J Cross-Cult Psychol, **42(5)**, 759-776 (2011)
8. R. Bonney, et al., BioSci, **59(11)**, 977-984 (2009)
9. K. Esau, Anal Kritik, **40(1)**, 55-72 (2018)
10. M.A. Ferrario, et al., *Software engineering for 'social good': integrating action research, participatory design, and agile development Companion*, in Proceedings of the 36th International Conference on Software Engineering – ICSE Companion 2014, 520-523 (2014)
11. K. Kalimeri, et al., Comput Hum Behav, **92**, 428-445 (2019)
12. A.V. Aksyukhin, A.V. Vitsen, Zh.V. Meksheneva, Mod High-Tech Techn, **11**, 50-52 (2009)
13. V.V. Laptev, Metodologiya vizualizatsii [Visualization methodology] (Mir, Moscow, 2011)
14. V.G. Gorokhov, Epistem Philos Sci, **2**, 14-32 (2008)
15. L.V. Skvortsov, Informatsionnaya kultura i integralnye znaniya: Izbrannye raboty [Information Culture and Integral Knowledge: Selected Works] (INION RAS, Moscow, 2001)
16. T.V. Nikulina, E.B. Starchenko, Pedag Edu Rus, **8**, 109 (2018)
17. Digitalization. Accessed on: October 14, 2021. [Online]. Available: https://en.wikipedia.org/wiki/Digitization
18. A. Marei, Digitalization as a paradigm shift. Accessed on: October 14, 2021. [Online]. Available: https://www.beg.com/ru-ru/about/beg-review/digitalization.aspx

19. E.L. Vartanova, et al., Industriya rossiiskikh SMI: tsifrovoe budushchee [The industry of Russian media: the digital future] (MediaMir, Moscow, 2017)
20. V.G. Khalin, G.V. Chernova, Manag Consult, **10**, 46-63 (2018)
21. Frank E.V., Mashevskaya O.V. & Ermolina L.V. (2016). Innovational mechanism of implementation of cluster initiatives in business//European Research Studies Journal. 2016. T. 19.№ 1. P. 179-190
22. A. Kohll, How to build a positive company culture (2018). Accessed on: October 14, 2021. [Online]. Available: https://www.forbes.com/ sites/alankohll/2018/08/14/how-to-build-a-positive-company-culture/#24410a1849b5
23. J.M. Palma-Oliveira, et al., Risk Anal, **38(3)**, 620-634 (2017). https://doi.org/10.1111/risa.12860
24. Y. Atif, C. Chou, J Educ Technol Soc, **21**, 152-154 (2018)
25. J. Hemerling, D. Dosik, Sh. Rizvi, A Leader's Guide to "Always-On" Transformation (2015). Accessed on: October 14, 2021. [Online]. Available: https://www.bcg.com/publications/2015/people-organization-leaders-guide-to-always-on-transformation
26. M.U. Scherer, Harv J, Law Technol, **29**, 353-400 (2016)
27. V.I. Panov, E.V. Patrakov, Tsifrovizatsiya informatsionnoi sredy: riski, vospriyatie, vzaimodeistvie [Digitalization of the information environment: risks, perceptions, interactions] (FSBI " Psychological Institute of RAO", Moscow, 2020)
28. K. Popper, Objective Knowledge: An Evolutionary Approach (1979)
29. K. Popper, The Logic of Scientific Discovery (Hutchinson, London, 1959)
30. K. Lorenz, Evolution and Modification of Behaviour (Methuen, London, 1966)

Education in Turkey in terms of gender asymmetry

Fatih Düzgün[1*], *Eda Havva* Tan Metresh[1], *Alina Aleksandrovna* Pozdnyakova[2], *Tatiana Stanislavovna* Grafkova[3], and *Marina Evgenievna* Kalita[4]

[1]Akdeniz University, Department of Russian Language and Literature, Antalya, Turkey
[2]The Kosygin State University of Russia, Institute of International Education, Moscow, Russia
[3]Moscow Pedagogical State University (MPGU), Department of German, Moscow, Russia
[4]Moscow Pedagogical State University (MPGU), Department of English and Digital Educational Technologies, Moscow, Russia

Abstract. The wide representation of women at different levels of the education system is not only an indicator of the prestige and status of the teaching profession, but also an important indicator of the socio-cultural and political development of society. The purpose of this article is to analyze the Turkish education system in terms of gender asymmetry. The research problems include: 1) the study of the historical reasons that led to the phenomena of gender asymmetry in certain "zones" and at certain levels of education; 2) the study of the motives for choosing a teaching profession by women of different age groups; 3) determination of the social profile of a Turkish female teacher; 4) the formation of recommendations for overcoming gender asymmetry in the field of Turkish education. Contrary to prejudice, Turkey is not a "gender bounded" country; moreover, its educational system is subject to problems common to the systems of most developed countries. The main ones are: 1) underrepresentation of female teachers in higher education (with a tendency to reduce the gender gap) and 2) underrepresentation of women in leadership positions. Despite the fact that all conditions have been created in Turkey for the professional development of a female teacher, the mentality of the average Turkish woman does not allow her to put her career interests above the interests of her family. This, in our opinion, largely explains the "restriction" of female education at the level of secondary specialized and insufficient "popularity" of higher education. For further study of the manifestations of gender asymmetry, a more detailed study of all stages of the Turkish education system is required, using data from different regions, in particular, from the southeast region. The situation there may differ significantly from the situation in the central regions, and the problem of adapting the education system to regional conditions will be extremely relevant.
Keywords: gender asymmetry, education system, education of women, Republic of Turkey

* Corresponding author: apozdnyakova@live.ru

1 Introduction

The concept of gender asymmetry means "a social phenomenon that reflects the objective regularity of the quantitative distribution of men and women in the social structure of society and in all spheres of its life, as well as the qualitative social consequences of this phenomenon" [1]. Such features of this phenomenon as "objectivity" and "regularity" are considered to be important in the definition of Sillaste, since they reflect the natural character of the development of disproportionate relations in a particular sphere of human interaction. It is from these positions that authors would like to consider the situation observed today in the Turkish education system.

2 Methods

To determine the range of reasons causing gender asymmetry in the education system of the Turkish Republic at certain stages of its development, an analysis of psychological, pedagogical, sociological, historical and philosophical literature was carried out. The following works of Turkish authors should be noted: Akyüz [2-4], Yaşar [5], Çakmak [6], Eşkinat [7, 8], Aydın [9], Tor [10], Güven [11], Sargın [12]; the research of Russian author Sillaste is highly relevant [1, 13, 14].

The information base for the analysis was data from open sources: 1) analytical reviews on the problems of economic and social development of regions [15-17]; 2) summarized statistical data related to the study of the teaching staff on teaching and learning [18-22]; 3) materials of websites of Turkish educational institutions.

To confirm the assumptions put forward regarding the nature of gender asymmetry in Turkish education, a survey of teachers in Turkish institutions of primary, general secondary and higher education, as well as women leaders who organize and supervise education was conducted.

3 Results

The wide representation of women at different levels of the education system is not only an indicator of the prestige and status of the teaching profession, but also an important indicator of the socio-cultural and political development of society.

For modern Turkey, these indicators are quite positive and comparable to those of many European countries (see Table 1). According to 2016 data, the proportion of women in the total number of teachers (HEI teachers) is at the border of 60% for primary and basic general education – the levels of education that are extremely feminized in Europe and traditionally defeminized in Muslim countries [22].

Table 1. Comparative representation of women in education.

Country / education levels	Primary general ISCED 1, %	Basic general ISCED 2, %	Secondary general ISCED 3, %	Secondary vocational and higher ISCED 5, 6, 7, 8, %
Turkey	59.5	58.9	50.4	43.3
The Czech Republic	94.4	77.8	62.7	38.4
The Netherlands	86.7	52.5	52.6	45.2
Russia	99.4	-	83.7	63.4

Remembering the history of female education in Turkey, the authors note that formerly primary schools were the only educational institution where girls could receive education. With the creation of the Turkish Republic in 1924, emphasis was placed on the democratization of education. The founder of the Turkish Republic, Mustafa Kemal Ataturk, believed that by providing women with equal right to education, it would be possible to find ways to solve social problems, however, according to researchers, "inequality in the application of laws was a clear sign of the latent patriarchy of modernism against the open patriarchy of traditionalism" [7]. This duality in the implementation and perception of socio-cultural reforms influenced the formation of the historical consciousness of the Turkish woman and determined the social profile of the modern teacher.

The modern Turkish female teacher is:

1) a woman with a secondary vocational education (ISCED 5), since a higher level of education for women "is not associated with increased employment opportunities" [23]. The researchers note that employment in the system of preschool (ISCED 0), primary general (ISCED 1) and secondary general (ISCED 2) education require the specified level of teacher's education is sufficient, and the unemployment rate is minimal, while in the higher education system (ISCED 5, 6, 7) the unemployment rate is constantly increasing;

2) a quite young woman; according to the OECD, the percentage of teachers under 30 working in the primary and secondary school system is particularly high (18%, rating 3/36, 2018), while the percentage of teachers over 50 in this area is 12.4% (rating 36/36, 2018);

3) a woman who is more oriented towards work in the public sector. In our opinion, this is due to two reasons: first, the fact that in the public sector the ratio of women's wages to men's wages is slightly higher than in the private sector – 0.783 and 0.767 [15]; secondly, the fact that this work is better combined with family responsibilities (shorter working hours, normalized working hours, etc.) and allows a woman to be in demand both at work and at home, which is extremely important for a Turkish woman;

4) a woman with a traditional family, although in recent years the number of unmarried women has also increased, especially in the field of higher education. Moreover, Turkey is a country where "the type of egalitarian family is recognized and supported by national legislation" [14];

5) a woman who does not actively strive for a career, but cares about improving the quality of the existing education, therefore, participating in various additional training programs;

6) a woman who is more focused on research in the natural sciences; the number of women focused on social sciences, economics is less [23]. This fact is confirmed by statistics on the number of women who completed doctoral studies [18-21]. Thus, the proportion of women who have completed doctoral studies in natural sciences, mathematics and statistics is quite high and amounts to 57.7% (rating 4/43, 2018).

The main reasons for choosing the profession of a teacher (HEI teacher) by Turkish women include the following:

1) The high status of the teaching profession in society, respect for the people of this profession. According to surveys of teachers in Turkish schools of different levels of education, the attractiveness of the profession is significant for 68% of respondents.

2) Sufficient, in the opinion of the respondents, the level of state support for education: 27% of teachers indicated satisfaction with this parameter, 43% indicated partial satisfaction, 7% of respondents expressed dissatisfaction with the level of state support, 23% found it difficult to answer the question. The data from related studies show that

among those financed from the state budget, the teaching profession is one of the most well-paid and socially protected.

3) Sufficient level of education. As we indicated above, secondary vocational education is sufficient for the implementation of pedagogical activity at all levels of the education system, except for higher education. This allows a woman, after graduating from college, to start working in her specialty, while the lack of higher education is not an obstacle to certain career growth. Turkey has one of the lowest proportions of women completing tertiary programs (52.8%, 33/36 in 2018) among OECD and partner countries for which data are available.

4) Lack of gender "confrontation" in this area. Up to half of the teachers in Turkish schools are men, and this parity largely determines the quality of professional interaction in the education system. In addition, in education, women and men receive almost the same wages. According to statistics, women's earnings as a percentage of men's earnings (among people aged 25 to 64 with higher education and income from work) is one of the highest among countries for which data are available (83.6%, rating 3 / 37, 2017).

The results of the survey confirm the assumption that the choice of the profession of a teacher (HEI teacher) for a Turkish woman in most cases is deliberate and largely based on deep motivation: a woman correctly evaluates her abilities in terms of combining work and family responsibilities, she knows well the content of the activities that she has to exercise, and feels confident enough in this area.

4 Discussion

According to the classification of Sillaste, Turkey belongs to the type of countries "gender symmetric, with parity of males and females in the structure of the population (50 to 50%)" [14]. In addition to Turkey, this group includes countries of "European orientation" – Australia, Greece, Denmark, Canada, Norway, Sweden, as well as the Republic of Korea, Tajikistan, Turkmenistan, Uzbekistan. Statistical data and analytical reviews make it possible to form an idea of the degree of women's involvement in education (see Table 2). The share of women at different levels of the educational system (from ISCED 1 to ISCED 5, 6, 7, 8) in the countries of the first subgroup is consistently decreasing [22].

Table 2. The share of women in education.

Country / education levels	Primary general ISCED 1, %	Basic general ISCED 2, %	Secondary general ISCED 3, %	Secondary vocational and higher ISCED 5, 6, 7, 8, %
Greece	71.4	66.5	55.7	33.8
Canada	74.5	-	74.5	49.0
Norway	74.6	74.6	53.2	45.7
The Republic of Korea	78.1	69.9	52.8	34.9
Sweden	76.9	76.8	53.4	44.5
Turkey	59.5	58.9	50.4	43.3

If we consider the countries of the second subgroup (Tajikistan, Turkmenistan, Uzbekistan), which became part of the Russian Empire in the middle of the 19th century, and then into the USSR, and therefore experienced a strong influence of the Russian educational tradition, then there is a similar trend: the share of women at all levels of education is about 70% with a significant decrease in higher education [24].

An interesting observation is that, despite different cultural, historical and educational traditions, countries are converging at the point of belonging to a "gender symmetric" group, while maintaining the essential characteristics of the ethnic education system. However, the large area of Turkey and its wide length from west to east implies the heterogeneity of the distribution of human resources and the different quality of education and educational services: the poor and backward southeastern regions are characterized by a strong influence of patriarchal religious and cultural traditions and significant manifestations gender asymmetry. According to Sillaste, "gender inequality inhibits human development; the average level of achievement in a particular country cannot be adequately assessed without taking into account the inequality between men and women" [14], so the development of recommendations to overcome gender asymmetry in the field of Turkish education is of high importance.

5 Conclusion

In our opinion, the balance between education and employment should be revised taking into account new programs targeting women in specific regions, including in the private sector. While the public sector is the main platform for employment of women, the private sector requires support by creating the conditions for hiring more women, reorienting tax policies, insurance conditions and employment schedules for specific levels of education. Moreover, conditions must be created to attract more women to higher education. The priority task of the government of the country is to improve the quality of life of women in poor regions, as well as the position of women in the education system of these regions in order to overcome the "gender asymmetry".

References

1. G.G. Sillaste, Higher Edu Rus, **3**, 122-133 (2004)
2. Y. Akyüz, Osmanlı Son Döneminde Kızların Eğitimi ve Öğretmen Faika Ünlüer'in Yetişmesi ve Meslek Hayatı [The Education of Girls in the Last Ottoman Period and the Upbringing and Professional Life of Teacher Faika Ünler] (1999). Accessed on: October 14, 2021. [Online]. Available: http://dhgm.meb.gov.tr/yayimlar/dergiler/Milli_Egitim_Dergisi/143/1.htm
3. Y. Akyüz, *Tanzimat'tan Cumhuriyet'e Türkiye'de Toplumsal Değişme ve Kadın Eğitimi* [Social Change and Education of Woman in Turkey from Tanzimat (Reformation) to Republic], in A.M. Tuncoku (ed), Toplumsal Gelişmede Türk ve Japon Kadınının Eğitimi [Women's Education in Turkey and Japan for Social Development], 1-59 (Ankara, 2011)
4. Y. Akyüz, Türk Eğitim Tarihi M.Ö. 1000 – M.S. 2016 [History of Turkish Education B.C. 1000 – A.D. 2016] (Pegen Akademi, Ankara, 2018)
5. O. Yaşar, Int J Human Sci, **4(1)**, 1-35 (2007)
6. B. Çakmak, J Theatre Critic Drama, **1(18)**, 51-88 (2011)
7. R. Eşkinat, *Türkiye'de Sivil Toplum Örgütlerinin Kadınların Eğitimine Katkısı* [Contribution of Civil Society Organizations in Turkey to Women's Education], in A.M. Tuncoku (ed), Toplumsal Gelişmede Türk ve Japon Kadınının Eğitimi [Women's Education in Turkey and Japan for Social Development], 259-279 (Ankara, 2011)
8. R. Eşkinat, J Faculty Law Anadolu Univ, **2(3)**, 267-282 (2016)

9. H. Aydın, Curr Res Soc Sci, **1(3)**, 84-96 (2015)
10. H. Tor, E. Ağlı, J Res Edu Teach, **5(9)**, 67-74 (2016)
11. Z.Z. Güven, *Eğitim ve Kadın İşgücü* [Education and Women's Labor Force], in B. Afşar, B. Büyükdoğan (eds), Dünyada Kadın İşgücüKadın İşgücü ve Disiplinlerarası Bir Bakış [A Comprehensive and Interdisciplinary Overview of the Women's Labor Force in the World], 115-124 (Gazi Kitabevi, Ankara, 2016)
12. N. Sargın, *Eğitim ve Kadın İşgücü* [Education and Women's Labor Force], in B. Afşar, B. Büyükdoğan (eds), Türkiye'de Kadın ve Eğitim. Kadın İşgücü ve Disiplinlerarası Bir Bakış [A Comprehensive and Interdisciplinary Overview of the Women's Labor Force in the World], 199-208 (Gazi Kitabevi, Ankara, 2016)
13. G.G. Sillaste, Gendernaya sotsiologiya i rossiiskaya deistvitelnost [Gender Sociology and Russian Reality] (Alfa-M, NIC INFRA-M, Moscow, 2016)
14. G.G. Sillaste, Human South Rus, **6(5)**, 48-62 (2017)
15. World Bank, Europe and Central Asia Economic Update, Spring 2021: Data, Digitalization, and Governance (Washington, 2021). https://doi.org/10.1596/978-1-4648-1698-7
16. Reiting stran mira po urovnyu gendernogo neravenstva [Gender Inequality Index] (2021). Accessed on: October 14, 2021. [Online]. Available: https://gtmarket.ru/ratings/gender-inequality-index
17. Reiting stran mira po indeksu urovnya obrazovaniya [Ranking of countries in the world by education level] (2021). Accessed on: October 14, 2021. [Online]. Available: https://gtmarket.ru/ratings/education-index
18. OECD, TALIS 2018 Results (Volume I): Teachers and School Leaders as Lifelong Learners (OECD Publishing, Paris, 2019). https://doi.org/10.1787/1d0bc92a-en
19. OECD, TALIS 2018 Results (Volume II): Teachers and School Leaders as Valued Professionals (OECD Publishing, Paris, 2020). https://doi.org/10.1787/19cf08df-en
20. OECD. Positive, High-achieving Students? What Schools and Teachers Can Do (OECD Publishing, Paris, 2021). https://doi.org/10.1787/3b9551db-en
21. Turkey. Overview of the education system (2020). Accessed on: October 14, 2021. [Online]. Available: https://gpseducation.oecd.org/CountryProfile?primaryCountry=TUR&treshold=10&topic=EO
22. N.V. Bondarenko et al., Pokazateli obrazovaniya: 2020: Statisticheskii sbornik [Education indicators: 2020: Statistical collection] (NRU HSE, Moscow, 2020)
23. E.E. Dinçsoy, O.M. Dinçsoy, *Japonya ve Türkiye'de Kadının Eğitimi ve İstihdamı* [Women's Education and Employment in Japan and Turkey], in A.M. Tuncoku (ed), Toplumsal Gelişmede Türk ve Japon Kadınının Eğitimi [Women's Education in Turkey and Japan for Social Development], 75-89 (Ankara, 2011)
24. Gender ko'rsatkichlari raqamlarda [Gender Statistics of Uzbekistan] (2021). Accessed on: October 14, 2021. [Online]. Available: https://gender.stat.uz

Development of the digital competence of students through the electronic course

Natalia Ivanovna Svetlova[1*], *Lydia Nikolaevna* Vasilieva[2], and *Irina Igorevna* Ilyina[3]

[1]Financial University under the Government of the Russian Federation, Department of Mathematics, Moscow, Russia
[2]Chuvash State University named after I.N. Ulyanov, Department of Automation and Control in Technical Systems, Cheboksary, Russia
[3]Bauman Moscow State Technical University, Department of Fundamentals of Mathematics and Informatics (SUNTs-1), Moscow, Russia

Abstract. The article is concerned with the role of an electronic course in higher mathematics in the development of digital competence of students of technical faculties. The principle of organizing the educational process with the help of an electronic course in higher mathematics, focused on the development of students' digital competence, is considered. The experience of testing an electronic notebook in higher mathematics, which is part of an electronic course, is presented. The experience of the implementation of an electronic notebook in higher mathematics in teaching students at all its stages is considered: the examples of problems solved in seminars are given; the design of classroom lessons is described taking into account the specifics of the academic subject "Mathematical Analysis". The structure of the electronic notebook, including test tasks and cases for intermediate certification of students based on the results of the seminar, is given. The effectiveness of the organization of the educational process using an electronic course in higher mathematics to develop the digital competence of students in a technical direction is shown. The work examines the attitude of students to teaching higher mathematics using an electronic course, their self-assessment of involvement in the educational process. Moreover, the educational results of the development of digital competence were demonstrated. A reliable relationship was confirmed between the level of development of digital competence and the positive attitude of students to a new format of teaching mathematics: an interactive electronic course with a component base in the form of an electronic notebook. Based on the results of the experimental work, it can be argued that the students positively assessed the use of the electronic course in teaching, because have constant access to training materials and to information about the individual trajectory of the course. This interactive development is convenient to use when preparing yourself for classes.
Keywords: e-course, e-notebook, digital learning, competence

* Corresponding author: svetlovani@mail.ru

1 Introduction

Currently, the development of information technology, the interaction of educational organizations on the network, e-learning, the use of the Internet, necessitate the use of qualitatively new opportunities in the field of education. The issue of increasing the level of development of digital competence is becoming relevant. According to the state program of the Russian Federation "Development of Education" for 2019-2025, the program "Digital Economy of the Russian Federation" (until 2024), the Federal State Educational Standards of the third generation (FSES 3 ++), the process of student learning must be built taking into account priority areas of informatization of society, focused on increasing the intensity of the use of digital technologies, electronic resources and products in the educational process.

The digital transformation of production makes rather high demands on the purposeful development of the professional competence of future engineers, the level of which depends on their professional activity [1]. An engineer should be fluent in digital skills associated with regularly solving complex professional problems in a digital environment, be creative as well as be able to work in a team and do critical thinking.

2 Materials and methods

The study aims to substantiate the development of digital competence of students of technical departments with the help of an electronic course in higher mathematics and checking the effectiveness of its application in experimental work.

The object of the research is the process of modern professional training of students of technical faculties in the context of the digital age.

The subject of the research is the level of development of the digital competence of students of technical faculties, as the basis of professional competence.

The materials for this study were both foreign scientific sources and domestic research. In the process of research, the authors used theoretical, diagnostic, empirical, experimental methods and methods of mathematical statistics.

The study [2] analyzes the international experience in the development and implementation of electronic educational resources and offers a practical description of the organization of e-learning for teachers in undergraduate programs.

The authors in [3] examine the trends in e-learning and analyze its impact from various points of view.

The article [4] proposes an algorithm for automatic recommendation of resources for training, based on a convolutional neural network. Automated learning resource recommendations are an increasingly pressing issue: they allow learners to discover new learning resources that suit their tastes, and enable eLearning to target their learning resources to the right learners.

E-learning recommendation systems play an important role in helping learners find useful and relevant learning materials that meet their learning needs. Personalized intelligent agents and recommendation systems are widely recognized as solutions to overcome information retrieval problems resulting from information overload in learners. An overview of ontology-based recommender systems for e-learning is given in the work [5].

In the Federal study [6], conducted in 2013, digital competence is understood as "based on the continuous mastery of competencies (knowledge, skills, motivation, responsibility), the ability of an individual to confidently, effectively, critically and safely select and use infocommunication technologies in different spheres of life (information environment, communications, consumption, technosphere), as well as his readiness for such activities".

Yachin and Fernandez believe that the concept of "digital competence" includes "confident and critical use of computers, mobile phones, tablets, and interactive whiteboards by students. This competence is based on logical thinking, a high level of mastery of information management and highly developed mastery of digital technology" [7].

Prikhodko considers digital competence as "mastery of methods of searching, structuring, systematizing and critical assessment of information using digital technologies and the global Internet for solving practical, educational and professional problems" [8].

In [9], digital competence is defined as "the readiness and ability to effectively and systematically use infocommunication technologies in different spheres of life based on the possession of information competencies as a system of knowledge".

However, in these works, the definition of the term digital competence of students of technical specialties is not considered.

The research of various aspects of e-learning is the subject of the works of Russian researchers: Vlasova [10], Zhozhikov, Tretyakova, Noskov, and others. Foreign authors such as Akaslan, Allen, Laanpere, Richardson, Seaman, Sheninger, Pratt, Whitby, Corbett, Skoulikari, et al. Scientists pay special attention to the platform on which the training takes place, and the form of organization of the educational process [11-13], the interaction of students with the teacher and ways of solving them.

However, to date, the problem of developing the digital competence of students of technical universities when teaching using an electronic course has not been sufficiently developed.

The analysis of the research showed the need to develop an electronic didactic tool (an electronic course in higher mathematics, including an electronic notebook) to increase the level of development of digital competence of students of technical specialties.

The digital competence of students of engineering specialties is considered by the authors as a set of general professional (GPC-8, GPC-7) and professional competencies (PC-10, PC-12), separated from the federal state standard of higher education for a specialist in the specialty 11.05.01 "Radio electronic systems and complexes" [14].

3 Results

One of the ways to solve the problem of developing the digital competence of students of technical specialties is to create conditions for the implementation of effective educational activities in higher education institutions [15].

The widespread use of digital technologies in interactive learning makes it possible to switch to an electronic course that promotes the integration of theory and practice through the visualization of the material [16]. Such developments help to develop students' critical thinking, form skills, identify and solve digitalization problems, and develop students' personal professional qualities [17].

An important component of the electronic course is an electronic workbook for the course of higher mathematics. This development allows not only to present and demonstrate the material, but also to interest and awaken the student's motivational mood, to show the necessity and importance of mastering the studied material for their further professional activities.

The structure of the electronic notebook is shown in Fig. 1.

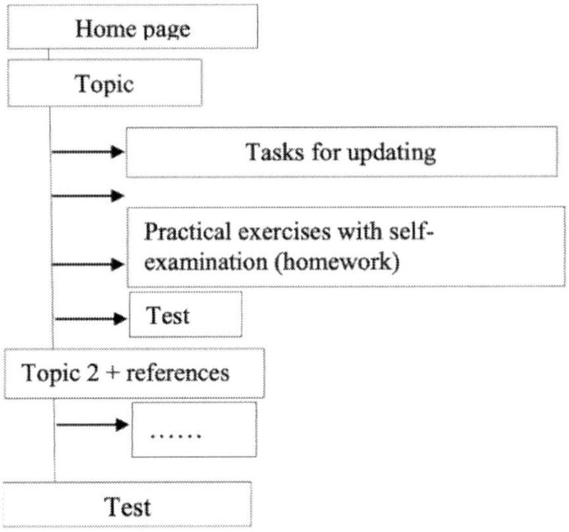

Fig. 1. The structure of the workbook.

The tool for diagnosing the level of digital competence in the study is the result of using an electronic notebook when teaching students mathematical disciplines in the form of an examination assessment, as well as the attitude of students to teaching mathematics using an electronic notebook, their self-assessment of involvement in the educational process [18].

To assess the level of development of digital competence, the following levels were identified: high, advanced, basic.

The high level of development of digital competence of the respondents includes: knowledge of modern principles of working with information (GPC 7), the ability to analyze it and use it to solve problems related to professional activities (GPC 8).

The indicators of the advanced level include: the ability to solve data processing problems using modern information technologies (GPC 7) by means of automation, as well as to have an idea of the digital needs in professional activities (GPC 8).

The basic level assumes: the ability to search and present up-to-date information about the state of the subject area (GPC 7), master the skills of using information and communication tools (GPC 8).

To analyze the level of development of digital competence in the discipline "Mathematical Analysis", the Pearson's criterion of agreement (chi-square) is used. Students of the experimental and control subgroups were divided into three levels in accordance with the level of training (Table 1).

Table 1. Levels of development of digital competence of students in the experimental and control groups.

Sample	Levels, %		
	Basic	**Advanced**	**High**
TG	34	51	15
RG	55	36	9

According to calculations, the value of the χ^2 criterion is 9.041, which is greater than the critical value of χ^2 at a significance level of 5.991. In accordance with the decision-making rule, the data obtained allowed us to state that, at the level of significance

$\alpha = 0{,}05$, the level of development of digital competence in the discipline "Mathematical Analysis" among students of technical faculties of the experimental group studying the discipline using an electronic notebook is higher than among students of the control group studying the discipline without using an electronic notebook.

4 Discussion

The authors paid special attention to the issues of student involvement in the educational process and the effectiveness of organizing interaction with the teacher. The independent work of students was analyzed. To assess these indicators, the reflexivity questionnaire of Karandashev (Table 2). The questionnaire consists of 33 questions. Testing time is 15 minutes. The reflexivity indicator is defined as the sum of the numerical indices of the respondents' answers according to the key. Up to 49 points characterize low reflexivity, from 50 to 59 points – average reflexivity, from 60 and above – high reflexivity.

Table 2. Results of the level of development of reflection before and after the experiment.

Group	Level of reflexivity, %					
	Low		Medium		High	
	before	after	before	after	before	after
RG	85.6	62.3	14.4	36.4	-	-
TG	82.7	41.5	15.3	50.5	-	8

Analysis of statistical data shows that there is an increase in the level of reflection in both the experimental and control groups. Thus, the level of reflection increases at all stages of training, but in the experimental group, it approaches a high value, which allows us to assert the success of the experimental work.

To diagnose the level of digital competence, in addition to analyzing the examination score, the authors tried to choose a methodology for assessing not only the cognitive component, but also the activity component. The work used a test developed by Microsoft, some of the questions of which were updated in accordance with modern requirements, based on the criteria of the All-Russian study "Index of digital literacy of citizens of the Russian Federation-2017" – a project of the Regional Public Organization "Center for Internet Technologies" (ROCIT), directed to measure the level of knowledge and skills of the population necessary for the safe and effective use of digital technologies and Internet resources [19] (Table 3).

Table 3. Analysis of the level of digital competence according to the criteria of the All-Russian study "Index of digital literacy of citizens of the Russian Federation-2017".

Group	Digital Competence Level, %									
	The digital age		Internet, cloud services		Standard programs		Privacy and security		Basic information	
	before	after	before	after	before	after	before	after	before	after
RG	54.4	53.3	38.7	39.5	46.5	50.1	35.3	38.1	50.2	49.3
TG	45	64.6	34.2	70.3	45.6	77.2	38.6	45.1	52.3	80.8

Thus, the pilot study and the analysis conducted (Table 4) make it possible to conclude that the use of an electronic course and its component – an electronic notebook, contributes to an increase in the level of development of digital competence of students of technical faculties, since there is a positive dynamics in the development of the studied competence.

5 Conclusion

Issues related to assessing the level of development of digital competence of students in technical areas, formed when working with e-courses, have been insufficiently studied, since the technology of introducing e-learning in the educational process is currently not widespread enough.

A further prospect of researching the problem will be aimed at solving issues related to the intensification of existing methods and technologies that are relevant for the development of digital competence of students.

References

1. X. Shen et al., *Automatic recommendation technology for learning resources with convolutional neural network*, in International Symposium on Educational Technology, 30-34 (ISET, 2016). https://doi.org/10.1109/ISET.2016.12
2. I.I. Lysova, O.A.Vitokhina, O.V. Volkova, Iss Journal, Edu, Linguist, **39(3)**, 408-418 (2020)
3. A.S. Imran, et al., *An Analysis of Social Collaboration and Networking Tools*, in P. Zaphiris, A. Ioannou (eds), Learning and Collaboration Technologies. LCT 2016. Lecture Notes in Computer Science, **9753** (2016). https://doi.org/10.1007/978-3-319-39483-1_31
4. I. Maslova, G. Burdina, I. Krapotkina, Int J Emerg Tech Learn, **15(16)**, 68-79 (2020)
5. V.A. Shershneva, et al., Europ J Contemp Edu, **17**, 357-367 (2016). https://doi.org/10.13187/ ejced.2016.17.5357
6. G.U. Soldatova, E.I. Rasskazova, Nation Psych J, **2(14)**, 27-35 (2014)
7. N.P. Yachina, G.G. Fernandez, Bul VSU, Ser: Prob Higher Edu, **1**, 134-138 (2018)
8. O.V. Prikhodko, Azimuth Sci Res: Pedag Psych, **9(1(30))**, 235-238 (2020)
9. S.M. Kosenok, T.N. Kurenkova, Human Sci Bul, **11**, 25-31 (2020)
10. E.A. Barakhsanova et al., IEJME-Math Edu, **11(10)**, 3447-3456 (2016)
11. V.A. Gaeva, D.Yu. Zakharov, Open Edu, **1**, 65-69 (2014)
12. J.K. Tarus, Artif Intell Rev **50**, 21-48 (2018). https://doi.org/10.1007/s10462-017-9539-5
13. C.Y. Yang, et al., *Learning performance evaluation in e-learning with the webbased assessment*, in K. Kim, N. Joukov (eds), Information Science and Applications 2017. ICISA 2017. Lecture Notes in Electrical Engineering, **424** (2017). https://doi.org/10.1007/978-981-10-4154-9_74
14. Ob Utverzhdenii Federalnogo Gosudarstvennogo Obrazovatelnogo Standarta Vysshego Obrazovaniya – Spetsialitet po Spetsialnosti 11.05.01 Radioelektronnye Sistemy i Kompleksy [On the approval of the federal state educational standard of higher education – specialty in the specialty 11.05.01 Radio electronic systems and complexes]. Accessed on: October 14, 2021. [Online]. Available: http://www.fgosvo.ru/news/21/3570
15. I.I. Ilyina, L.N. Vasilyeva, *Napravleniya ispolzovaniya sovremennykh IKT pri podgotovke bakalavrov tekhnicheskikh napravlenii* [Directions of the use of modern

ICTs in the preparation of bachelors of technical directions], in A.N. Matveeva (ed), Actual problems of methods of teaching mathematics. Collection of reports and scientific articles based on the materials of the regional scientific and practical conference, 98-101 (2019)

16. G.J. Hwang, H. Xie, Knowl Manag E-Learn: Int J, **10(4)**, 371-374 (2018). https://doi.org/10.34105/j.kmel.2018.10.022
17. L.N. Vasilyeva, et al., Sci Edu Today, **10(2)**, 124-137 (2020)
18. N.I. Svetlova, N.A. Efimova, *Elektronnaya rabochaya tetrad kak dopolnitelnoe sredstvo formirovaniya tsifrovoi kompetentnosti subektov obrazovatelnogo protsessa* [Electronic workbook as an additional means of forming digital competence of subjects of the educational process, in Socio-humanitarian and legal problems of modern society: digital reality], 10-19, Materials of the XX interuniversity scientific conference on general humanitarian, legal and economic issues. Cheboksary, MGEU CHI (f) February 27, 2020 (LLC "Publishing House "Pegas", Cheboksary, 2020)
19. Vserossiiskoe issledovanie. Indeks tsifrovoi gramotnosti 2017 [All-Russian research. Index of digital literacy 2017]. Accessed on: October 14, 2021. [Online]. Available: https://rocit.ru/uploads/769c4df4bc6f0bd6ab0fbe57a056e769b8be6bcf.pdf

Problems of assessing students' knowledge in mathematics using LMS Moodle

Larisa Petrovna Konnova, *Larisa Vladimirovna* Lipagina, *Elena Fedorovna* Olekhova, *Alexander Arkadievich* Rylov, and *Irina Kimovna* Stepanyan[1]

Financial University under the Government of the Russian Federation, Department of Mathematics, Moscow, Russia

Abstract. The widespread introduction of the LMS Moodle educational environment into the educational process in the framework of distance learning and recommendations to use the abovementioned resource for full-time education served as the prerequisites for this study. The authors believe that the knowledge assessment system for distance learning cannot be automatically applied to assess students of other forms of education, but should be adapted to them. The purpose of the research was to study the issues of assessment in the Moodle environment, to develop a methodology for using it to assess the knowledge of students of economic bachelor's degree in Mathematics. To achieve the goal, scientific literature on this topic was studied, pedagogical design, experiment and student survey were used as research methods. As a result, the authors of the article designed a control and measuring system for assessing knowledge in mathematics in LMS Moodle and the procedure for intermediate certification using this system. The experience of introducing a knowledge assessment system in the first year of the Financial University under the Government of the Russian Federation and a questionnaire survey showed the specificity of approaches to assessing students' knowledge in an electronic environment in Mathematics. To control the academic achievements of students, the authors consider it necessary to check additionally the performance of the teacher with the subsequent correction of points. The novelty of the research lies in the development of a methodology for using LMS Moodle for assessing students' knowledge in mathematics in the course of full-time education.
Keywords: LMS Moodle, testing, measurement control system, teaching mathematics

1 Introduction

Digital transformation is actively involved in the educational processes of higher education institutions. One of the indicators of the introduction of digital technologies into the educational process is the use of electronic learning systems, LMS Moodle in particular. The platform developed for virtual communication between a student and a teacher in the

[1] Corresponding author: IKStepanyan@fa.ru

course of distance learning becomes increasingly popular when working with full-time students.

In the 2020-2021 academic year, the authors have developed electronic training courses (ETC) in the Moodle environment for communication with first-year undergraduate economics students at the Financial University. The practical evaluation made it possible to single out a number of problems related to the assessment of students' knowledge, which in turn led to the research hypothesis: the knowledge assessment system used in LMS Moodle for distance learning cannot be automatically applied to assess the knowledge of full-time students and those, who study in blended learning mode.

The purpose of the study was to develop a methodology for using the LMS Moodle platform for assessing students' knowledge in mathematics in a blended learning mode. To achieve the goal, the following tasks were set:

- to explore the experience of LMS Moodle electronic testing in teaching mathematical disciplines;
- to create a base of test items in mathematics for assessing first-year undergraduate economics students at the Financial University;
- to test the created base for monitoring the academic achievements of students;
- to study the opinion of students on the effectiveness of the introduction of electronic assessment tools.

2 Materials and methods

LMS Moodle has long established itself as an environment for creation and implementation of the ETCs with a pedagogical support system. Consider the recent works associated with this research.

A recent review of e-learning in mathematical content was carried out by the Spanish [1]. A comparative analysis of blended, distance, e-learning and virtual learning using the example of students from the Philippines is given in [2], while the scientific papers [3] of the authors from Portugal describe the possibilities of LMS Moodle as an online assessment tool for a number of educational programs in mathematics.

The experience of designing the ETCs in mathematics for blended learning of future teachers is reflected in the work [4], the article [5] is devoted to the problem of testing. The works [6, 7] related to the process of creating adaptive e-learning courses in mathematics require special attention. The authors also would like to note the article [8], which characterizes the capabilities of the Moodle platform for automatically checking the course assignments with computer support.

The problem of assessment of the learning outcomes within the framework of the ETCs is relevant for the research being conducted. In a recent expert study [9], the instrumental capabilities of assessment of the course elements are revealed, the authors of the article [10] define the technology for test items development. Note the useful experience of developing an assessment system in the humanities [11, 12] and technical [13] disciplines. In a survey work on the assessment capabilities of the LMS Moodle [14], the organization of the main forms of assessment was presented: cumulative (Summative assessment) and formative (Formative assessment).

As a summary, the authors emphasize that the existing publications mainly describe the use of cumulative assessment of students' knowledge in LMS Moodle, while inadequate attention is paid to the formative assessment aspect. This prompts us to focus specifically on the formative assessment associated with the face-to-face learning regime.

In addition to the review of publications on assessment tools in the Moodle environment, the authors note the following methods of the proposed research:

- pedagogical design of a system of assessment tools for current and intermediate control;
- practical evaluation of the base of test problems in mathematics as part of the creation of the individual ETCs;
- questioning – a survey of students who have completed test tasks in the Moodle environment during the year to get feedback.

3 Results

3.1 Pedagogical design of the assessment tool system

According to the "Mathematics" work program the course is divided into two main sections "Mathematical Analysis" and "Linear Algebra", each of which has its own technical features when creating tests.

In the first semester of the academic year, a bank of test questions of 10 categories in the mathematical analysis was created: a total of about 1200 questions, about 60% of which are associated with the generation of parameters. In the second semester, 9 categories in mathematical analysis, 11 categories in linear algebra and 2 categories in linear programming were created: several thousand test questions were obtained, 95% of which are generated.

When developing test questions in mathematical disciplines, the following problems arise:

- complex problem statement;
- dependence on many parameters;
- analytical solution;
- a large number of options;
- approximate or numerical solutions to problems;
- ambiguity in the presentation of answers;
- complex structure of the answer;
- receipt of an answer in a convenient form (whole numbers, ordinary fractions);
- control of solution steps.

Let us focus on one of the many types of test questions: Formulas – a plug-in for Moodle [15], which is an optional component. This type of question can use random parameters and complex answer structures. Answer fields, placed anywhere in the question, make it possible to create assignments with different types of answers (coordinates, polynomial, matrix, etc.) Thus, the Formulas type is a powerful and functional tool for generating mathematical problems taking into account their peculiarities and for forming a more flexible variable assessment.

The formation of a bank of tasks for the subject includes a very important aspect – the assessment system. The solution of tasks in mathematics consists of many logical and computational operations therefore it would be incorrect to evaluate such tasks by the final answer.

The authors believe that a partial solution to this problem involves the development of test questions with step-by-step assessment. In this case, the student receives points from LMS Moodle for each correct step of the solution, depending on the significance and complexity of the actions. However, this approach does not allow to entrust the assessment of students' knowledge completely to the electronic platform. The authors concluded that in addition to electronic verification of the answer, the solution of the task should be verified by the teacher.

3.2 Experience in using the control and measuring system for assessing knowledge on the Moodle platform

To evaluate the experimental test work performed by students on the Moodle platform, it was decided to take into account the results of the teacher's verification of the written decision and, after adjusting, set the final score. The verification experience showed a significant discrepancy between the electronic assessment and the assessment of the written decision. See Table 1 for examples of such discrepancies. 1. The presence of a sufficiently large percentage of discrepancies (on average 30-35%) confirmed the hypothesis of the need for a final revised assessment.

Table 1. Fragment of the table of checking the test work in mathematics.

Moodle	Assessment by the teacher	Revised assessment	Reason for revision
0	1	0.9	the solution is correct the answer was entered in a format violation
0	1	0.8	the solution is correct, no answer entered
0	0.7	0.7	arithmetic error with the right solution
0	0.4-0.7	0.4-0.7	resolved partially
1	0.5	0.5	insufficient justification
1	0	0	no solution

Based on the analysis of the test results on the Moodle platform, the Regulations for the examination in Mathematics were developed. The document reflects the mandatory verification by the teacher of the written solutions given by the students and subsequent correction of scores from Moodle.

According to the Regulations, the assessment of examination papers is carried out according to the criteria:

1. Each correct answer to a question is entered in electronic form, if there is a correct solution in the examination paper, it is estimated at 10 points.

2. For a correctly entered answer 0 points are given if:

- the decision of the corresponding item of the assignment has not been submitted;
- the entered solution is incorrect.

3. For the absence of an answer in electronic form to the question of the examination paper, if there is a correct substantiated written solution, a penalty of 5 points out of 10 is removed.

To organize control over students' academic achievements in the Moodle environment, it is necessary, first, to acquaint them with the specifics of testing and assessment. In this regard, it is recommended to conduct short tests and trainings, some recommendations for the types of assessment are presented in Table 2.

Table 2. Distribution of assignment types by assessment type.

Assessment type	Types of recommended assignments
Moodle	theoretical express survey, training
Moodle + check by the teacher	independent and control work,

	test, exam
check by the teacher	homework, case, project

According to the results of the survey of 205 first-year students who had the experience of testing on the Moodle platform for two semesters, it can be stated that students also prefer combined verification by the electronic system and the teacher (Fig. 1).

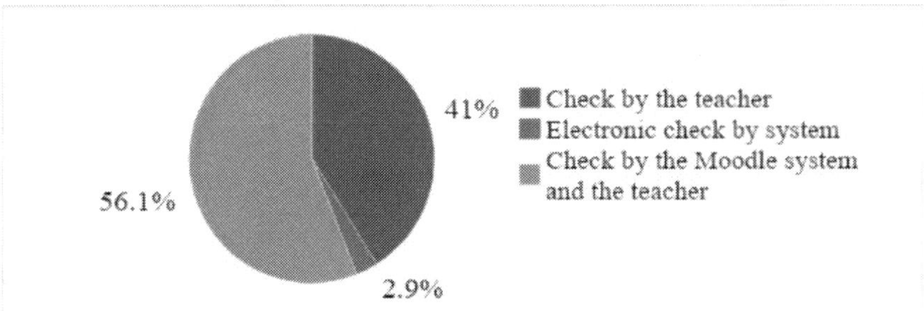

Fig. 1. What assessment of works do you consider the most objective? *Source:* compiled by the authors.

4 Discussion

The specificity of the mathematics makes it necessary to fix the sources and consequences of various mathematical errors, to check the consistency and validity of the methods used to obtain the correct result. Effective tracking of these aspects is possible only by checking the written solution to the problem and, as a result, summarizing the result in a cumulative assessment. This creates a balance between cumulative and formative assessment. We also agree with the authors of the review [14], who assign the leading role to tracking the student's individual progress by means of formative assessment.

We share the opinion of the authors [16], who consider the combination of automated assessment with other species to be the most effective. Interestingly, in addition to the teacher's assessment, the importance of self-assessment and mutual assessment is noted.

Working with electronic tests is one of the important communicative digital competence skills related to digital communication [17]. In their future professional activities, students will face the need to communicate in the format of strict digital regulations, and during their studies, it is necessary to prepare them for this.

5 Conclusion

The experience of testing in the Moodle environment and the research conducted allow us to conclude that LMS Moodle has a positive impact on the educational process of full-time students.

The important benefits of using Moodle when teaching math are:

- the ability to generate a large number of tasks;
- the variety of types of test items;
- the development of students' digital communication skills.

The formulated hypothesis of ineffectiveness of the automatic application of Moodle work algorithms in terms of evaluating work for full-time students was confirmed. The same conclusion follows from a survey of students with experience working on the platform. For the widespread introduction of LMS Moodle into the knowledge assessment process, specially developed regulations and guidelines are required, and they should not duplicate similar regulations for distance learning. For the subject "Mathematics", the authors generally recommend using a combined check: Moodle electronic assessment for effective student feedback and teacher check. The different types of verification are justified by the content and structure of the proposed assignments.

To create the most objective system for assessing the knowledge of students of any form of education, a megabase of test items is needed, focused on increasing the variety of types of tasks. The development of such a megabase is a laborious and time-consuming process that requires serious testing and adjustment. This is a prospect for the further work of the group of authors.

References

1. A-J. Moreno-Guerrero, I. Aznar-Díaz, P. Cáceres-Reche, S. Alonso-García, Mathematics, **8(5)**, 840 (2020). https://doi.org/10.3390/math8050840
2. C.P. Cortez, Europ J Interact Multim Edu, **1(1)**, e02001 (2020). https://doi.org/10.30935/ejimed/8276
3. F. Soares, A.P. Lopes, *Online assessment through Moodle platform*, in Proceedings from ICERI2018: 11th annual International Conference of Education, Research and Innovation, 4952-4960 (2018). http://dx.doi.org/10.21125/iceri.2018.2124
4. E.I. Deza, Sci Schl, **6**, 160-169 (2016).
5. M.D. Vasiliev, O.I. Matveeva, Soc: Soc, Psych, Pedag, **10(66)**, 122-127 (2019). https://doi.org/10.24158/spp.2019.10.18
6. Y. Vainshtein, et al., Proc ERPA 2018: SHS Web Conf, **48**, 0101. (2018). https://doi.org/10.1051/shsconf/20184801010
7. T.V. Zykova, et al., Persp Sci Edu, **4(34)**, 58-65 (2018)
8. Yu.V. Peryazeva, Mod Inf Techn IT Edu, **15(4)**, 876-885 (2019). https://doi.org/10.25559/SITITO.15.201904.876-885
9. O. Vaganova, et al., Int J Adv Trends Comp Sci Eng, **9(3)**, 2488–2492 (2020). https://doi.org/10.30534/ijatcse/2020/01932020
10. Zh.V. Smirnova, M.V. Mukhina, Mod Probl Sci Edu, **2**, 182-189 (2017)
11. S. Kapsargina, Zh. Shmeleva, Ju. Olentsova, *The use of LMS Moodle in the implementation of point-rating system of evaluation in the discipline Foreign Language*, in Proceedings from SGEM 2019: 19th International Multidisciplinary Scientific GeoConference, 361-368. (2019). https://doi.org/10.5593/sgem2019/5.4/S22.049
12. I.Yu. Luchenetskaya-Burdina, A.A. Fedotova, Yaroslavl Pedag Bul, **3**, 131-135 (2017)
13. S.H. Gamage, et al., Int J STEM Edu, **6**, 27 (2019). https://doi.org/10.1186/s40594-019-0181-4
14. V.A. Tunda, E.A. Tunda, Manag Probl Soc Sys, **8(12)**, 38-75 (2015). https://doi.org/10.17223/2077-6160/12/4
15. Question types: Formulas question type. Accessed on: October 3, 2021. [Online]. Available: https://moodle.org/plugins/qtype-formulas
16. N. Floratos, T. Guasch, A. Espasa, Open Praxis, **7(2)**, 141-152 (2015)
17. L.P. Konnova, I.K. Stepanyan, Stand Monit Edu, **6**, 21-27 (2020). https://doi.org/10.12737/1998-1740-2020-21-27

Raising the quality of designers' professional training through STEAM and CLIL

Ekaterina Sergeevna Marnitcyna[1] and *Ksenia Vladimirovna* Kaisheva

Saint Petersburg State University of Industrial Technologies and Design, Foreign Languages Department, Saint Petersburg, Russia

Abstract. The issues of quality assurance in the process of professional training have always been widely discussed. The current research presents the English for Academic Purposes section of the English language course design based on the STEAM approach and the feedback model for its evaluation. The study aims to assess the role of the English language course in building professional competencies with the help of the STEAM instructional design. Qualitative and quantitative methods, a formative pedagogical experiment, and an online survey were conducted. The study shows that interdisciplinary course evaluation can be based on feedback and outcomes indicators. The content and language integrated learning methods are perceived by the students as relevant. STEAM-based course design motivates students to perform fundraising activities for further professional development and continue improving their English language skills. Thus the English language can become a valid and stable base for further interdisciplinary projects in university education. The research found that the majority of the first-year art and design students learned about MOOCs for the first time and took courses to improve their professional skills during the English Language course. The authors conclude that STEAM and CLIL methodologies have much in common and are positively evaluated by both students and academics.
Keywords: educational quality, foreign language, STEAM, CLIL, art and design students

1 Introduction

Large-scale reforms implemented in the Russian educational system recently have become an integral part of the educational process and have already made some impact on the real studying and teaching practices. We have definitely come to the stage when we can discuss and analyze some of them.

The research focuses on the ways of improving the quality of education based on STEAM approach and highlights the role of the foreign language (the English language) as an auxiliary discipline when teaching art and design students. Even though the English language cannot be underestimated today and is acknowledged as lingua franca in scientific

[1] Corresponding author: ekaterinamarn@gmail.com

and academic spheres [1] according to new trends in higher education in Russia, its role (i.e. contact hours) is diminishing. However the linguists are trying to do their best to improve the English syllabus and further develop and hone students' skills, thus the new recommended syllabus includes not only traditional modules like General English, Business English and English for Specific Purposes (for professional communication) but English for Academic purposes [2], which is part of the international trend and is marked and researched in many countries, for instance, need analysis show that English for academic purposes is one of students' priorities in other countries as well [3]. Some linguists even claim that teaching ESP (English for specific purposes) at university level will soon be divided into EAP (English for academic purposes) and EOP (English for occupational/professional purposes) [4], which have already happened in some higher educational establishments. New textbooks, teaching materials and research results have been published on the topic [5-7].

2 Methods and materials

The current experiment was designed to prove that the English language could become an integral part of university curriculum, could be widely used in interdisciplinary tasks and thus it can contribute to developing both soft and hard skills. The experiment was based on the modern STEAM approach which emphasizes promotion of creativity with rationalization, this is interdisciplinary approach which focuses on creativity and problem-solving skills [8-10].

The forming pedagogical experiment was conducted in 2017-2018 at Saint Petersburg State University of Industrial Technologies and Design. The Academic Mobility section (the name used in the university as an equivalent to the term English for Academic Purposes) of the English syllabus taught to undergraduate students of the university was supported by a special textbook and online course designed on the university MOODLE platform, both being developed by the academics of the foreign languages department of the above-mentioned university.

Key areas of exploration included the ability of students to learn and study, new opportunities for acquiring professional skills, chances to study the chosen profession (design) abroad or online. The aim was to encourage students to study and develop individually not only within the English language curriculum but professionally and personally as well and to share the skills within and across art and design (their major) and auxiliary disciplines (such as a foreign language).

The experiment consisted of three phases. In the first stage, the content analysis of the best STEAM practices and methods was done, and the syllabus was designed. In the second stage, the course General English together with Academic Mobility section was delivered to the first-year students majoring in art and design. Face-to-face classes took place twice a month (totally 17 contact hours) and each face-to-face class was supported by online session which included the lesson plan (covering both the contact hours for students who missed the class to be able to catch up and online activities description). Online activities included exercises to revise new vocabulary, a text with pre- and post-reading activities and minimum one extra activity such as working out personal development plan, writing a CV, recording a video with self-presentation, writing an essay, etc. All tasks had a deadline to encourage students to build self-discipline. The final assessment was organized as a board game and group presentations (for oral practice) and as a final written test. In the third stage, the learning outcomes were assessed and the feedback using an online survey was gathered. Qualitative and quantitative research methods were employed to interpret the survey results.

Students' feedback is acknowledged to be an effective tool to correct and improve the educational process and is prescribed to be used in all countries implementing the Bologna process [11]. Feedback is a foundation to provide any process with quality assurance and basing on it, it is possible to find a reasonable balance between teachers' control and students' independent work.

Moreover, the desktop analysis of scientific literature, i.e. publications in peer-reviewed scientific journals cited in Web of Science research intelligence platform was conducted. The analysis of papers published in 2015-2020 showed rapid growth in the number of publications proposing tools and strategies of raising the quality of professional training based on STEAM approaches [12, 13]. Being a widely discussed in scientific discourse phenomena, education quality has a great variety of valid methods. The current research uses UNESCO's conceptualization of quality: a framework for understanding, monitoring and improving education quality [14]. UNESCO's methodology for quality assessment has been applied for more than 50 years and matches the requirements of interdisciplinary research. UNESCO proposes 4 dimensions that influence the core processes of teaching and learning: contextual; enabling inputs; teaching and learning; outcomes. STEAM practices on a course level can be measured with the help of the last two dimensions. The outcomes and feedback indicators for evaluating the quality of the designer's professional training were used to assess the effectiveness of STEAM-CLIL approaches embedded into the Academic Mobility section of the English syllabus.

The survey was conducted with the assistance of the university Quality Assurance Department which is in line with the University policy and meets the demands of Bologna process. The survey was published on the University's official website and consisted of 22 questions to cover different aspects of the English for academic purposes course, e.g. online support, opinion on new types of activities, etc., and included Likert scale question types and open-ended questions. 93 students participated in the experiment. 87 survey responses were gathered. The survey aimed at collecting students' feedback and investigating opinion of the participants on the new part of the English syllabus and if it contributed to both their personal and professional development. This article considers six questions logically connected with the aim of the research and with each other.

3 Results

Taking into consideration that traditional and STEAM-based English courses have the same learning outcomes it is reasonable to analyze the feedback parameters to evaluate the quality. The results are presented in two diagrams in accordance with the question type and the method.

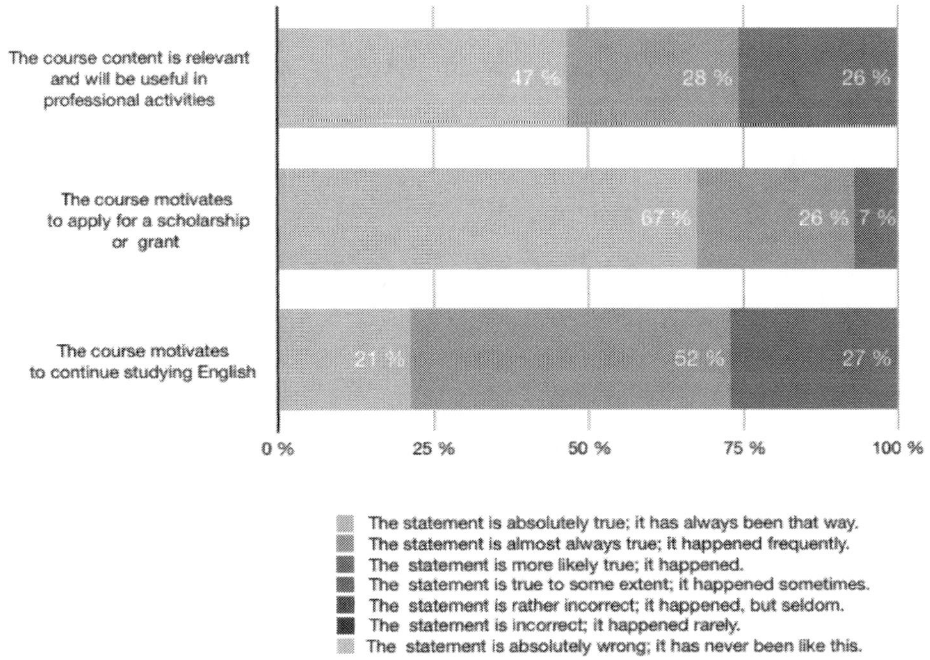

Fig. 1. Diagram 1.

Being an interdisciplinary approach STEAM focuses on problem-solving activities and innovations. Survey results presented in diagram 1 show that this project-based teaching and learning strategy motivates students to apply for a scholarship or any other funding source. More than half of respondents started considering the funding opportunities for their project. The traditional instructional design of English courses does not motivate students to perform fundraising activities.

Despite (or rather to say thanks to) the interdisciplinary character of the course it motivates respondents to continue studying the English language.

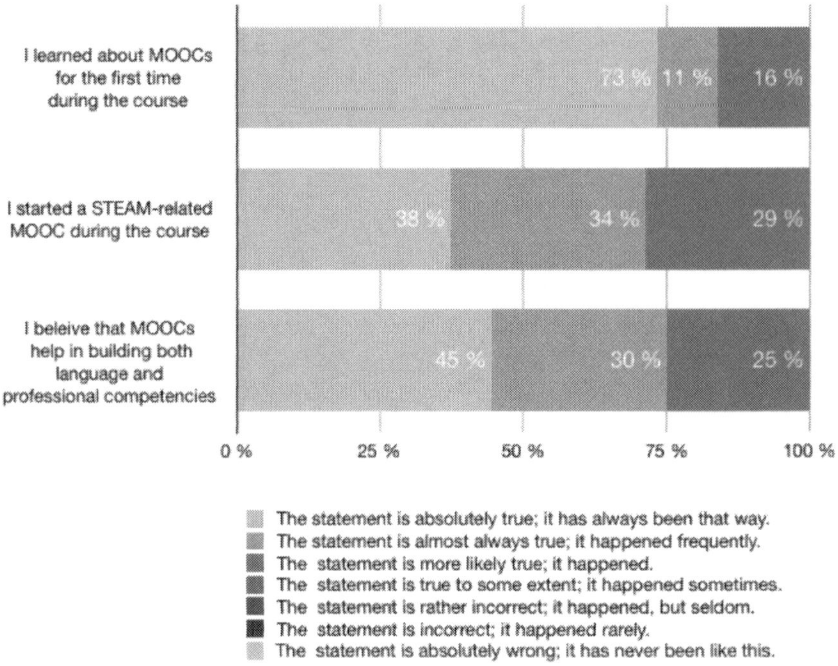

Fig. 2. Diagram 2.

The STEAM-based approach requires wide interdisciplinary knowledge. It is almost impossible to achieve the required level of STEAM subjects within a regular English course. To solve this problem, students were encouraged to choose and take any MOOC connected with their professional field or any other field where they lack skills. Surprisingly, as we see from Diagram 2, the majority of the first-year students never heard about MOOCs and the educational opportunities they provide. As a result, most of the students chose the course and started at least one of the MOOCs. Students found MOOC's to be an effective tool in studying both English and STEAM-subject or profession.

4 Discussion

The outcomes of the specially designed short course are very optimistic, you can see that students used only the three most positive answers of the Likert scale, which shows not only high level of students' satisfaction but good perspectives for further integrating the English language into university curriculum even at undergraduate level. Attempts to involve students into more profound independent work in their professional spheres turned out to be very successful by means of the foreign language; it is confirmed by the amount of students who agree that this course is useful for their occupation and by those, who really started a MOOC course. Three questions about MOOC do prove that English teachers managed to increase students' awareness of learning opportunities to develop any type of skill including professional skills. This brings us to new conclusions. Doing any course in the frame of MOOC in English (if it is not a mother tongue for a student) as a part of English for Academic Purposes with the assistance of an English teacher the process becomes a real content and language integrated learning (CLIL) practice. The STEAM methodology has much in common with CLIL and deals with building up both language and professional competencies and skills. CLIL is defined by its founder D. Marsh as a dual-focused educational approach when a foreign language is used for learning and teaching a new subject and the language itself with the aim to master both (the language and the subject) [15, 16]. Developing the ideas of D. Marsh linguists teaching languages and applying to CLIL methods concluded that the content here is the base [17]. This is exactly what happened when we forwarded students to MOOC in the frame of the English course, students concentrated on the content using the English language as a tool but still learning it. To go further we can involve a teaching strategy called 'collaborative teaching', when two teachers, a language teacher and a content teacher, cooperate [18], even though this type of cooperation is more reasonable at a later stage.

5 Conclusion

Even though this connection was confirmed by the experiment, attempts to marry these two approaches are already studied, encouraged and funded, for example, projects developed in the frame of the European project CLIL for STEAM funded by the Erasmus+ Programme of the European Union. Moreover, current interest in STEM/STEAM education generates a lot of interesting research, real practice description, their analysis and systematization [19] which could be applied to education process at different levels. This approach is perceived by both students and educator as relevant and it should not be missed by curriculum, syllabus and module template designers at university level to further develop professional skills and to save some auxiliary disciplined in the educational process.

References

1. A. Mauranen, Helsinki Eng Stud, **6(6)**, 28 (2010)
2. S.G. Ter-Minasova, E.N. Solovova, Inostrannyi yazyk dlya neyazykovykh vuzov i fakul'tetov, primernaya programma [Foreign language for non-linguistic universities and faculties, sample program] (Moscow, 2009)
3. T.A. Parnawati, A. Ulinuha, Premise J, **8(2)**, 226-240 (2019). https://doi.org/10.24127/pj.v8i2.2296
4. A. Kusni, *Reformulating English for Specific Purposes (ESP) in Indonesia: Current issues and future prospects*, in Proceedings of National Seminar on English Language Teaching, 36-48 (Universitas Negeri Padang, Padang 2013)
5. S.V. Bogolepova, Bul Omsk State Pedag Univ. Human Stud, **1(2)**, 36-39 (2014)
6. I.B. Korotkina, Domestic Foreign Pedagogy, **4(52)** (2018)
7. E.V. Dvoretskaya, Questions Mod Sci Prac, **1**, 147-152 (2016). https://doi.org/10.17277/voprosy.201601.pp.147-152
8. A. Harris, L.R. de Bruin, Aust. Art Educ, **38**, 54–75 (2017)
9. J.C. Hong, C.H. Chang, C.R. Tsai, K.H. Tai, Int. J. Sci. Edu, **41**, 1667–1681 (2019). https://doi.org/10.1080/09500693.2019.1624992
10. C. Lin, J. Huang, R. Lin, Edu Sci, **11**, 171 (2021). https://doi.org/10.3390/educsci11040171
11. J. Kettunen, M. Kantola, Edu Manag, **12:3**, 257-267 (2006). https://doi.org/10.1080/13583883.2006.9967172

12. D. Herro, et al., Int J STEM Edu, **4(1)**, 1-12 (2017). https://doi.org/10.1186/s40594-017-0094-z
13. N.H. Kang, Asia Pac. Sci. Edu, **5**, 6 (2019). https://doi.org/10.1186/s41029-019-0034-y
14. A New generation of external quality assurance dynamics of change and innovative approaches. International Institute for Educational Planning (2021). Accessed on: September 24, 2021. [Online]. Available: https://unesdoc.unesco.org/ark:/48223/pf0000377497?posInSet=1&queryId=dbddbb76-396c-4e27-a915-aadc20cbe41d
15. D. Marsh, P. Mehisto, D.Wolff, M. Frigols-Martin The European Framework for CLIL Teacher Education (European Centre for Modern Languages (ECML), Graz, 2010)
16. D. Marsh, Content and Language Integrated Learning (CLIL). A Development Trajectory (University of Cordoba, 2012)
17. L.P. Khalyapina, Bul PNRPU. Prob Ling Pedag, **2**, 149-155 (2017)
18. D. Poedjiastutie, Edu Res Rev, **12(6)**, 338–349 (2017). https://doi.org/10.5897/ERR2016.3125
19. C. Liao, *Creating a STEAM Map: A Content Analysis of Visual Art Practices in STEAM Education*, in Khine M., Areepattamannil S. (eds) (STEAM Education. Springer, Cham 2019). https://doi.org/10.1007/978-3-030-04003-1_3

Immigration situation in Russia (2016–2020)

Elena Viktorovna Ganina[1*], *Nina Aleksandrovna* Kozlovtseva[1], *Veronika Eduardovna* Matveenko[2], *Ayyub Sevdim oglu* Kerimov[3], and *Daria Valerievna* Sukhorukova[1]

[1]Financial University under the Government of the Russian Federation, Department of Foreign Languages and Intercultural Communication Moscow, Russia
[2]Saint Petersburg University, Center of Additional Educational Programmes in the Field of Russian as a Foreign Language, Saint Petersburg, Russia
[3]Azerbaijan Technical University, Department of Social Sciences, Baku, Azerbaijan

Abstract. The article overviews the peculiarities of migration in the Russian Federation in the modern period and analyzes the situation faced by people with a migrant background in Russia. The study provides a detailed description of the regions that are popular among immigrants and the underlying reasons for their choices. The main directions of migration flows in Russia are studied (Moscow, Moscow Region, St. Petersburg, Leningrad Region, Krasnodar Region, Novosibirsk and Tyumen Regions); the reasons for the popularity of these regions among migrants are revealed. The goals of the arrival of foreign migrants in Russia, as well as the information about their level of education and age category, are presented. It was revealed that more than 80% of migrants who arrived in Russia were males of working age. The article also presents statistical data on refugees and stateless persons in Russia, as well as information on national diasporas organized in the territory of the Russian Federation.
Keywords: foreign immigrants, reasons for migration

1 Introduction

The Russian Federation is the second country in the world ranking in terms of the number of migrants. Migration processes significantly affect the socio-economic and demographic situation of the Russian Federation. The migration attractiveness of Russia in comparison, for example, with the countries of the European Union, is low for a number of reasons: climatic, economic, and social. Migrants arriving in Russia are predominantly citizens from the countries of the former USSR. The high level of foreign migration to Russia is the reason for the ethnocultural, economic, and national security problems that are completely new for Russia and specific regions (for example, Siberia, Moscow, and the Moscow Region). One of the issues of foreign migration in Russia is the problem of statistical reporting of migration. Some experts [1] note that the volume of illegal migration is many times higher than the officially registered one. In this regard, the issue of collecting migration statistics in the Russian Federation and the need for a thorough analysis of the

* Corresponding author: eganina@fa.ru

problems that arise as a result of a large influx of migrants to Russia become more relevant. Nowadays "more than ever before, migration touches many countries and people of all walks of life and background" [2].

Objective: analyze the immigration situation in Russia (2016–2020), identify the causes of migration and the main directions of migration flows, and determine the impact of migration movements on the economic and social spheres in the country.

2 Materials and methods

The article relies on data from official websites of the Federal Migration Service, the Border Guard Service of the Federal Security Service, the Federal State Statistics Service, and the Central Bank of the Russian Federation. According to the Border Guard Service of the Federal Security Service data, posted on the website of the Federal State Statistics Service, in 2016 the total number of foreign migrants in Russia was 27,811,917 [3]. According to the Federal Migration Service of the Russian Federation, the total number of foreign migrants who arrived in Russia in 2016 amounted to 575,158 people, most of which were males of working age. The number of migrants from the countries of the former USSR (Azerbaijan, Armenia, Belarus, Kazakhstan, Kyrgyzstan, Tajikistan, Moldova, Uzbekistan, Ukraine, etc.) who arrived in 2016 was 511,773 people, which is 89% of the total number of foreign migrants arriving. According to the data of the Main Directorate for Migration of the Ministry of Internal Affairs of the Russian Federation (statistics as of April 2016), one of the main purposes of entry of foreign migrants is to work for hire (40% of migrants). The next most important purpose is a private one (39%), transit – 2%, business – 2%, education – 3%, business trip – 1%, tourism – 7%, commercial purpose – 1%, other reasons account for 6% [4]. The bulk of migrant employees work in the food, construction, marketing, and cleaning spheres, and also assist in households. As Lisitsyn and Rezaev have stated: "The employer prefers foreign labor migrants to local (Russian) workers due to the ability to quickly and easily resolve overtime issues beyond the time specified by law. A migrant's working day per week averages 10–12 hours" [5]. "The discrimination experienced by migrants and the other ethnic minorities through both formal policies and processes as well as informal practices or cultures is well-established" [6].

3 Results

The main flow of foreign migrants in 2016 is attributed to the Central Federal District (especially Moscow and the Moscow Region) – 190,414 foreign migrants arrived here, to the North-West Federal District (in particular, the Leningrad Region and St. Petersburg), which welcomed 59,587 foreigners in 2016; Southern Federal District (69,951 migrants), Siberian Federal District (76,511 migrants), and Volga Federal District (78,603 foreign migrants arrived). In detail, the most attractive regions for foreign migrants in 2016 were Moscow and the Moscow Region (25,788 and 35,267 people arrived, respectively), St. Petersburg (22,391 people) and the Leningrad Region (10,890 people in 2016), the Krasnodar Region (23,550 foreigners), and the Voronezh Region (18,580 migrants). These regions are characterized by a large number of jobs and a fairly high level of wages, as well as optimal climatic conditions and tourist attractions. In these regions, various nationalities and representatives of different religions live together for a long time, usually without acute ethnic conflicts. The Siberian part of Russia receives a large flow of foreign migrant workers due to the rapid development of industry in this region: in 2016, 22,360 foreigners arrived in the Novosibirsk Region, 14,471 – in the Krasnoyarsk Region, and 11,853 – in the Tyumen Region. As a rule, the Far East region of Russia is attractive mainly for Chinese workers – 11,458 foreign migrants arrived in the Khabarovsk Territory in 2016, and 12,139 people – in the Primorye Territory Chinese of working age accounted for 65% of the total number of migrants who arrived in the Far East region [3].

"The workplace and labor market experiences of migrants are highly diverse" [7]. Migrants with higher education and a doctorate/candidate of science degree tend to work in the kindergartens, schools and higher education, medicine, economics, and law; they also occupy managerial positions in various companies, etc. As for migrants from other countries (non-CIS), the largest number of them with higher education comes from Germany (1069 people), Georgia (876 people), Abkhazia (395 people), China (391 people), Latvia (403 people), Syria (304 people), Turkey (324 people), and Estonia (381 people) [8]. In 201, 5 migrants from Georgia and 4 migrants from Syria with a doctorate came to Russia, and 6 migrants from Syria and 5 migrants from Georgia had a candidate of science degree (the highest rates) [8].

"Migration is always the result of a complex combination of macro-, meso-, and micro-factors" [9]. The reasons for migration to the Russian Federation are:

1. Economic reason. The presence of friendly relations between Russia and many countries of the former USSR and a visa-free entry-exit regime create conditions for relatively free movement of the population for employment purposes. The main flow of labor migrants comes from Tajikistan, Uzbekistan, Kazakhstan, Kyrgyzstan, Azerbaijan, Armenia, Moldova, and Ukraine. In comparison with Russia, in these countries, wages are low.

2. Return of ethnic Russians to Russia (mainly from the countries of the former USSR).

3. Political reason (including military) – one of the reasons for migration to Russia is people's disagreement with the political course of the government of their native state or the military situation in the country. In 2016, there was a large influx of population from Syria (1,107 people), Ukraine (including from its western part – a total of 178,274 people), Iraq (955 people), and Afghanistan (847 people).

4) Education (academic mobility).

"There are … a number of problems related to comparing the impact of the migration dimension on acculturation, as this often entails comparisons of culturally very different groups" [9]. Migrants in Russia create their own national diasporas, public organizations, associations, communities, and cultural centers in different regions. The largest diasporas are:

1. Tajik – represented by the All-Russian social movement "Tajik Migrant Workers" (Moscow), non-governmental organization "Tajik Union of Russia" (Moscow), regional non-governmental organization "Vatan", which means "Motherland" (Moscow), non-governmental support organization for Tajikistan citizens "Compatriots" (Moscow), non-governmental organizations "Tajik Community" (Moscow) and "Tajik Culture Society" (Moscow), Tajik national cultural center "Samanids" (Yaroslavl), Tajik culture community center "Sino" (Ryazan), Tajik cultural center "Navruz" (Lipetsk, Voronezh), non-governmental organization "Friendship of Russian-Tajik People "Somonien", "Ajam" (St. Petersburg), Union of assistance in cultural and national traditions preservation, in family consolidation, in women and children protection, women involvement in social and economic activity (St. Petersburg), Charity Support Fund for Tajik migrants (St. Petersburg), Information and Culture Centre "Tajikistan" (Kaliningrad), non-governmental organization for the protection of Tajik rights and interests "Pamir" (Astrakhan), non-governmental organizations for assistance in cultural restoration of Central Asia people

"Assistance" and "Mekhri Somonien" (Krasnodar), non-governmental organizations of Tajikistan migrants "Umed" (Nizhny Novgorod) and "Tajik Union of Russia" (Perm), non-governmental organization of Tajiks "Vakhdat" (which means "Unity") (Orenburg), Tajik culture national centers "Somoni" (Kazan and Ufa), and others.

2. Uzbek – represented by the non-governmental organization Uzbek Cultural Center "Uzbekistan" (Astrakhan), Union of Uzbek Compatriots (Volgograd), Uzbek community "Farishta" (Voronezh), Uzbek national and cultural autonomies (Moscow and St. Petersburg), Council of Uzbek Diasporas (Novosibirsk), Uzbek National and Cultural Center (Bashkortostan, Russia), Uzbek National and Cultural Center "Aziz" (Rostov-on-Don), and others.

3. There are also Ukrainian, Kazakh, Moldavian, Armenian, Georgian, Syrian, Kyrgyz, Azerbaijani, Jewish, and other diasporas in Russia that organize their own cultural centers, compatriot support centers, youth (student) organizations, etc.

According to the Central Bank of the Russian Federation, in 2018, individuals located in Russia, but not citizens of the Russian Federation, transferred $7,491 million to the former Soviet republics, and $3,285 million to other countries. The average amount of one transfer to the CIS countries was $336, far abroad – $3,661. Therefore, more than twice as much money was transferred to the CIS countries in 2018 than to other countries of the world.

Transfers from Russia made in 2018 by non-resident individuals (foreign citizens) are presented in Table 1 (the figures are given for 20 countries – counterparties of the Central Bank of the Russian Federation) [11].

Table 1. Geography of transfers from Russia.

Recipient country	Transfer amount (USD mln)	Amount of one transfer (USD)	Recipient Country	Transfer Amount (USD mln)	Amount of one transfer (USD)
Former Soviet Union	7491	336	15. USA	307	15806
1. Uzbekistan	2940	427	16. UK	185	13243
2. Tajikistan	1510	195	17. China	208	7048
3. Kirghizia	1401	416	18. Spain	52	8998
4. Ukraine	124	276	19. Germany	117	10843
5. Azerbaijan	400	344	20. Cyprus	42	24287
6. Armenia	521	355	21. Austria	68	19985
7. Kazakhstan	273	1,413	22. Luxembourg	153	396922
8. Moldova	236	329	23. Italy	96	7962
9. Belarus	85	284	24. Netherlands	61	12973
10. Turkmenistan	0,5	219	25. Turkey	38	2251
Others countries	3,285	3,661	26. Monaco	32	122,585
11. Switzerland	306	91658	27. Georgia	159	544
12. Latvia	36	5640	28. France	136	10072
13. Israel	30	6903	29. Singapore	15	22179
14. Czech Republic	51	10385	30. United Arab Emirates	66	19425

4 Discussion

Currently, the issue of foreign migration to the Russian Federation is actively being discussed not only in the Russian media but also by academia – the papers by Gadelshina and Zubaidullina [12], Postavnin and Vlasova [1], Lisitsyn and Rezaev (2015), Florinskaya and Mkrtchyan, and others. Gadelshina and Zubaidullina determine the following main problems of foreign labor migration to Russia: "1. All rubles earned by migrants are exported from the country and spent outside its borders. 2. As a result of migration processes, the unemployment rate among the natives is increasing. 3. Most of these workers are illegal immigrants. This means that employers profit from their work and do not make any contributions to the budget. In addition, if there is a cheap labor force, there is no need to introduce new technologies and develop" [12]. The data of the Prosecutor General's Office of the Russian Federation on the state of crime for January-March 2019 are alarming: "Foreign citizens and stateless persons in the territory of the Russian Federation committed 9,288 crimes, of which 8,228 crimes were committed by migrants from the countries of the former USSR" [13].

According to Florinskaya and Mkrtchyan, "the number of documents issued by migrants for work in the Russian Federation does not in any way correspond to the number of those staying for the purpose of employment. The problem of legalization does not lose its relevance; moreover, in 2016 it became even more acute. In total, at the end of September 2016, 1.8 million foreigners had valid documents giving the right to work in the Russian Federation (and 4 million were for the purpose of 'work for hire')". "Companies play a central role in structuring the opportunities offered to migrants" [14]. There are 1.3 million completely illegal migrants in Russia. There is still a fairly significant share of those who entered the Russian Federation stating a "private purpose", but at the same time working without registration. A small number of foreign workers enter into official relations with Russian employers: out of 900 thousand citizens of the EAEU countries coming to "work for hire", more than 60% worked unofficially [15]. "States should redouble their efforts to reconcile national border security concerns and the basic human rights of migrants, refugees, and asylum seekers" [16].

Also, one of the most important problems is the health of migrants. "Despite the many health challenges that migrants face, data on their health status and needs are limited and fragmented" [17].

5 Conclusion

It should be noted that in recent years, the inflow of migrants from foreign countries (especially from the CIS) to the Russian Federation has sharply increased, the main purpose of which is work, since the standard of living in Russia is quite high compared to most of the CIS countries and some other neighboring countries, which serves as an "attraction" factor for migrants. By now, certain practices of resettlement of immigrants, as well as directions for the integration of migrants into the economic sphere of the regions, have already been formed. The main areas with a high concentration of migrants are Moscow and the Moscow Region, St. Petersburg and the Leningrad Region, the Krasnodar Region, the Novosibirsk Region, and the Khabarovsk Region. As a result, new large elements of the ethnocultural structure are being formed in the Russian context.

References

1. V.A. Postavnin, N.I. Vlasova, Demogr Overview, **4(7)**, 196-202 (2017)
2. D. Hughes, F. Akkök, G. Arulmani, H. Zelloth, British J Guid Couns, **47**, 1-5 (2019). https://doi.org/10.1080/03069885.2018.1564898

3. The Federal State Statistics Service. The size and migration of population of the Russian Federation in 2016. Accessed on: October 1, 2021. [Online]. Available: http://www.gks.ru/bgd/regl/b17_107/Main.html
4. The Federal State Statistics Service. The allocation of foreign migrants according to the purpose. 2016. Accessed on: October 1, 2021. [Online]. Available: http://www.gks.ru/bgd/regl/b17_107/IssWWW.exe/Stg/tab2-11-16.xls
5. P.P. Lisitsyn, A.V. Rezaev, Trudovye migranty v Rossii: strategii sotsialnoi adaptatsii [Labour migrants in Russia: strategies of social adaptation]. St. Petersburg. 2015
6. G. Netto, G. Craig, Soc Pol Society, **16(4)**, 613-622 (2017). https://doi.org/10.1017/S1474746417000203
7. C.F. Wright, S. Clibborn, J Indus Rel, **61(2)**, 157-175 (2019). https://doi.org/10.1177/0022185618824137
8. The Federal State Statistics Service. Allocation of migrants aged 14 and older according to the level of education and the home / host country in the Russian Federation in 2016. Accessed on: October 1, 2021. [Online]. Available: http://www.gks.ru/bgd/regl/b17_107/IssWWW.exe/Stg/tab2-14-16.xls
9. F. Castelli, J Travel Med, **25(1)**, tay040 (2018). https://doi.org/10.1093/jtm/tay040
10. T.F. Hedegaard, H. Bekhuis, CMS **6**, 9 (2018). https://doi.org/10.1186/s40878-018-0072-8
11. Central Bank of the Russian Federation. Statistics of external sector. Statistical data. Transborder transfers of individuals in 2018. Transfers from Russia. Accessed on: October 1, 2021. [Online]. Available: http://www.cbr.ru/statistics/?PrtId=svs
12. L.R. Gadelshina, D.V. Zubaidullina, NovaInfo.Ru, **62**, 112-116 (2017)
13. The Prosecutor General's Office of the Russian Federation. Level of crime in Russia. January-March 2019. Accessed on: October 1, 2021. [Online]. Available: http://crimestat.ru/analytics
14. L. Sandoz, J Ethnic Mig Stud, **46(1)**, 222-241 (2018). https://doi.org/10.1080/1369183X.2018.1502657
15. Y.F. Florinskaya, N.V. Mkrtchyan, Monit Econ Sit, **19(37)**, 37-41 (2016). https://doi.org/10.31737/2221-2264-2018-37-1-9
16. J. Chamie, J Mig Human Sec, **8(3)**, 230-245 (2020). https://doi.org/10.1177/2331502420948796
17. M. Norredam, Ch. Agyemang, The Lancet Glob Health, **7(7)**, E813-E814 (2019). https://doi.org/10.1016/S2214-109X(19)30224-4

Labour migration: legal means of regulation in the context of contemporary challenges

Ludmila Vasilievna Andrichenko and *Inna Vladimirovna* Plyugina[1]

Institute of Legislation and Comparative Law under the Government of the Russian Federation, Center for Public Law Research, Moscow, Russia

Abstract. For a long time, the Russian Federation has been one of the world's leaders in terms of the number of international migrants, a significant proportion of which are labour migrants. This required the development of a legal framework that would make it possible, on the one hand, to ensure the security of the state and the host society, create conditions for a prompt and flexible response to the situation in the labour market, and, on the other hand, protect the rights and freedoms of foreign nationals working in Russia, and create an attractive migration climate. The article highlights the current developments in the statutory regulation of labour migration, the legal means used to respond to modern challenges and formulates the major problems facing government bodies. Purpose: on the basis of an analysis of the development of the migration legislation of the Russian Federation, to identify the features of the regulation of labour migration and determine the areas of modernisation of the legal framework in that field. Basic methods: formal-logical, system-structural, special-legal, comparative-legal; content analysis, etc. Results: by evaluating the dynamics of legal regulation of labour migration in the Russian Federation in recent decades, the following features can be distinguished: detailed and differentiated legal status of various categories of foreign nationals involved in labour activities in Russia; the focus of legislation on the use of flexible mechanisms for foreign labour intake, which presuppose taking into account professional qualities and the level of qualifications; increased role of supranational regulators in the legal regulation of migration relations (CIS, EAEU, etc.); introduction of the technological advance achievements into the public administration mechanism which made it possible to improve control measures and migration registration etc. The developed legal means for regulating migration processes need interval updating in order to ensure an adequate response to changes in political, social, economic and other conditions, effective impact on the migration situation by taking into account the purpose and intent of the state migration policy.

Keywords: labour migration, labour migrants, foreign nationals, labour activities, migration legislation

[1] Corresponding author: inna_wp@mail.ru

1 Introduction

For several decades, Russia has been at the forefront of global migration processes; in 2020 it ranked fourth in the world in terms of the number of international migrants [1]. The COVID-19 pandemic situation and the associated restrictions had a direct impact on migration processes and reduced the Russian positive migration balance. In Quarter 1 of 2020, it corresponded to the average level in recent years, but was 38% lower than in the same period of 2019 [2]; however, similar processes can be recorded in other countries, where the pandemic has also seriously disrupted all forms of human mobility including international migration. Although the influx of labour migrants has decreased due to the COVID-19 pandemic, the areas and composition of labour migration as a whole remain stable being shaped in the overwhelming majority of citizens of the Commonwealth of Independent States.

Foreign nationals constitute a significant component of the labour resources of economic sectors of a number of regions of the country (construction, agriculture, housing and communal services, the provision of transport services, etc.). This determines the intensive enhancement of the Russian regulatory framework in the field of migration in general and, above all, in the field of labour migration. As the law enforcement practice results show, the legislative measures adopted in recent years have generated a tangible positive effect, and the foundations of a balanced migration policy of the Russian Federation have been laid. As a result of regulatory changes, the state has received tools for more flexible regulation of migration processes including optimisation of many administrative procedures which has increased the predictability of migration flows; the effectiveness of measures of state control (supervision) and the provision of public services has increased; the public-private partnership possibilities in the implementation of programmes for adaptation and integration of migrants have been expanded. At the same time, updating the strategic goals and objectives of the state migration policy, the development in 2020 of a new concept of legal regulation of its key areas requires an additional scientific assessment of the existing legal tools used to regulate labour migration, substantiation of migration legislation development forecasts.

The research community has been strongly focused on issues of the legal regulation of labour migration; there are both comprehensive studies [3, 4] and those devoted to its individual aspects [5, 6]; a significant number of works are focused on the study of those issues in the aspect of international law [7-9], although in general, the share of legal research is not dominant in the total array of works. Researchers critically assess the foreign labour intake mechanisms, the licensing procedures used to regulate labour migration, they are looking for an optimal balance of public and private interests.

It is noted that migration research, for a long time, has suffered from a divided approach based on a simplified opposition of forced and voluntary migration: "Research in sociology, anthropology and political science has discovered the inadequacy of this binary categorisation, given that the movement of people across borders is caused by a combination of many reasons, aspirations, motivations and factors that overlap and are not mutually exclusive" [7]. It is stated that it is difficult to talk about forced and voluntary migration as distinct categories as they probably exist rather on a continuum [10].

Migration issues are regulated by various branches of law including labour, administrative, civil ones, etc. When assessing the correlation of migration legislation with other branches and institutions, it is noted that in some countries, in practice, compliance with migration legislation is a priority, in particular, in relation to legislation in the area of work and employment [11].

A number of publications criticise the labour migration management paradigm, which assumes dependence on short-term economic needs and demand from local employers; justifies the feasibility of transition to a "hybrid system", according to which labour migrants are selected taking into account various criteria (language, age, previous experience in the country of destination, professional experience, etc.), and not only on the basis of their immediate needs of the employer [12].

An increase in the number of legal statuses of labour migrants, the emergence in this regard of various legal configurations, is recorded [13]. At the same time, some researchers point to the fact that the categories of migrants are artificial without often meeting the realities of life [14].

A promising area of research is the research into the limits of administrative discretion in the implementation of public administration in the field of labour migration. It is noted that law-making always implies forcing a general solution onto specific and often extremely differentiated individual situations [15]. It is emphasised that discretionary powers are a tool that ensures flexibility of migration regimes, which must stand the test of time and, in general, be able to meet the individual needs of both workers and employers; at the same time, it is emphasised that in the case when discretionary power holders are not held accountable, this power may become arbitrary [16].

2 Materials and methods

The issues raised were addressed with a systemic and structural analysis of Russian legislation and international regulatory documents, a critical study of the key outcomes of research and analysis related to the labour migration regulation. In the course of the research, the latest initiatives were taken into account involving the systematisation of migration laws and the modernisation of the overall model for regulating labour migration (this refers to the concept of the draft federal law "Conditions of Entry (Exit) and Stay (Residence) in the Russian Federation of Foreign Nationals and Stateless People"). The use of formal-logical and special-legal research methods contributed to ensuring consistency of conclusions and judgments, identifying the main trends in the development of labour migration legislation, assessing the existing tools used to achieve the goals and objectives of the state migration policy. Comparative legal and systemic and structural methods were used to determine possibilities for improving legal regulation in that area.

3 Results

1. Almost every recipient state has to resolve issues related to the protection of the national labour market and the prevention of illegal labour activities of foreign nationals.

To ensure the request of business entities to involve those particular foreign specialists that are in demand, foreign labour selection tools are used. Some states use a point system (for example, Canada, Australia), others, as the main means of state influence, use quota mechanisms that establish prohibitions and restrictions on engaging foreign nationals in certain areas or professions, etc.

The migration policy implemented by the Russian Federation pursues the regulation of the number of foreign workers involved, also based on the real situation and trends in the Russian labour market taking into account the needs of the economy and the citizens' interests. In this regard, the Russian Federation Government annually determines the general need for foreign workers arriving in the country with a visa including by priority professional qualification groups and taking into account the political, economic, social and demographic situation. In addition, the permissible share of foreign workers employed in various sectors of the economy by business entities operating in Russia is being changed. At the same time, the need for further differentiation of the conditions and procedure for involving foreign labour and expanding the practice of using selective mechanisms has become apparent for a long time.

The subsequent development of the national model for regulating labour migration can be seen in two possible areas, namely by improving the existing mechanisms and further differentiating the procedure for attracting foreign labour, or through a gradual transition to a point system, the feasibility of which is named by many researchers [17, 18]. At the same time, switching to a point-based system will require significant organisational and time costs, and will also entail a reform of the entire migration legislation and the existing institutional framework, since it will lead to the transfer to the state, that currently sets only general parameters, of the principal risks associated with assessing the level of professional qualifications from employers and customers of work (services).

2. In assessing the dynamics of legal regulation of labour migration in Russia, it is essential to state that the selective approach is increasingly being introduced into the general procedure for involving foreign labour, as manifested in stimulating the entry of certain categories of foreign nationals. The differentiation of the legal statuses of labour migrants is still in progress. In addition to such well-known categories as foreign workers, foreign nationals registered as individual entrepreneurs, foreign journalists, etc., new categories of labour migrants have appeared, in particular highly qualified specialists; qualified specialists; those invited to the Russian Federation as health workers, teaching staff or research employees and others. This "fragmentation" of legal statuses generally corresponds to the global trend, it makes it possible to more flexibly adjust the labour market in the state, solve issues associated with involving labour migrants for the needs of the economy and social sector. But if there are no any clear criteria for differentiation or any labour migration development areas determined in strategic planning documents of the Russian Federation, this approach leads to excessive complication and unjustified overloading of migration legislation.

In recent years, in the Russian legislation, a tendency to establish the specifics of carrying out labour activities by certain categories of foreign nationals in relation to specific territories has increasingly developed. Such special features are established in relation to foreign nationals working for residents of the free port of Vladivostok, in the Skolkovo innovation centre, the international medical cluster as well as the areas of advanced social and economic development, etc. At the same time, the regulation of such features is only growing, which increases the already accumulated array of reference provisions in the federal law "Legal Status of Foreign Citizens in the Russian Federation", although many such regimes for attracting foreign nationals duplicate each other. This requires the categorisation of legal regulation of the peculiarities of the labour activities of foreign nationals in special economic areas.

3. The experience of such regional interstate organisations as the Commonwealth of Independent States, and now the Eurasian Economic Union, testifies that the international legal regulation of labour migration today is becoming a unifying motive; it is aimed at solving the problems that are generated in connection with the intensification in that region of migration. In particular, the 2014 Eurasian Economic Union Treaty (EAEU) includes, among other forms of cooperation, the creation of a single labour market and migration management in the context of the free movement of goods, services, capital and workers in the common economic space. The adoption of the Treaty led to the allocation of a new

category of migrants within the EAEU with an independent status, workers of the member states of the Union, who are granted the most favoured nation treatment as an element of the international legal cooperation mechanism. By establishing a simplified legal mode for the stay and employment of those persons, the EAEU member states create regional advantages in the global competitive environment for labour and thus isolate their regional labour market. All this proves that within the framework of the EAEU, being created is their own migration system characterised by stable migration flows that have developed thanks to strong historical, economic, cultural and other ties, as well as the presence of a special system of legal regulation of migration relations supported by the regulation of adjacent relations with the involvement of migrants (in the field of labour relations, social security, education, health care, etc.).

The weak side is that within the framework of the EAEU, a unified concept for managing migration processes, which would ensure effective interstate interaction between countries of destination and origin of migrants, has not yet been developed. There is also no interstate migration registration system in the EAEU, which would make it possible to carry out a comprehensive analysis of migration processes and ensure that they are managed. This indicates the presence of constraining factors for the development of the migration system within the EAEU in the interests of all its members. It is important to focus on one more factor, that hinders the formation of a full-fledged migration system within the EAEU, the EAEU migration policy is focused only on labour migration flows within the Union by regulating the status of only migrant workers who are citizens of the Union. It does not in any way affect the provisions of the national legislation addressed to migrants from other countries of the EAEU countries. On the one hand, this preserves the independence of these states in the choice of areas of their migration policy development, on the other hand, this does not ensure the unity of this migration system, restricts its integration properties.

4. In recent years, a noticeable emphasis in the regulation of migration relations in Russia, as in many other countries, has been on the determination of the IT procedure for the regulation of migration processes including labour migration. In particular, the point at issue is about the introduction of an electronic visa system, the use of digital and information technologies in order to simplify and optimise border control, the expansion of the use of modern biometric technologies for personal identification in order to increase the efficiency of state control (supervision) in the field of migration, ensure public safety and provide government services in the field of migration; improving administrative procedures in the field of migration through the introduction of electronic forms of interaction, as well as the creation, maintenance and development of state information systems etc. Nevertheless, the level of application of information technologies in order to identify foreign nationals when providing them with public services and exercising state control in the field of migration is still not high enough. Thus, experts note that it is necessary to further expand the use of up-to-date digital technologies in order to study, assess and predict changes in the migration situation; the use of biometric information in the implementation of state control (supervision) and the provision of public services in the field of migration [19].

In this regard, the initiative to issue ID-cards with personal data for foreign nationals staying in the Russian Federation for a long time, as well as for persons who have received the right to permanent residence, seems promising. It is assumed that such a card will be an identity document; it will be possible to issue it subject to fingerprint registration, photographing, medical examination, and confirmation of the existence of a voluntary medical insurance policy.

5. The shift in the focus of the problems of regulatory policy in the field of labour migration to the socio-cultural sphere is due to the special importance of issues of the migrant socialisation. The influence of the sociocultural context is determined by the fact that migration policy has features that distinguish it from policy in other social areas. In this context, to a greater extent, certain sociocultural restrictions remain, which are due to the practice of interaction between the host population and immigrants. The "blurring" of the boundaries between the host community and migrants largely depends on the focus of the latter towards adaptation in the new social environment.

The provision of the legal basis for social and cultural adaptation and integration has not yet been completed, which is primarily due to the difficulties of the final choice of a common integration model. The creation of an integrated system for the provision of information services and legal support, including in the national languages of those states, from which the largest number of foreign nationals arrive, could contribute to solving the adaptation and integration problems. Measures aimed at adaptation and integration should be aimed not only at the labour migrants themselves but also their family members, who enter with them, and also involve working with the host society. In addition, it is necessary to more clearly delineate the powers of public authorities at various territorial levels.

4 Discussion

One of the debatable issues is the assessment of the possibility and feasibility of the transition to a point system. Despite the fact that this idea has a significant number of supporters and can be generally implemented in practice by taking into account the peculiarities of the modern legal system of the Russian Federation; nevertheless, it can hardly be taken unconditionally, given the significant number of foreign workers in Russia, as well as the visa-free border crossing regime established for many countries, as already indicated in earlier studies [20]. At the present stage, one of the regulatory options is the selective use of a point system to attract certain categories of foreign specialists, for example, those invited to implement large investment projects. This will make it possible to go through its approbation, identify the chances of further using it for the selection of specialists on a larger scale.

The employment differentiation trend by certain categories of foreign nationals recorded in recent decades will continue and will be associated with a layering of regulatory impact in the context of various legal frameworks applied in certain areas.

In the scientific literature, various migration systems are distinguished, among them North American, European, Asia-Pacific, Middle East, South American [21]; Russian scientist I.V. Ivakhnyuk substantiated that a Eurasian migration system has developed in the post-Soviet space, namely a group of post-Soviet countries linked by strong and numerous migration flows, which are caused by the interaction of a number of factors, i.e. historical, economic, political, demographic, socio-ethnic, geographical [22]. It can be argued that the migration sub-system is currently being generated within the EAEU.

The national integration model, which is not finalised in its final form, raises questions as to which social and cultural adaptation and integration measures should be applied to labour migrants, at whose expense they should be financed, over what period of time the use of them is relevant. The proposed combination of funding sources for such events should be supported; it can be carried out both at the expense of a foreign national himself/herself, the inviting party (in particular, the employer), civil society institutions, and at the expense of budgetary funds. An additional scientific assessment needs to determine the role of employers and/or work (service) customers during the adaptation and integration.

5 Conclusion

The identified trends in the development of migration legislation are due to the peculiarities of the political, social, economic, cultural and other conditions of the Russian Federation as a recipient state, in the territory of which there are a lot of foreign nationals. Currently, the Russian Federation has developed a toolkit that makes it possible to regulate labour migration and respond to factors affecting the labour market and migration situation; however, its regular updating takes place taking into account the adjustment of strategic goals and objectives, changes in the structure of migration flows, the emergence of new challenges and threats.

References

1. International Organization for Migration. World migration report 2020. Accessed on: October 2, 2021. [Online]. Available: https://www.un.org/sites/un2.un.org/files/wmr_2020.pdf
2. Y.F .Florinskaya, N.V. Mkrtchyan, Econ Devel Rus, **28(1)**, 50-54 (2021)
3. P.V. Pankin, Regulirovanie rynka truda s uchetom osobennostei trudovoi migratsii [Labor market regulation taking into account the specifics of labor migration]. (Yelets State University named after I.A. Bunin, Yelets, 2019)
4. A.V. Topilin, O.A. Parfentseva, Perspektivy trudovoi migratsii v Rossii: ot kolichestvennykh k kachestvennym parametram [Prospects for labor migration in Russia: from quantitative to qualitative parameters]. (MAKS Press, Moscow, 2008)
5. E.A. Malyshev, Regulirovanie vneshnei trudovoi migratsii administrativno-prinuditelnymi merami [Regulation of external labor migration by administrative and compulsory measures]. (INFRA-M, Moscow, 2016). https://doi.org/10.12737/18210
6. I.N. Scherer, Problemy molodezhnoi trudovoi migratsii Rossii [Problems of youth labor migration in Russia]. (Non-Governmental Educational Institution of Higher Professional Education, Volgograd, 2010)
7. V. Chetail, International Migration Law. (Oxford University Press, 2019)
8. D.K. Bekyashev, D.V. Ivanov, Mezhdunarodno-pravovoe regulirovanie vynuzhdennoi i trudovoi migratsii [International legal regulation of forced and labour migration]. (Prospect, Moscow, 2014)
9. S.E. Metelev, Mezhdunarodnaya trudovaya migratsiya v usloviyakh globalizatsii i nelegalnaya migratsiya v Rossii [International labor migration in the context of globalization and illegal migration in Russia]. (UNITI: Law and Law, Moscow, 2006)
10. E. Snel, Ö. Bilgili, R. Staring, J Ethnic Mig Stud, **47(14)**, 3209-3225 (2020). https://doi.org/10.1080/1369183X.2020.1804189
11. W. Palmer, A. Missbach, Third World Quart, **40(5)**, 908-925 (2019). https://doi.org/10.1080/01436597.2018.1522586
12. S. Sarolea, Legal migration in the "New Pact": modesty or unease in the Berlaymont? Accessed on: October 2, 2021. [Online]. Available:

https://eumigrationlawblog.eu/legal-migration-in-the-new-pact-modesty-or-unease-in-the-berlaymont/

13. J. Könönen, Work, Empl Soc, **33(5)**, 777-793 (2019). https://doi.org/10.1177/0950017019835133
14. L. Sandoz, J Ethnic Mig Stud, **46(1)**, 222-241 (2020). https://doi.org/10.1080/1369183X.2018.1502657
15. M. Zamboni, The Theory Pract Leg, **7(2)**, 101-133 (2019). https://doi.org/10.1080/20508840.2020.1729599
16. T. De Lange, P. De Sena, The Theory Pract Leg, **7(2)**, 135-151 (2019). https://doi.org/10.1080/20508840.2020.1729559
17. A.G. Grishanova, E.S. Krasinets, Immig Law, **1**, 26-30 (2014)
18. O.T. Suyunchaliyeva, Proc Acad Manag Ministry Int Aff Rus, **2(22)**, 123-125 (2012)
19. V.E. Vasiliev, K.V. Vlasova, Sci J "Epomen", **44**, 62-73 (2020)
20. L.V. Andrichenko, J Rus Law, **7**, 5-16 (2011)
21. St. Castles, M.J. Miller, The Age of Migration: International Population Movements in the Modern World. (Palgrave Macmillan, New York, 2009)
22. I.V. Ivakhnyuk, RUDN Newsl, Int Rel Ser, **2**, 21-28 (2008)

The selectivity of attracting foreign nationals to work in recipient countries

Marina Leonidovna Galas[1]

Financial University under the Government of the Russian Federation, Department of Political Science of the Faculty of Social Sciences and Mass Communications, Moscow, Russia

Abstract. The article provides a comparative analysis of practices on regulating the employment of foreign nationals, used in large recipient states (Germany, USA, Canada, Russia, Republic of Korea). The subject of the study is the process of external labor migration, taking into account country regulatory approaches to attracting foreign labour. The study aims to develop theoretical principles and methodology for regulating the process of external labor migration. The methodology is based on the polyparadigmal concept of the study, using theoretical experience in the study of migration processes and the practice of influencing their regulation. The scientific novelty of the study is to assess the effectiveness of a selective approach to attracting foreign workers to national labour markets. As a result of the study, a methodology has been developed to assess the impact of migration processes on labour market regulation in recipient countries. Promising for countries hosting external migrant workers is the mechanism of targeted organized recruitment of foreign nationals to carry out work activities. At the heart of this mechanism is the selective selection of job seekers on the basis of professional, sociocultural, educational and personal criteria that correspond to the interests of the host external migrants of society and the recipient state. To understand the basic administrative approaches and the system of economic, legal and social instruments for regulating the process of external labour migration, it is important to study the practice of adaptation activities of large recipient States, which have a systematic and proven mechanism of inclusion of foreign job seekers. Measures to regulate external labour migration needed to be balanced, since increased restrictive measures could lead to an increase in illegal external labour migration, and excessive preferences for foreign workers could restrict the rights of recipient citizens.
Keywords: external labour migration, adaptation of migrants, selective employment

1 Introduction

It is known that migrants, including labour, are socially vulnerable because of objective factors related, especially to their presence in a foreign country. The pandemic of the new

[1] Corresponding author: MLGalas@fa.ru

coronavirus infection, which has affected all the states of the world community without exception, and its limitations have multiplied this vulnerability many times, putting migrant workers in a very difficult position that has proved much more problematic than that of its own citizens. The situation is also aggravated by its very uniqueness – our civilization has not previously had to face a crisis of such a global scale, and accordingly, the nations of the world community have no experience in such difficult conditions.

Analysis of the adaptive policy of recipient states towards foreign workers shows the risk of developing conflicts of interest between citizens and external migrant workers [1-4], identifies the problems of integrating foreign workers into national systems of economic relations and labour market regulation [5], as well as the social and economic inclusion of refugees [6]. Adaptation issues are addressed to human rights and freedoms [7]. Recipient states in the situation of increasing crisis migration are increasingly taking selective (restricting entry of undesirable categories of external migrants) in addition to adaptive preferences for foreign workers in demand in the labour market [8-10]. International migration is transforming the structure of the host society [11]. One of the most modern and most promising mechanisms for the formation of an effective labour market to date is the scheme of selectively organized recruitment of foreign labor (or managed employment). The Russian Federation has only just begun to implement this mechanism and its full potential has yet to be discovered and assessed. In this regard, it is interesting to see the practice of implementing the measures of an organized set of countries where this mechanism has been used for a long time, tested and well-established.

2 Materials and methods

The comparative analysis of the most representative and comparable country criteria for regulating external labour migration processes, including the adaptation of foreign workers, is indicative. These criteria include: the competence of public authorities and governments that have the power to regulate migration; The level of legislative framework for the administration of external labour migration; a system of rights and responsibilities for foreign job seekers; The level of involvement of civil society structures in the adaptation of foreign workers; mechanisms for regulating labour migration need to be updated. Since external migration has demographic and economic potential, it contains a number of risks that could escalate into threats to public stability and the interests of recipient countries. In this regard, it is important to develop a system of effective mechanisms for selective regulation of external labour migration. The risks of labor migration are considered by the author as a factor in the occurrence of conflict situations. The challenge is to prevent such risks through regulatory measures, applying the principle of selectivity in the selection of foreign labour. The author proposes to correlate the process of external labor migration with the level of social development of recipient states, with measures of social inclusion of foreign workers. A comparative analysis of national approaches to selective selection of foreign labour is viewed in conjunction with external and domestic socio-economic and political processes. Methods have been developed: socio-political forecasting of the risks of external labor migration; Assessing the impact of labour migration on the inclusivity of the economy; the use of social entrepreneurship tools to encrust for foreign workers.

3 Results

The Republic of Korea is a leader in the Asian region in implementing and implementing temporary labour migration programmes based on the principle of organized recruitment. The object of the Korean organized recruitment program, Employment Permit System (EPS), is an unskilled foreign workforce. Such foreign workers are afforded protection of labor and social rights at exactly the same level as their own citizens. The EPS program, launched in 2003, is implemented through bilateral intergovernmental agreements, with full exclusion from the recruitment and selection process of private recruitment organizations or agencies. The treaties define the reciprocal responsibilities and responsibilities of the signatory States, the Republic of Korea and the sending countries, and establish the manner in which the parties work together to hire, select, locate, protect and benefit migrant workers. State control is a key element of the system; EPS was originally designed to root out corruption, with human rights violations. In 2011, EPS was awarded the UN Public Service Award for its contribution to transparency and anti-corruption.

The Russian Concept of State Migration Policy for 2019-2025 one of the key areas of work on labour migration regulation is the development of mechanisms for the organized recruitment and recruitment of foreign nationals for work; qualitatively improving the professional and qualification level of foreign workers involved [12]. In the short term, the Russian Federation needs to move fully to the organized attraction of external migrant workers who are in a manner that does not require a visa (citizens of the member states of the Eurasian Economic Union).

The Eurasian Economic Union Treaty of May 29, 2014, states that member states are cooperating in harmonizing labour migration policies, as well as facilitating the organized recruitment of workers from member states to carry out work activities. Social guarantees and equal access to work for workers from the EAEU member states with Russian citizens are enshrined in the legislation. The ability to ensure a legally enshrined set of rights depends on the economic opportunities of the host regions.

The first example of the development of the organized recruitment system was the Memorandum between the Government of St. Petersburg and the Executive Authority of Dushanbe (Republic of Tajikistan) on cooperation on labor migration. The agreement established cooperation in the issues of attracting foreign labor between the Committee on Labor and Employment of the population of St. Petersburg and the Office of Migration Service of the Ministry of Labor, Migration and Employment of the Republic of Tajikistan in Dushanbe. In 2017, an intergovernmental agreement was signed between Russia and the Republic of Uzbekistan on the organized recruitment of citizens of Uzbekistan to carry out temporary work in Russia [13]. Federal law No. 366-Fz on ratification of this agreement came into force on 16.12.2017. The agreement provides for the training of Uzbek workers in the country of residence to carry out work in Russia. Uzbekistan's authorised authorities can select candidates for employment in Russia from among Uzbek nationals according to the professional qualifications and experience required for employers in the relevant field of work; to check the law-abidingness of candidates. In order to implement the Agreement, a system of training and professional certification of job seekers in accordance with the requirements of professional standards in Russia will be organized; legislated to establish in Russia and Uzbekistan the legal responsibility of Uzbek workers for violations of the employment contract without good reason; to establish a supervisory structure.

In addressing the selection of foreign workers, Russia has formed a system of preferences for sociocultural and economically promising categories of foreign nationals: the right of visa-free entry on the basis of a patent; Permission to operate during the holidays and in their spare time to persons studying full-time in professional educational organizations and educational organizations of higher education; invitation to the country as scientific or pedagogical and medical workers. The Government of the Russian Federation sets a limit on each branch of the economy to attract foreign workers on the basis of the All-Russian Classifier of Economic Activities.

The Government of the Russian Federation recommends that actors monitor the labour market in the light of the total number of foreign workers and the level of total unemployment (based on the methodology of the International Labour Organization). The decision of the Russian Trilateral Commission for the Regulation of Social and Labour Relations is to check the proportion of foreign workers employed in the number of economically active people and the number of employed in the economy, which should not exceed the level of tension in the regional labor market.

The German Labour External Migration Act should be considered a systemic element of the national legal framework for the employment of external migration. By 2035, the working-age population is expected to shrink by 4 million to 6 million. The German government intends to attract more foreign workers and has already made it easier for asylum seekers to access training programmes. In 2016, at the legislative level, asylum seekers were allowed to complete a three-year training program and continue to work for the same company for another 2 years. Until the start of employment, a refugee must have lived in Germany for at least 3 months and apply for an initial work permit, which is issued on the condition that the employment of the applicant will not have a negative impact on the employment prospects of German citizens, EU citizens or persons entitled to permanent residence in the country [14].

Organized involvement of foreign nationals in the workforce is an important indicator of social inclusion in Germany, and a tool for balancing the distribution of foreign labour, taking into account national and regional economic opportunities, and needs, taking into account the legitimate interests of the citizens of the host state and human rights.

The lack of high-skilled high-tech workers in Germany in the 2000s led to changes in legislation aimed at supporting and controlling, social preferences in obtaining the right to permanent residence, facilitating access to integration courses, and employment for refugees, foreign nationals with socio-economic potential. In Germany, the most favoured regime for highly qualified specialists.

Labour external migration for Canada is a source of progressive economic and social growth, a means of human development. The Canadian Multiculturalism Act 1988 regulated the sources of funding for integration activities, and allowed agreements between the executive branch and provincial governments to implement migration policy. The Immigration and Refugee Protection Act 2001 legitimized the categories of foreign nationals granted preference for permanent resident status; enshrined the host State's obligations with regard to foreign nationals, as well as the rights and responsibilities of immigrants. All levels of government and civil society institutions (employers, trade unions, educational and social institutions, health structures) are involved in regulating the social aspects of migration processes. Taking into account the socio-economic opportunities of the provinces, it is planned to use foreign human capital in the labor and employment markets. Unlike Germany and Russia in Canada, municipalities play a key role in the immigration partnership, which provides access to partnerships to leading economic market participants, addressing the issues of social inclusion of foreign nationals.

For the United States, an effective system for regulating external migration involves addressing the problem of employers using foreign workers without a certain legal status. In November 1986, Congress passed the Federal Immigration and Nationality Act, 2020 [15], which tightened sanctions on employers who hired "undocumented" "undocumented" migrants, making it easier to obtain U.S. citizenship for such foreign nationals who were in the U.S. at the time of the law's passage. For the first time, U.S. federal law has placed employers in charge of verifying and keeping records of the work permits they employ foreign workers. States and municipalities also have the power to regulate migration. In 2001, a draft Law on the Development, Assistance and Education for Foreign Minors was submitted to the U.S. Congress. The Dream Act proposed the introduction of a conditional temporary residence institution with the right to work for skilled migrants who arrived in the U.S. under the age of 16 and met the required qualities to obtain the right to permanent residence [16], known as the Innovation and Leadership Knowledge Act, aims to increase the attraction in the U.S. of legitimized highly skilled immigrants engaged in science, technology, mathematics and engineering by increasing H-1B visa quotas, lifting green card restrictions for those with degrees. The Comprehensive Immigration Reform Act of 2006 is interesting for foreign workers who have been living in the U.S. for a long time (more than 5 years) and have not been documented for a long time or continuously living in the United States[17]. Comprehensive Immigration Reform Act 2007 [18] simplified the citizenship process for highly qualified U.S. employees. Law S 744 [19] on border security, 2013 is indicative of the introduction of a ball system of qualification, educational and loyal assessment of immigrants, as well as a new differentiated visa program that grads the statuses of highly skilled, low-skilled and unskilled immigrants, foreign entrepreneurs. Green cards have been upgraded for immigrants studying at U.S. universities and immigrants with a degree in high technology, engineering and mathematics, protected in the United States. The law approved a program to open jobs for young immigrants.

4 Discussion

In the author's previous works, the problem of selective organized recruitment of foreign workers was not addressed on purpose. The focus was on the issues of world and national factors that cause the increase in migration, on migration risks and threats, as well as on the mechanisms of social and economic inclusion of foreign nationals in Russia and in the recipient countries [1, 2].

Borneman J., Ghassem-Fachandi P. [4] considered the German concept of Stimmung, which compares the internal state of the migrant-receiving society, and objective forms of response to the growth of migration flows. The impact of the migration process on the political situation in the country is shown. This author concentrates the study on the psychological adaptation of migrants, leaving aside the problem of organized involvement of foreign nationals in the German economy and social environment.

Bygnes S. correlates on the basis of the data of the sociological poros of the financial and economic world crisis and migration flows to the EU countries. The author focuses on private migration, which does not allow us to see the general patterns of the migration crisis and to propose a mechanism for economic and social adaptation of external migrants [3].

Mackenzie R., Lucio M. Public and private methods of regulating the labour market at the macro, meso- and micro-levels are characterized. These authors propose models of the socio-economic transformation of labor relations. A significant role is played in the models of trade union structures, forms of employment contract and professional compliance. These researchers are interested in the social experience of external migrant workers [5].

The genesis of international regulation of refugee and international migrants is theoretically and methodologically investigated by Braithwaite A., Salehyan I. & Savun B. These scientists pay more attention to the motivation of international migration These scientists pay more attention to the motivation of international migration [6]. National, regional and international priorities in the regulation of international migration re explored by Akanle O. This author analyses international migration regarding the activities of international organizations. More researchers are interested in unresolved migration and measures to facilitate the return of international migrants to countries of departure [11]. Global, regional and national migration flows can be interpreted from a security and human rights perspective. Модели регулирования миграции определенных групп населения при таком рассмотрении строится только на вопросах правам человека и безопасности мигрантов [7].

Strategies to reform international and regional migration policies in terms of asylum and humanitarian protection, resolution of border conflicts and forms of humanitarian assistance are considered [8]. The complex problems of organizing the entry of international migrants, depending on the capacity of the recipient states, are not taken into account. structures knowledge about migration and shapes policy decisions and governance structures.

The example of Mediterranean migration explores the process of political decision-making in the EU countries and mechanisms for managing migration flows [9]. Armed conflict is a major factor in modern international migration. The intra-migration selection of foreign nationals forced to leave their countries for relatively prosperous EU states is analyzed. This selection is based on the social status and financial capacity of the migrants themselves [10].

5 Conclusion

Modern breeding vectors in attracting foreign workers to recipient countries: helping migrant workers to master sought-after professions and professions; creating organized forms of migration of tour resources to minimize risks for both migrants and employers. The model of selectively organizing the recruitment of foreign nationals for temporary employment in the recipient country may include steps:

- determining the needs of regional labour markets to attract foreign workers;
- selection of applicants for employment in the recipient country in accordance with the needs of the labour market, the request of employers, professional and qualifications;
- conducting a set of training activities and organizing the entry of job seekers in the recipient country;
- organizational support of job seekers in the recipient country in terms of paperwork authorizing the implementation of work activities, conclusion of employment contracts; monitoring and monitoring the activities of foreign workers.

Two ways can be used to enhance the process of organized recruitment of foreign workers: 1) mantaining mandatory standards for work/service customers (employers) and foreign workers; 2) develop a legal system to encourage customers of work and services in the development and development of a mechanism for the organized attraction of foreign workers to the labor market. Incentives can be:

- a simplified procedure for obtaining a patent or work permit to carry out the work of foreign nationals participating in an organized recruitment;
- State support for the establishment of a national system for assessing the professional qualifications of foreign professionals in demand in the labour market;
- introduction of the institution of targeted migration amnesty in respect of foreign workers in demand on the labour market, who have undergone the procedure of national assessment of professional qualifications under the obligation of the customer of works and services;
- the introduction of legal liability for the uncertainty of the legal status of job seekers in relation to employment agencies engaged in the organized recruitment of foreign workers;

- tightening the liability of a foreign worker employed under the targeted recruitment programme for non-compliance with obligations under an employment contract for reasons not provided by the recipient country's labour laws;

- in order to encourage the entry of prospective foreign youth, it is advisable to apply the mechanisms of targeted admission to higher education programs at the expense of regional and local budgets, taking into account the specialties and training areas of specialists most sought after in the sectors of the recipient country's economy.

Recipient States are tasked with defining for the medium and long term those categories of migrants that are important for national socio-economic interests are ready to adapt and integrate into the country of entry without conflict.

References

1. M.L. Galas, G.G. Sillaste, K.D. Shevchenko, Migratsiya v usloviyakh krizisnogo razvitiya mirovogo soobshchestva i ee riski dlya Rossii [Migration in the face of the crisis development of the world community and its risks for Russia] (Prometheus, Moscow, 2020)
2. M.L. Galas, T.A. Gorochnikova, K.D. Shevchenko, Integratsiya inostrannykh grazhdan: realnost i prognozirovanie migratsionnykh protsessov [Inclusion of foreign citizens: reality and forecasting of migration processes] (Prometheus, Moscow, 2021)
3. S. Bygnes, Acta Soc, XX(X), 1–13 (2019). https://doi.org/10.1177/0001699319833143
4. J. Borneman, P. Ghassem-Fachandi, J Ethn Theory, 7(3), 105–135 (2017). https://doi.org/10.14318/hau7.3.006.
5. R. Mackenzie, M.M. Lucio, J Indust Rel, 61(2), 176-197 (2019). https://doi.org/10.1177/0022185618814280
6. A. Braithwaite, I. Salehyan, B. Savun, J Peace Res, 56(1), 5-11 (2018). https://doi.org/10.1177/0022343318814128
7. B. Purkayastha, Curr Soc, 66(2), 167-191 (2017). https://doi.org/10.1177/0011392117736302
8. V. Squire, The British J Pol Int Rel, 20(2), 441-458 (2018). https://doi.org/10.1177/1369148118759616
9. N. Dines, N. Montagna, E. Vacchelli, Sociology, 52(3), 439-447 (2018). https://doi.org/10.1177/0038038518767372
10. J. Schon, J Peace Res, 56(1), 12-27 (2018). https://doi.org/10.1177/0022343318806044
11. O. Akanle, Int Soc, 33(2), 161-170 (2018). https://doi.org/10.1177/0268580918757105
12. Ukaz "O Kontseptsii gosudarstvennoi migratsionnoi politiki Rossiiskoi Federatsii na 2019–2025 gody" [Decree on the Concept of State Migration Policy of the Russian Federation for 2019 – 2025]. Accessed on: September 26, 2021. [Online]. Available: http://kremlin.ru/events/president/news/58986
13. Soglashenie mezhdu Pravitelstvom Rossiiskoi Federatsii i Pravitelstvom Respubliki Uzbekistan ob organizovannom nabore i privlechenii grazhdan Respubliki Uzbekistan dlya osushchestvleniya vremennoi trudovoi deyatelnosti na territorii Rossiiskoi Federatsii [An agreement between the Government of the Russian Federation and the Government of the Republic of Uzbekistan on the organized recruitment and involvement of citizens of the Republic of Uzbekistan for temporary work in the

russian Federation]. Accessed on: September 26, 2021. [Online]. Available: https://docs.cntd.ru/document/542614408

14. The State of Refugee Integration in Germany in 2019. Accessed on: September 26, 2021. [Online]. Available: https://wenr.wes.org/2019/08/the-state-of-refugee-integration-in-germany-in-2019
15. Immigration and Nationality Act, 2020. Accessed on: September 26, 2021. [Online]. Available: https://www.uscis.gov/legal-resources/immigration-and-nationality-act#top
16. Dream Act 2017. Accessed on: September 26, 2021. [Online]. Available:: https://www.congress.gov/bill/115th-congress/house-bill/3440/text
17. Comprehensive immigration reform act, 2006. Accessed on: September 26, 2021. [Online]. Available: https://www.congress.gov/bill/109th-congress/senate-bill/2611
18. Comprehensive immigration reform act S1348. Accessed on: September 26, 2021. [Online]. Available: https://www.congress.gov/bill/110th-congress/senate-bill/1348
19. Law S 744 on border security, economic opportunities, immigration and modernization. Accessed on: September 26, 2021. [Online]. Available: https://www.congress.gov/bill/113th-congress/senate-bill/744

Trends in the settlement of migrant workers in the context of social and economic processes in contemporary Russia

Mikhail Yuryevich Zelenkov[1*], *Aleksander Aleksandrovich* Nikolaev[1], *Pavel Viktorovich* Razov[1], *Tatiana Innokentyevna* Pak[1], and *Julia Olegovna* Sulyagina[2]

[1]Financial University under the Government of the Russian Federation, Department of Sociology, Moscow, Russia
[2]Russian State Social University, Department of Management and Administrative Management, Moscow, Russia

Abstract. The purpose of the article is to identify the basic trends in the settlement of migrant workers in the Russian Federation. The methodology of the research was based on the humanistic approach, which allowed focusing on the problems of labor migrants' life activity and their search for value-based orientations. The systemic and comprehensive approach allowed identifying and substantiating the directions of the trends in the settlement of labor migrants by regions and districts of Russia. The comparative analysis of scientific works helped to identify the causes of settlement of labor migrants by place of their labor activity. Two identified opposing trends are the main results of the study. The first is dispersed settlement, which objectively promotes the integration process of representatives of immigrant ethnic groups into the urban environment of the area of residence, at the expense of a simultaneous weakening of intra-ethnic social ties and an increase in the depth and density of contacts with the local population. The second is the opposite trend – the self-isolation of immigrant groups within the framework of compact settlements, leading to the emergence of large ethnic diasporas of migrants in certain areas of the hosting country. The novelty of the study comes from an original approach to the identification and justification of migrants' settlement trends in the socio-economic conditions of modern Russia.
Keywords: tendency, adaptation, integration, migration, labor migrants

1 Introduction

International migration has become a global phenomenon in the second decade of the 21st century. More and more people are looking for employment opportunities outside their countries. Before the global pandemic, according to the World Bank, the number of

* Corresponding author: mz60@mail.ru

migrants (not including refugees) reached about 240 million, but even more impressive is the amount of remittances sent home by foreign workers – $483 million [1]. In recent years, labor migration has tended to change in terms of settlement by regions and districts of the country of their stay. Moreover, the changing geography of labor migration was first noted in scientific papers in the United States in the 1990s [2, 3], followed by parallel developments concerning the southern parts of Europe at the turn of the millennium [4-6]. Studies on the settlement of labor migrants in the northern parts of Europe have recently emerged and contain the EU enlargement process that began in 2004 in their origins [7-10].

As for Russia, studies show that the increased pace of development of its economy, economic restructuring, together with a low inflow of replenishment of the national labor market due to the low birth rate among Russians and the increased duration of education for young people has resulted in relatively high demand for labor resources in the Russian Federation [11]. A comparative analysis of scientific works has revealed a wide range of methodological approaches to the study of labor migration: demographic [12], political science [13], historical [14], sociological [15-17], psychological [18, 19], economic [20]. The problem of labor migration is also considered on the basis of the network approach [21], the new institutional approach [22-24], risk theory [25], the concept of "human capital" [26-28].

Such a wide range of methodological approaches can be explained by the fact that the Russian population, for the most part, has a relatively high level of general education and professional qualifications, and mostly has the job that corresponds to its level of capabilities, needs, and interests. In addition, the domestic representatives of the working population avoid taking low-skilled, low-paid, and non-prestigious jobs, even despite the unemployment (5.8% as of January 2021, according to Rosstat). All this combined has led to the fact that Russia has a shortage of low-skilled labor force, which started to be eliminated by immigration of labor force mainly from the CIS countries, where there is an excess of labor population, primarily Uzbekistan, Moldova, Tajikistan, and Kyrgyzstan. Under these conditions, the process of labor migration has become so important that it already affects the formation of certain sectors of the labor market in Russia (construction and utilities).

In such an environment, a dilemma for the executive branch is on the agenda. On the one hand, the Russian economy is experiencing an acute shortage of human resources, which is eliminated, to a large extent, with the help of foreign hired workers. On the other hand, for the successful use of such personnel, it is necessary to create conditions for their adaptation and safe labor activity. Adaptation should be understood as a process of adjustment of incoming workers to different living conditions. As a result, new behavioral stereotypes are formed, and a new way of life is actually developed, which implies a certain structure, including the main areas of people's life: financial, economic, social, family, religious, household, security. An important role in this process is played by the area and region of foreign workers' settlement, which today is already subject to certain unwritten laws. This study is devoted to identifying the main trends in the settlement of migrants in Russia.

2 Materials and methods

The methodological basis of the study consists of scientific works that describe the problem of labor migration [1-12, 15, 17, 21, 26, 27, 29-34] and trends in the settlement of migration in different regions of the world and in Russia [13, 16, 23, 35-39]. The methodology of the study included a humanistic approach, which allowed focusing on the problems of migrant workers' life activity and their search for value-based orientations, a systemic and comprehensive approach, which allowed identifying and substantiating the trends in the settlement of migrant workers in the regions of Russia, and a comparative analysis of scientific works in order to determine the causes of migrant workers' settlement in the regions of their labor activity.

3 Results

The results of the study led to the following conclusions:

1. Labor exporting countries (Kyrgyzstan, Tajikistan, Uzbekistan, Ukraine, Moldova, Belarus) are interested in labor migration of their citizens to Russia because it reduces social tension in these countries and helps labor migrants to survive in conditions of low level of economy of those countries.

2. Generally the migration process in Russia has acquired a recurring character: some foreign workers return to their homes, others acquire the status of permanent residents. As a result, the internal structure of labor migrants begins to change.

3. The areas of employment of migrant workers are vegetable and clothing markets, construction sites, the service sector, and the housing and utilities sector, where locals do not want to work as long as they consider such work to be non-prestigious, unskilled, and low-paid.

4. The process of settlement of migrant workers in Russia is characterized by two opposite trends.

The first is dispersed settlement, which objectively contributes to the integration process of ethnic groups in the urban environment by simultaneously weakening intra-ethnic ties and increasing the depth and density of contacts with the local population. This trend is characteristic of rural and sparsely populated areas where labor migrants stay, since the process of living in this area requires the integration of residents, regardless of ethnicity and citizenship of the population.

The second trend is opposite. It is self-isolation of immigrant groups within compact settlements. Such settlements are formed along ethnic lines and are controlled by the respective ethnic diaspora. In these areas, migrants live in accordance with their ethnic culture, traditions, and customs of family and household life. For example, in Moscow, the best-known places for such localization of migrant workers are Ochakovo, Lianozovo, Tekstilshchiki, Koptevo, Lyublino, Biryulevo. It is extremely difficult to estimate the number of their inhabitants since many of them are illegal migrants who are not eager to have contact with the official authorities.

4 Discussion

Historically, international labor migration has been an urban phenomenon. Most immigrants settled in large cities and industrial regions, where the demand for labor is the highest [31]. This is since the central issue of the migration process is employment and earning financial resources. Some experts express the opinion that "guest workers" contribute to the increase of unemployment in the country because they take the jobs of Russians. However, as studies show, the incoming labor force, for the most part, does not compete in the Russian labor market with locals, because it occupies its own niche and performs such types of work that are not attractive to Russian citizens [36], especially in such large cities as Moscow and St. Petersburg, where fundamental investment projects of Russia are concentrated.

The main reason for this situation is that jobs with a predominance of unskilled and semi-skilled labor are not in demand in the Russian labor market due to low wages and labor content that does not meet the needs of the local population [38]. In addition, it has become obvious that foreign labor is needed even in situations of economic crisis and unemployment. It is no longer possible to replace migrant workers with locals in monotonous and tedious, dirty and hard, low-paid and non-prestigious jobs that do not require high qualifications [30].

At the same time, it should be noted that the complicated procedure for obtaining a work permit and strict registration rules lead to informal labor activity. Labor migrants, in conditions of legal uncertainty, tend to agree to work without guarantees from the employer. But this form of employment, in practice, often means low wages, no protection from being dismissed at any time, increased risk of injury at work, unregulated working hours, lack of collective agreement, trade union, vacations, any social benefits, etc.

In such a situation, guest workers try to get a job with their fellow countrymen, since they have some guaranteed minimum for accommodation and payment. Also, there is a new lower level being formed among foreign workers, which seems to be between two worlds. On the one hand, being in a foreign country, a migrant is largely alienated from his or her culture and, at the same time, remains for a long period excluded from the social life of the host country. Finding themselves in a different socio-cultural environment, guest workers begin to feel the difficulties associated with differences in customs and traditions, social attitudes, and value-based orientations in the process of interacting with the local population. At the same time, the social and cultural component of "guest workers" often forms misunderstanding and causes irritation among the locals [29].

This sometimes discriminatory and sometimes unfriendly attitude towards "guest workers" on the part of some local residents has a socio-psychological property [37], which prevails over the personal one – "I don't like the "newcomers", our electrician is a nice guy though". At the same time, the presence of discriminatory attitudes towards migrants is closely connected with the socio-economic situation in which the local population finds itself and reflects the uncertainty and unreliability of their positions in society, as well as the uncertainty about the future. Some Russians seek to compensate for this in the form of the superiority demonstration for their national group and thus consolidating their position.

Practice shows that not only migrant workers, but also their families with their natural needs begin to exist in the Russian social infrastructure, and the age and sex structure of ethnic diasporas become close to that of the local population [39]. This leads to a significant "visibility" of migrants in some cities. They begin to be competitors for some of the locals in the urban housing market and, as a result, the intracity structure changes. As a consequence, the tendency of national concentrations is formed, which leads to the emergence of a parallel social subsystem with its own specific needs, interests, and most importantly, requirements for the social medium of migrants' stay.

5 Conclusion

Thus, the study has shown that for the stable existence of the market in Russia, the executive power should timely identify the prevailing trend of the settlement of labor migrants in its region and implement a migration policy considering the identified trend. Competent formulation of migration policy, that reflects the prevailing trend of the settlement of migrant workers in a municipal or urban area will help to timely uncover the threats that contribute to the radicalization of migrant workers and encourage their behavior toward social society.

The authors note that this study is only one of the possible vector approaches to the problem of labor migration. One should not limit oneself to identifying the trends outlined, but should further substantiate the factors influencing this process. Given that the results of this study are based on macro-level scientific documents, it would be useful to study this issue at the micro-level as well. Such a study might allow for a better understanding of the subject matter under analysis.

References

1. Migration and Development Brief. Accessed on: October 18, 2021. [Online]. Available:https://www.knomad.org/publication/migration-and-development-brief-31
2. D.S. Massey, New Faces in New Places: the Changing Geography of American Immigration (Russell Sage Foundation, New York, 2008)
3. V. Zuniga, Hernandez-Leon R. New Destinations: Mexican Immigration in the United States (Russell Sage Foundation, New York, 2005)
4. K. Hoggart, C. Mendoza, Sociol Rural, **39(4)**, 538-562 (1999)
5. J. Gertel, S.R. Sippel, Seasonal Workers in Mediterranean Agriculture (Routledge, Abingdon, 2014)
6. A. Corrado, C. de Castro, D. Perrotta, Migration and Agriculture Mobility and Change in the Mediterranean (Area Routledge, London, 2017)
7. B. Jentsch, Int J Multicult Soc, **9**, 1-12 (2007)
8. R. McAreavey, Sociol Rural, **52**, 488-507 (2012)
9. S. Scott, P. Brindley, Geography, **97**, 29-38 (2012)
10. S. Scott, Popul Space Place, **19(5)**, 459-471 (2013)
11. I. Polevoy, A. Shunyaev, Man Labor, **11**, 55-61 (2012)
12. P.P. Lisitsyn, Sravnitelnyi analiz sotsialnoi adaptatsii legalnykh trudovykh migrantov iz Vyetnama i Uzbekistana v Sankt-Peterburge [Comparative analysis of the social adaptation of legal labor migrants from Vietnam and Uzbekistan in St. Petersburg], PhD thesis in Sociology (Moscow, 2009)
13. T. Regent, Econ Sci Mod Rus, **3**, 96-107 (1998)
14. V.P. Budakov, Rossiya v poiskakh sebya [Russia in quest of itself] (ROSSPAN, Moscow, 2007)
15. L.M. Drobizheva, Sotsialnye problemy mezhnatsionalnykh otnoshenii v postsovetskoi Rossii [Social problems of interethnic relations in post-Soviet Russia] (Center for Universal Values, Moscow, 2003)
16. L.L. Rybakovsky, Migratsiya naseleniya (teoreticheskie voprosy) [Population migration (theoretical questions)] (ISPII RAN, Moscow, 2003)
17. V.V. Petrov, Etnicheskie migranty v prinimayushchem obshchestve. Metodologiya i teoriya issledovaniya tolerantnosti i migrantofobii [Ethnic migrants in the host society. Methodology and theory of research of tolerance and migrant-phobia] (Krasnodar, 2005)
18. V.N. Pavlenko, *Psikhologicheskaya adaptatsiya i reabilitatsiya vynuzhdennykh migrantov: podkhody i problemy. Akkulturatsionnye strategii i modeli transformatsii*

identichnosti migrantov [Psychological adaptation and rehabilitation of forced migrants: approaches and problems. Accultural strategies and models of identity transformation in migrants], in G.U. Soldatova (ed), Psychology of refugees and forced migrants: the experience of research and practical work (Smysl, Moscow, 2001)

19. T.G. Stepanenko, *Adaptatsiya k novoi kulturnoi srede* [Adaptation to the new cultural environment], in Ethnopsychology (Aspect-Press, Moscow, 2003)
20. V.V. Radaev, J Sociol Soc Anthrop, **IV(3)**, 109-130 (2001)
21. G.V. Gradoselskaya, Setevye izmereniya v sotsiologii [Network measurements in sociology] (Novyi Uchebnik, Moscow, 2004)
22. S.G. Kirdina, Sociol Res, **12**, 23-33 (2002)
23. V.I. Mukomel, Migratsionnaya politika Rossii: postsovetskii kontekst [Migration policy of Russia: Post-Soviet context] (Dipol-T, Moscow, 2005)
24. J.M. Hodgson, *Viability of institutional economics*, in Evolutionary economics at the threshold of the XXI Century. Papers and speeches of the International Symposium, 29-74 Pushchino, September 23-25, 1996 (Japan Today, Moscow, 1997)
25. V.P. Buyanov, K.A. Kirsanov, D.M. Mikhailov, Riskologiya. Upravlenie Riskami [Riskology. Risk Management] (Ekzamen, Moscow, 2006)
26. G. Bekker, Povedenie cheloveka: ekonomicheskii podkhod [Human behavior: economical approach] (GU – Higher school of economics, Moscow, 2003)
27. R.I. Kapelyushnikov, Rossiiskii rynok truda: adaptatsiya bez restrukturizatsii [Russian labor market: adaptation without restructuring] (GU – Higher school of economics, Moscow, 2001)
28. Y.A. Korchagin, Chelovecheskii kapital kak intensivnyi sotsialno-ekonomicheskii faktor razvitiya lichnosti, ekonomiki, obshchestva i gosudarstvennosti [Human capital as an intensive socio-economic factor in the development of personality, economy, society, and statehood] (GU – Higher school of economics, Moscow, 2011)
29. S.V. Dementieva, Bul Tomsk Polytech Univ, **308(5)**, 195-199 (2005)
30. S.A. Stepanov, SGZ, **2**, 28-47 (2012)
31. S. Castles, H. de Haas, M.J. Miller, The Age of Migration. International Population Movements in the Modern World (Palgrave Macmillan, Basingstoke, 2014)
32. M.Y. Zelenkov, et al., Acad Strat Manag J, **20(1)** (2021)
33. M.Y. Zelenkov, et al., Acad Strat Manag J, **20(5)** (2021)
34. M.Y. Zelenkov, et al., Europ J Sci Theol, **17(3)**, 91-101 (2021)
35. V.A. Bogatyrev, P.V. Golubev, V.B. Aksinin, *Sovremennye tendentsii trudovoi* migratsii [Modern trends of labor migration], in Market. Entrepreneurship. Quality, 26-28 (Press, N. Novgorod, 1996)
36. N.N. Zinchenko, SGZ, **1**, 138-147 (2012)
37. E.A. Ivanova, Postgrad Bul Volga Reg, **10(5-6)**, 139-142 (2010). https://doi.org/10.17816/2072-2354.2010.0.5-6.139-142
38. I.V. Ivakhnyuk, Employ Serv, **3**, 47 (2015)
39. A.A. Nikolaev, Bul Finan Univ, **4**, 90-94 (2018). https://doi.org/10.26794/2226-7867--2018-8-4-90-94

AUTHOR INDEX

EDP Sciences – Web of Conferences
17, Avenue du Hoggar
Parc d'Activité de Courtabouf
BP 112
F-91944 Les Ulis Cedex A
France

ISBN 978-1-7138-4078-7

CURRAN ASSOCIATES INC.
proceedings.com